Ecotoxicological Testing of Marine and Freshwater Ecosystems

Emerging Techniques, Trends, and Strategies

ECOVISION WORLD MONOGRAPH SERIES

Series Editor

M. Munawar

Managing Editor

I.F. Munawar

Ecotoxicological Testing of Marine and Freshwater Ecosystems

Emerging Techniques, Trends, and Strategies

Edited by

P.J. den Besten and M. Munawar

Taylor & Francis
Taylor & Francis Group

Boca Raton   London   New York   Singapore

A CRC title, part of the Taylor & Francis imprint, a member of the
Taylor & Francis Group, the academic division of T&F Informa plc.

Published in 2005 by
CRC Press
Taylor & Francis Group
6000 Broken Sound Parkway NW, Suite 300
Boca Raton, FL 33487-2742

Library of Congress Cataloging-in-Publication Data

Ecotoxicological testing of marine and freshwater ecosystems : emerging techniques, trends, and
   strategies/ [edited by] P.J. den Besten, M. Munawar.
       p. cm.
   Includes bibliographical references and index.
   ISBN 0-8493-3526-4 (/05/$0.00+$1.50)
   1. Water quality bioassay. 2. Toxicity testing. 3. Marine ecology. 4. Freshwater ecology. I. Besten,
   P. J. den. II. Munawar, M. III. Title.

   QH90.57.B5E29 2005
   577.6'275--dc22

                                                                                2004022548

Taylor & Francis Group
is the Academic Division of T&F Informa plc.

Visit the Taylor & Francis Web site at
http://www.taylorandfrancis.com

and the CRC Press Web site at
http://www.crcpress.com

Editor's Note

M. Munawar

Within the past decade, the Aquatic Ecosystem Health and Management Society (AEHMS) has been actively engaged in organizing ecotoxicological symposia and conferences on a variety of themes and topics. The papers originating from these well-attended scientific gatherings have been published by the AEHMS in its journal, *Aquatic Ecosystem Health and Management*, or via its Ecovision World Monograph Series (Munawar et al. 1995a, 1995b; Munawar and Luotola 1995). The AEHMS also took a lead by focusing on sediment toxicity issues and established a Sediment Quality Assessment (SQA) working group. The SQA working group was charged with organizing and facilitating integrated and in-depth publications on the discipline. So far six SQA symposia have been organized across the world in a series of biennial meetings. The SQA meetings are highly successful, productive, and have resulted in the publication of several special issues and books (AEHMS, 1995; 1999a; 1999b; 2000; 2004; Munawar and Dave 1996; Munawar 2003).

Participants in various AEHMS symposia and conferences have asked for a comprehensive and concise compendium of modern techniques of aquatic ecosystem health-assessment strategies for professionals who deal with environmental issues, either in general or within specific fields. An opportunity to gather material on the current status of ecotoxicological techniques was offered by the 6th International Conference of the AEHMS, "Aquatic Ecosystem Health: Barometer of Integrity and Sustainable Development" (November 4–7, 2001, in Amsterdam), sponsored by the AEHMS, the Institute for Inland Water Management and Waste Water Treatment, and the Netherlands Society of Toxicology.

The concept of sustainable development necessitates the integration of ecotoxicological sciences with environmental management, legislation, and policy making. Aquatic ecosystem health assessment is a broad and integrated field of disciplines made up of structural and functional assessments in the field and laboratory. The field plays a key role in achieving sustainability since water and sediment quality are important prerequisites for the protection of the environment and human health. There have been several attempts to publish books on this subject. The AEHMS published a large compendium of environmental bioassay techniques in 1989 (Munawar et al. 1989). Most of these books, however, focused either on the scientific basis of ecosystem health assessment or on case studies in which risk-assessment strategies were demonstrated.

This monograph documents recent innovations and developments, listed below, in the fields of water and sediment quality assessments. These fields have integrated considerable advancement in ecotoxicology as well as in environmental chemistry:

- Chemical assessment of bioavailability
- Biosensor techniques to detect specific groups of contaminants
- Bioassays more relevant to species diversity or exposure routes
- Integrative approaches
- Modeling of bioaccumulation and consequences of sediment or water toxicity at higher trophic levels
- Communication strategies that focus on risk perception by the public, investigators, policy makers, and government agencies

All papers included in this monograph were invited and peer reviewed by a panel of international referees, using standard AEHMS publication guidelines. Accepted manuscripts were meticulously revised by authors, reviewed by the coeditors, and edited for technical and linguistic issues by the technical editor. We hope that this collection of papers provides a holistic and timely picture of the fast-changing field of ecotoxicological testing and is useful to toxicologists, environmentalists, researchers, managers, and policy makers across the world.

I sincerely thank Dr. P.J. den Besten of the Institute for Inland Water Management and Waste Water Treatment for his devotion, hard work, and cooperation that resulted in the preparation and publication of this landmark book. I also thank Nabila F. Munawar, Sharon Lawrence, Iftekhar F. Munawar, Susan Blunt, and Calais Irwin for their assistance in the processing of this book. Thanks also to Randi Cohen for her interest, encouragement, and assistance in the publication of this book with Taylor & Francis/CRC Press.

References

AEHMS (Aquatic Ecosystem Health and Management Society). J. Aquat. Ecosyst. Health 4(3), 133-216, 1995.

AEHMS. *Sediment Quality Assessment: Tools, Criteria and Strategies (special. issue).* Aquat. Ecosyst. Health Mgmt. 2(4), 345-484, 1999a.

AEHMS. *Integrated Toxicology (special issue).* Aquat. Ecosyst. Health Mgmt. 2(1), 1-71, 1999b.

AEHMS. Aquat. Ecosyst. Health Mgmt. 3(3), 277-430, 2000.

AEHMS. *Assessing Risks and Impacts of Contaminants in Sediments (special issue).* Aquat. Ecosyst. Health Mgmt. 7(3), 335-432, 2004.

Munawar, M. (Ed.). *Sediment Quality Assessment and Management: Insight and Progress.* Ecovision World Monograph Series. Aquatic Ecosystem Health and Management Society, Canada, 361 pp. 2003.

Munawar, M., Dave, G. (Eds.). *Development and Progress in Sediment Quality Assessment: Rationale, Challenges, Techniques and Strategies.* Ecovision World Monograph Series. SPB Academic Publishers, the Netherlands, 255 pp. 1996.

Munawar, M., Luotola, M. (Eds.). *The Contaminants in the Nordic Ecosystem: the Dynamics, Progress and Fate.* Ecovision World Monograph Series. SPB Academic Publishing, the Netherlands, 276 pp. 1995.

Munawar, M., Dixon, G., Mayfield, C.I., Reynoldson, T, Sadar, M.H., (Eds.). *Environmental Bioassay Techniques and their Application.* Hydrobiologia, 188/189, 680pp. 1989.

Munawar, M., Chang, P., Dave, G., Malley, D., Munawar, S., Xiu, R., (Eds.). *Aquatic Ecosystems of China: Environmental and Toxicological Assessment.* Ecovision World Monograph Series. SPB Academic Publishing, the Netherlands, 119 pp. 1995a.

Munawar, M., Hanninen, O., Roy, S., Munawar, N., Karenlampi. L., Brown, D., (Eds.). 1995b. *Bioindicators of Environmental Health.* Ecovision World Monograph Series. SPB Academic Publishing, the Netherlands, 265 pp. 1995b.

Foreword

G. Dave

During the last 50 years most of us have realized that the "the solution to pollution is not dilution." Books like *Silent Spring* and *The Frail Ocean* and TV programs by Jacques Cousteau have alerted scientists, decision-makers, and the public to the threat of chemicals to environmental health. We have added other threats like acidification, eutrophication, overexploitation of natural resources (biological as well as geophysical), and global warming. We have also realized that the environment is a very complex system in which unexpected events may occur, such as eggshell thinning caused by chlorinated hydrocarbons and imposex in gastropods caused by tributyl tin. These examples illustrate the need for precautionary principles.

Experience has shown that the majority of environmental problems are of global concern, and that we need international cooperation to solve them. This is certainly the case for the marine environment. In many parts of the world it is overexploited while it also suffers from pollution, illustrating the "tragedy of the commons." Cooperation does work, and has resulted in positive action at international, national, regional, and local levels. The unifying principle of the Rio conference in 1992, "think globally, act locally," and the acceptance of Agenda 21 have certainly affected the Aquatic Ecosystem Health and Management Society (AEHMS). The AEHMS has acted globally by organizing conferences and publishing the journal *Aquatic Ecosystem Health and Management*. The AEHMS has also produced numerous special issues and peer-reviewed books such as this monograph and the Ecovision World Monograph Series (http://www.aehms.org/).

This book is one of several important steps toward a better understanding of the effects of chemicals and assessment of ecosystem health. During the last decade there has been an increasing emphasis on monitoring of biological parameters in the aquatic environment. This may be seen as a shift in emphasis from laboratory studies and toxicity tests toward field studies and bioassays, and from measurements of concentrations of pollutants toward measurements of biological diversity and ecological function and interaction. However, these changes in focus should be complementary and not occur at the expense of each other. The complexity of aquatic ecosystems requires consideration of both exposure to chemicals and effects of chemicals, as well as the interaction between organisms and the influence of confounding factors such as weather and climate. We also need to communicate these matters to decision-makers and the public.

The chapters of this book present various methods that can be used to improve our understanding of the aquatic environment and its response to disturbances. The book as a whole promotes the understanding of the structure, function, and performance of healthy and damaged aquatic ecosystems (freshwater, marine, and estuarine) from integrated, multidisciplinary, and sustainable perspectives, and explores the complex interactions between human society, ecology, development, politics, and the environment. This makes the book a valuable contribution to the ideas and philosophy of our society and to the AEHMS in particular.

Preface

P.J. den Besten and M. Munawar

Over the past 25 years the discipline of ecotoxicology has undergone two major developments. Firstly, new assays have been developed, deploying organisms that bear added relevance to the specific environment under investigation. Several new procedures assess the effects on organisms after exposure to environmental samples rather than to spiked water or sediment samples. Also noteworthy is the considerable attention given to effects of chronic exposure to low levels of contaminants. These developments are of great importance for the application of ecotoxicological techniques in risk-assessment approaches. They create new possibilities for building lines of evidence as part of weight of evidence (WOE) approaches (Burton et al. 2002). Secondly, progress is apparent from the increased attention given to effecting measurements at different levels of biological organization. Including new endpoints in assays at the cellular, subcellular, or molecular level may increase the sensitivity, specificity, or throughput capacity of the assays. Such developments will prove to be crucial steps in the application of screening steps in water and sediment quality assessment. Furthermore, these techniques may help to build prognostic tools that can be used in early-warning systems (den Besten 1998).

Almost 15 years ago, a state-of-the-art assessment of environmental bioassays and their applications was published (Munawar et al. 1989). Since then several other books with different scopes about the scientific background of ecotoxicology and its application in environmental risk assessment have appeared. This book is intended to capture the progress and developments made in this field since 1989.

Most chapters focus on the impairment of aquatic ecosystem health due to the pollution of water and sediments. However, it is clear that there are many more stressors that can threaten aquatic ecosystems. Impacts by human activities can also be observed at different scales, from local to global. Direct impacts occur through catchment runoff, discharge of wastes, atmospheric deposition of pollutants, eutrophication, overexploitation, and habitat modification. Insidious impacts include the spread of introduced species and manifestations of global warming. A special chapter in this book deals with the role of remote sensing technologies in monitoring, predicting, and managing changes within coastal ecosystems. Important improvements in information technology and data processing make possible the assessment of spatial variability.

The information from ecotoxicological assessments is used to make recommendations to preserve, enhance, or restore ecosystem functions. Decisions regarding the commitment of political or resource expenditures necessary to implement these recommendations are often made by nontechnical experts such as elected officials in consultation with the public. These audiences are often unfamiliar with the data and techniques used to assess aquatic ecosystems. It is important that assessment results be effectively communicated in comprehensible terms and language to ensure that decision-makers and the public are making informed choices. Therefore, this book contains a chapter describing the background of risk perception and communication. This information should show scientists how to effectively communicate the outcome of their risk assessments.

Ecotoxicological testing of water and sediment implies that the quality of water and sediment is not only based on information from chemical analyses, but also (or as a first step) on effect measurements. Effect measurements are in this respect usually referred to as bioassays or toxicity tests. The terms effect-based water quality assessment and effect-based sediment quality assessment are used to underscore the change from the classical chemical approaches. Effect-based water and sediment quality assessments have been implemented in different countries to a variable degree. Generally speaking, effect monitoring is gaining importance in the following water and sediment management tasks:

- Surface water quality assessment
- Drinking water quality assessment
- Wastewater quality assessment (before and after treatment)
- Sediment quality assessment (decision frameworks for remediation)
- Dredged material quality assessment (for selecting disposal options)

The reason for the increasing importance of effect-based quality assessment is that we generally know the identity of just a small percentage of the chemicals that are released into the environment. Furthermore, it is obvious that the presence of chemical substances in the environment is important for the ecosystem because effects occur, and not just because the chemicals are present. For example, most chemical analyses do not include an evaluation of the biological availability, even though this is essential information for understanding the actual risks. When quality assessment is also based on effect measurements, important information about availability and about unknown toxic compounds is included in the evaluation.

The focus of this book is on ecotoxicological testing of water and sediment quality in both freshwater and marine waters. In many cases, effect-based quality assessment approaches include field surveys of pelagic or benthic invertebrates or wildlife populations (offspring size, bioaccumulation levels, and so on). The expertise involved in this work is partly from ecology and partly from ecotoxicology, and thus is not entirely outside the scope of this book. However, this book is primarily dedicated to recent developments in bioassays (toxicity tests with water or sediment samples) and new technologies such as gene-expression analysis and remote sensing. It also contains a description of techniques included as appendices at the end of some of the chapters, enabling the reader to understand and comprehend the strengths and limitations of various techniques and providing access to additional literature. An overview and synthesis of the current status of techniques and strategies is included in the last chapter.

This book focuses on the following topics:

- Emerging fields of research on biomarkers, genome expression, multispecies tests, and tiered approaches
- Experimentally oriented strategy (although the book does not contain information about ecology)
- Overview of methods for processing and integration of data, risk communication, and risk perception
- Use of information from biological testing in decision- and policy-making
- Selected and simple proven techniques that may be used for testing and training purposes (in the appendices)

The reader may find some inconsistencies in the terms and definitions used by the different authors for specific techniques, such as toxicity test, bioassay, biosensor, and so on. In the opinion of the editors, these differences reflect personal views on the roles these techniques may play in risk assessment. Tests can be chemically oriented, focusing on the mode of action of a toxic compound, or be ecologically oriented, aimed to link cause and effect observed in the field. Since this book is not intended to reach agreement in the definition of those terms and techniques, occasional differences among the chapters should be interpreted as the personal preferences of the authors.

References

Burton, G.A., Jr., Batley, G.E., Chapman, P.M., Forbes, V.E., Smith, E.P., Reynoldson, T., Schlekat, C.E., den Besten, P.J., Bailer, J., Green, A.S., and Dwyer, R.L., 2002. A weight-of-evidence framework for assessing ecosystem impairment: improving certainty in the decision-making process. *Human Ecol. Risk Assessment*, 8, 1675–1696.

den Besten, P.J., 1998. Concepts for the implementation of biomarkers in environmental monitoring, *Mar. Environ. Res.* 46, 253–256.

Munawar, M., Dixon, G., Mayfield, C.I., Reynoldson, T., and Sadar, M.H., (Eds.) 1989. Environmental bioassay techniques and their application. *Hydrobiologia*, 188/189.

Contributors

P.J. den Besten
Institute for Inland Water
  Management and Waste Water
  Treatment
Ministry of Transport, Public Works
  and Water Management
PO Box 17
8200 AA Lelystad
The Netherlands

N.W. van den Brink
Centre for Ecosystem Studies
PO Box 47
6700 AA Wageningen,
The Netherlands

A. Brouwer
BioDetection Systems BV and
Institute for Environmental Studies
  (IVM)
Badhuisweg 3
1031 CM Amsterdam,
The Netherlands
and
Institute for Environmental Studies
  (IVM)
Vrije Universiteit
De Boelelaan 1087
1081 HV Amsterdam
The Netherlands

B. van der Burg
BioDetection Systems BV
Badhuisweg 3
1031 CM Amsterdam
The Netherlands

W.M. De Coen
Laboratory for Ecophysiology,
  Biochemistry and Toxicology
University of Antwerp
Groenenborgerlaan 171, B-2020
  Antwerp
Belgium

G. Dave
Department of Applied
  Environmental Science
University of Goteborg
Goteborg
Sweden

K.T. Ho
Department of Applied
  Environmental Science
University of Goteborg
Box 464
405 30 Goteborg
Sweden

D.S. Ireland
U. S. Environmental Protection
  Agency
Chicago, Illinois
United States

K. Koop
New South Wales Department of
  Environment & Conservation
Sydney South, NSW
Australia

A. Lange
University of Antwerp
Laboratory for Ecophysiology,
  Biochemistry and Toxicology
Groenenborgerlaan 171, B-2020,
  Antwerp
Belgium

D. Leverett
Environment Agency, National
  Centre for Ecotoxicology and
  Hazardous Substances
4 The Meadows,
Waterberry Drive
Waterlooville, Hampshire
P07 7XX
United Kingdom

M. Maras
University of Antwerp
Laboratory for Ecophysiology,
  Biochemistry and Toxicology
Groenenborgerlaan 171, B-2020
  Antwerp
Belgium

M. Munawar
Fisheries & Oceans Canada
Burlington, Ontario
Canada

R. van der Oost
DWR, Institute for Water
  Management and Sewerage
Environmental Toxicology
PO Box 94370
1090 GJ Amsterdam
The Netherlands

L. Pelstring
Damage Assessment Center
National Oceanic and Atmospheric
  Administration
Silver Spring, Maryland
United States

T.R. Pritchard
University of Waikato
Hamilton
New Zealand

M.R. Reiss
U.S. Environmental Protection
  Agency
New York, New York
United States

M. Tonkes
Institute for Inland Water
  Management and Waste Water
  Treatment
Ministry of Transport, Public Works
  and Water Management
PO Box 17
8200 AA Lelystad
The Netherlands

C. Porte-Visa
Environmental Chemistry
  Department
IIQAB-CSIC
C/ Jordi Girona, 18
08034 Barcelona
Spain

Contents

Chapter one   Toxicity tests for sediment quality assessments 1
                  D.S. Ireland and K.T. Ho

Chapter two   Bioassays and tiered approaches for monitoring
                  surface water quality and effluents.....................................43
                  M. Tonkes, P.J. den Besten, and D. Leverett

Chapter three   Biomarkers in environmental assessment......................87
                  R. van der Oost, C. Porte-Visa, and N.W. van den Brink

Chapter four   Molecular methods for gene expression analysis:
                  ecotoxicological applications...153
                  A. Lange, M. Maras, and W.M. De Coen

Chapter five   Bioassays and biosensors: capturing biology in
                  a nutshell 177
                  B. van der Burg and A. Brouwer

Chapter six   Satellite remote sensing in marine ecosystem
                  assessments..195
                  T.R. Pritchard and K. Koop

Chapter seven   Risk perception and public communication of
                  aquatic ecosystem assessment information..................229
                  M.R. Reiss and L. Pelstring

Chapter eight   Ecotoxicological testing of marine and freshwater
                  ecosystems: synthesis and recommendations..............249
                  P.J. den Besten and M. Munawar

Index...261

chapter one

Toxicity tests for sediment quality assessments

D.S. Ireland and K.T. Ho

Contents

Introduction 2
The need for toxicity tests in sediment quality assessments 2
Assessment approaches 5
    Tiered testing approaches 5
    Applications of sediment toxicity tests 5
Sediment sampling 9
    Sample design 9
    Sample collection, processing, transport, and storage 10
    Sample manipulation 12
Recommended procedures for both freshwater and marine test
    organisms 14
Interpretation 17
    Laboratory versus field exposures: what is the ecological
        relevance? 17
    Future research recommendations 23
Summary 24
Acknowledgments 24
References..25
Appendix 36
Toxicity tests for sediment quality assessments 36
Freshwater test organisms 36
    Hyalella azteca 36
    Chironomus riparius 38
Marine test organisms 39
    Ampelisca abdita 39
Microtox 41
References..41

Introduction

Toxic sediments have contributed to a wide variety of environmental problems around the world. The observed effects include direct toxic effects to
aquatic life, biomagnification of toxicants in the food chain, and economic
impacts. This chapter discusses the use of toxicity tests as an integral part
of contaminated sediment assessments, and summarizes the use of sediment
toxicity testing in existing tiered regulatory guidance for addressing toxic
sediments and dredge spoils in several countries. Sampling design, collection, handling, and storage of sediments for toxicity testing are discussed in
relation to the project objectives.

A number of sediment toxicity tests exist for both fresh and marine
waters. A brief description of the type of test, collection method for the test
organism, volume of test material needed, suitable test matrix, level of standardization, and references where detailed methodology can be found are
also included in this chapter. Several studies are highlighted that discuss the
ecological significance of toxicity testing, and recommendations for future
research in the area are included.

The need for toxicity tests in sediment quality assessments

Sediment is an integral component of aquatic ecosystems, providing habitat,
feeding, spawning, and rearing areas for many aquatic organisms. In aquatic
systems, sediments accumulate anthropogenic (man-made) chemicals and
waste materials, particularly persistent organic and inorganic chemicals.
These accumulated chemicals are then reintroduced into waterways (USEPA
1998) and have contributed to a variety of environmental problems. Contaminated sediments may be directly toxic to sediment-dwelling organisms
or be a source of contaminants for bioaccumulation in the food chain. The
direct effects of contaminated sediments can be obvious or subtle. Evident
effects include loss of important fish and shellfish populations (USEPA 1998);
decreased survival, reduced growth, and impaired reproduction in benthic
invertebrates and fish (USEPA 2002); and fin rot and increased tumor frequency in fish (Van Veld et al. 1990). Adverse effects on organisms in or near
sediment can occur even when contaminant levels in the overlying water
are low (Chapman 1989).

More subtle effects resulting from contaminated sediments include
changes in composition of benthic invertebrate communities from sensitive
to pollution-tolerant species and decreases in aquatic system biodiversity
(USEPA 1998). Tolerant species may process contaminants in a variety of
ways, and the resulting novel metabolic pathways and products may affect
ecosystem functions such as energy flow, productivity, and decomposition
processes (Griffiths 1983).

Loss of any biological community in the ecosystem can indirectly affect
other components of the system. For example, if the benthic community is
significantly changed, nitrogen cycling might be altered such that forms of

nitrogen necessary for key phytoplankton species are lost and replaced with blue-green algae, capable of nitrogen fixation (Burton et al. 2002). Many examples of direct impacts of contaminated sediment on wildlife and humans have been noted. Bishop et al. (1995, 1999) found good correlations between a variety of chlorinated hydrocarbons in sediment and concentrations in bird eggs. These researchers found that this relationship indicated that the female contaminant body burden was obtained locally, just prior to egg-laying. Other studies by Bishop et al. indicated a link between exposure of snapping turtle (*Chelydra s. serpentina*) eggs to contaminants (including sediment exposure) and developmental success (Bishop et al. 1991, 1998).

Contaminated sediments can also be a source of chemicals for bioaccumulation in the food chain (USEPA 2000a; ASTM 2002a). Contaminants may be bioaccumulated by transport of dissolved contaminants in interstitial water (ITW — sometimes referred to as pore water) across biological membranes and/or the ingestion of contaminated food or sediment particles with subsequent transport across the gut. For upper-trophic–level species, ingestion of contaminated food is the predominant route of exposure, especially to hydrophobic chemicals; it is through the ingestion of contaminated fish and shellfish that human health can be impacted from contaminated sediments. Other investigations of environmentally persistent organic compounds (chlorinated hydrocarbons) have shown bioaccumulation and a range of effects in the mud puppy, *Necturus maculosus* (Bonin et al. 1995; Gendron et al. 1997). For humans, there is evidence that chronic exposure to significant quantities of polychlorinated biphenyls (PCBs) via consumption of freshwater fish results in low–birth-weight infants, reduced head circumference, and delays in developmental maturation at birth (Swain 1988). In fact, fish consumption represents the most significant route of aquatic exposure of humans to many metals and organic compounds (USEPA 1992a). In addition there is anecdotal evidence from cases like Monguagon Creek, a small tributary of the Detroit River, where incidental human contact with the sediment resulted in a skin rash (Zarull et al. 1999).

Consequently, contaminated sediments in aquatic ecosystems pose potential hazards to sediment-dwelling organisms (epibenthic and in-faunal invertebrate species), aquatic-dependent wildlife species (fish, amphibians, reptiles, birds, and mammals), and humans (USEPA 2002; MacDonald et al. 2002a, 2002b).

In addition to animal health, human health, and ecological impacts, contaminated sediments may cause severe economic effects. Economic impacts may be felt by the transportation, tourism, and fishing industries. In one Great Lakes harbor (the Indiana Harbor Ship Canal), navigational dredging has not been conducted since 1972 "due to the lack of an approved economically feasible and environmentally acceptable disposal facility for dredged materials" from the canal (USACE 1995). The accumulation of sediment in this canal has increased costs for industry. Ships carrying raw materials have difficulty navigating in the harbor and canal. In addition, ships come into the harbor loaded at less-than-optimum vessel drafts. The use of various docks is restricted, requiring unloading at alternative docks and double-handling of bulk commodities to the preferred dock. These problems are causing increased transportation costs of waterborne commerce in this canal, estimated in 1995 to be $12.4 million annually (USACE 1995).

Assessments of sediment quality commonly include the analyses of anthropogenic contaminants (sediment chemistry), geochemical factors that affect bioavailability, benthic community structure, and direct measures of toxicity (toxicity tests). All of these measures provide useful and unique information relating to the quality of the sediment. However, sediment chemistry measurements alone might not accurately reflect risk to the environment (USEPA 2000b). Bioavailability of chemicals in sediment is a function of the chemical class and of speciation and geochemical factors, as well as the behavior and physiology of the organism. In addition, complex chemical analyses are often impractical, expensive, and in many cases almost impossible due to the high number of unknown contaminants. Benthic community surveys may be inadequate because they can fail to discriminate between effects of contaminants and effects from noncontaminant factors (for example, physical parameters such as salinity and flow).

Sediment toxicity tests allow a direct measure of sediment toxicity or bioaccumulation by exposing surrogate organisms to sediments under controlled conditions (ASTM 2002b; USEPA 2000b, 2001a). These tests have evolved into standardized, effective tools providing direct, quantifiable evidence of biological consequences of sediment contamination that can only be inferred from chemical or benthic community analyses (ASTM 2002b; USEPA 2000b, 2001a). Some advantages of sediment toxicity tests are that they measure the bioavailable fraction of contaminants, they require limited special equipment, they can be applied to all chemicals of concern, and tests applied to field samples reflect cumulative effects of contaminants and contaminant interactions (ASTM 2002b; USEPA 2000b, 2001a). Some disadvantages of using sediment toxicity tests are that natural geochemical characteristics of sediment may affect the response of test organisms, indigenous animals may be present in field-collected sediments, tests applied to field samples may not discriminate effects of individual chemicals, and few comparisons have been made of methods or species (ASTM 2002b; USEPA 2000b, 2001a).

Traditionally, sediment toxicity test data have been expressed as a percentage of survival in comparison to a control or reference for indicator organisms exposed to the field-sampled sediment in laboratory toxicity tests (ASTM 2002b, 2002c, 2002d; USEPA 1994a, 1994b, 2000b, 2001a). Methods for testing the short- and long-term toxicity of sediment samples to benthic freshwater and marine organisms have been developed (see reviews in API 1994; Burton et al. 1992; Lamberson et al. 1992; USEPA 1994a, 1994b, 2000b, 2001a). More recently, sublethal measurements (reduction in survival, growth, and reproduction) are also being used (Ingersoll et al. 2001).

Assessment approaches

Tiered testing approaches

Tiered testing refers to a structured, hierarchical procedure for determining data needs relative to decision-making that consists of a series of tiers (levels or steps) of investigative intensity. Tiered testing represents a logical, technically sound approach for evaluating contaminated sediments and is used in a variety of regulatory programs throughout the world (including those described below). Typically, increasing tiers in a tiered testing framework involve increased information and decreased uncertainty (USEPA 1998). The objective of the tiered testing approach is to make optimal use of resources in generating the information necessary to make a contaminant determination, using an integrated chemical, physical, and biological approach. The initial tier uses available information that may be sufficient for completing the evaluation in some cases. Evaluation at successive tiers requires information from tests of increasing sophistication and cost. For example, some frameworks prescribe the use of short-term (acute) sediment toxicity tests in tier 2, and long-term (chronic) sediment toxicity tests as well as bioaccumulation tests in tier 3. If the information gathered in a tier is inadequate to make a decision, testing proceeds through subsequent tiers of more extensive and specific testing until sufficient information is generated to support a decision. The most logical and cost-efficient approach is to enter tier one and proceed as far as necessary to make a determination.

The general conclusions that are made at each of the tiers is that either the available information either is or is not sufficient to make a contaminant determination. With the tiered testing structure, it is not usually necessary to obtain data for all tiers to make a contaminant determination. It may also not be necessary to conduct every test described within a given tier to have enough information for a determination. The underlying philosophy is that only that data necessary for a determination should be acquired.

Applications of sediment toxicity tests

All sediment toxicity tests are undertaken for a specific reason and most are done for some type of regulatory purpose. This may include support for sediment remediation; dredged material disposal; sediment monitoring; and in the U.S., possible support for total maximum daily loads (TMDLs) and natural resource damage assessments (NRDAs). Numerous regulations exist throughout the world that authorize programs for addressing contaminated sediments. A few of these regulations and frameworks that use toxicity tests include dredged material disposal in the U.S., Canada, and Australia, and sediment remediation in the U.S.. This is not meant to be all-inclusive, but serves to provide some examples.

In most navigational dredging situations, the decision has been made that the material will be moved. The question is whether or not the material can be disposed of in an unrestricted fashion (no treatment of the material) in open water as opposed to in some type of confined disposal facility (either on land or in the water). In the U.S., the U.S. Environmental Protection Agency (USEPA) and U.S. Army Corps of Engineers (USACE) are responsible for governing the regulatory program concerned with evaluating navigation dredged material. About 400 million cubic yards (roughly 500 million tons) of sediment are dredged annually in the U.S. to maintain more than 400 ports and 25,000 miles of navigation channel. Dredged material transported for disposal at ocean sites is regulated by Section 103 of the Marine Protection, Research and Sanctuaries Act (MPRSA). Guidance for conducting evaluations for material being proposed for ocean disposal is described in *Evaluation of Dredged Material Proposed for Ocean Disposal — Testing Manual* (USACE/USEPA 1991), otherwise known as the Ocean Testing Manual (OTM). The dredged material unsuitable for ocean disposal is either placed in upland environments (confined disposal facilities) or is managed within the aquatic environment rather than disposed of in open water. Dredged material that is proposed to be managed within the aquatic environment landward of the baseline of the territorial sea is regulated under Section 404 of the Clean Water Act (CWA). Guidance for conducting evaluations under Section 404 is contained in *Evaluation of Dredged Material Proposed for Discharge in Waters of the U.S. — Testing Manual* (USACE/USEPA 1998), otherwise known as the Inland Testing Manual (ITM).

The same evaluative framework is used in the OTM and the ITM to characterize exposure and effects. The framework uses a tiered approach, as outlined above, proceeding through subsequent tiers until there is sufficient information to determine if the material would cause unacceptable impacts in the aquatic environment. Tier 1 involves the collection and analysis of existing information on the physical, chemical, and biological properties of the material in question. Tier 2 involves the collection and use of chemical data. In tier 3 and tier 4, sediment toxicity tests are conducted to assist in the decision-making process regarding the disposal of dredged material. Toxicity tests with whole sediments are designed to determine whether dredged material is likely to produce unacceptable adverse effects on benthic organisms. In these tests, the test animals are exposed to whole sediments, and the effects (lethality in tier 3 and sublethality in tier 4) are recorded. For whole-sediment toxicity tests, both the OTM and the ITM recommend the use of three sensitive species, representing a filter feeder, a deposit feeder, and a burrowing organism (where possible). If only two different species are tested they should, together, cover the above three life-history strategies. Additional information on the test requirements can be found in the OTM and the ITM.

The OTM and the ITM also provide information necessary to estimate the potential for bioaccumulation. A bioaccumulation test in tier 3 is normally conducted only when there is a reason to believe that specific chemicals of concern may be accumulated in the tissues of target organisms (USACE/USEPA 1998). Both the OTM and the ITM require two 28-day bioaccumulation tests utilizing species from two different tropic niches (where possible), representing a suspension-feeding/filter-feeding and a burrowing deposit-feeding organism (USACE/USEPA 1991, 1998). If results of the bioaccumulation test in tier 3 are indeterminate, further testing may be required in tier 4, recognizing that an exposure period of 28 days may not be sufficient for the selected test species to achieve a steady-state tissue concentration in the normal tier 3 bioaccumulation test. In a tier 4 bioaccumulation test, testing may be done in the lab or in rare cases in the field, and testing options may also include time-sequenced laboratory exposures in excess of the standard 28 days in order to reach a steady-state concentration (USACE/USEPA 1998).

The management of dredged material disposal in Canada for marine sediments follows a similar tiered structure as in the U.S.. Each year in Canada, 2 to 3 million tons of material are disposed of at sea. Most of this is for keeping shipping channels and harbors clear for navigation and commerce. Environment Canada administers the control of disposal at sea under the Canadian Environmental Protection Act, 1999 (CEPA). This permitting system applies to both marine and internal marine waters and lives up to the commitments made under the 1996 Protocol to the Convention on the Prevention of Marine Pollution by Dumping of Wastes and Other Matter (known as the London Convention). The assessment framework used for controlling material for open-water disposal mirrors the Waste Assessment Guidance of the 1996 Protocol and has been reproduced in the CEPA. Material not suitable for disposal at sea may be left in place, or disposed of or treated on land under various other jurisdictions. Similar to the U.S., evaluations are conducted in a tiered approach and proceed through subsequent tiers until there is sufficient information to determine if the material would cause unacceptable impacts in the aquatic environment. The exposure pathways to support this determination include both whole-sediment and ITW tests using three marine or estuary sediment bioassays, including an acute lethality test (Environment Canada 1998a) and two sublethal tests (Environment Canada 2001a); or one sublethal test and one bioaccumulation test (USEPA 1993a). Species were originally selected to be both representative of Canadian environments and ecologically important. Additional information on this framework can be found in the Canadian Disposal at Sea Regulations (Environment Canada 2001b).

In May 2002, Environment Australia released the National Ocean Disposal Guidelines for Dredged Material (Environment Australia 2002). Like Canada, Australia is party to the London Convention (ratified in December 2000), and under the Environment Protection (Sea Dumping) Act 1981 (the Sea Dumping Act), Australia implements the protocol of the London Convention by regulating the dumping of wastes and other matter into the sea. The Sea Dumping Act provides the basis for the permitting and ongoing management of such actions. These guidelines are intended to provide a comprehensive framework to assess potential environmental impacts from disposing dredged material at sea in accordance with the Sea Dumping Act and other environmental protection legislation, including the Environment

Protection and Biodiversity Conservation Act 1999 and Australia's international obligations (Environment Australia 2002). Under these guidelines, Australia has developed a tiered approach for assessing sediment contamination using four phases. Sediment toxicity testing is in phase three (acute toxicity) and phase four (subacute or chronic toxicity). Protocols for conducting these test are outlined in the Australian and New Zealand Guidelines for Fresh and Marine Water Quality (ANZECC/ARMCANZ 2000). The National Ocean Disposal Guidelines for Dredged Material states that sediment toxicity testing, using protocols based on those developed by USEPA (the OTM and the ITM outlined above) or by the American Society for Testing and Materials (ASTM), is considered the best available method for predicting the bioavailability and subsequent toxicity potential of contaminated sediments for open-sea disposal of dredged material. Whole-sediment tests are preferred, where available, because the water tests available are not necessarily on the most ecologically relevant species (Environment Australia 2002). As stated above, further details on toxicity testing are set out in ANZECC/ARMCANZ (2000).

The U.S. Comprehensive Environmental Response, Compensation and Liability Act of 1980 (CERCLA, often referred to as Superfund) as amended by the Superfund Amendments and Reauthorization Act of 1986 (SARA) provides one of the most comprehensive authorities available to the USEPA for obtaining sediment cleanup, reimbursement of USEPA cleanup costs, and compensation to natural resource trustees for damages by contaminated sediments. The USEPA Superfund program carries out the Agency's mandate under CERCLA/SARA. The primary regulation issued by the Superfund program is the National Oil and Hazardous Substances Pollution Contingency Plan (NCP). To date, about 300 sites (approximately 20%) on the Superfund National Priorities List (NPL) — the list of national priorities among the known releases or threatened releases of hazardous substances, pollutants, or contaminants throughout the U.S. based on a hazard-ranking system — appear to have some kind of contaminated sediment (USEPA 2004). To assist in identifying sites where the risks to human health or the environment are unacceptable due to sediment contamination, the USEPA has recently developed and published guidance for conducting Ecological Risk Assessments (ERAs) within the Superfund program (USEPA 1997). ERAs are most often conducted by the USEPA during the Remedial Investigation/Feasibility Study (RI/FS) phase of the Superfund response process and are composed of eight steps or phases. These ERAs are used to evaluate the likelihood of adverse ecological effects occurring as a result of exposure to any physical, chemical, or biological entities that can induce adverse responses at a site. Steps 1 and 2 involve the compilation of existing information, steps 3 through 6 are data collection, step 7 is risk characterization, and step 8 is risk management.

Sediment toxicity tests are commonly used in ERAs to assist in determining if there is an unacceptable risk from sediment contamination. During step 3 (problem formulation), assessment endpoints are selected. These endpoints are an explicit expression of the environmental value (species, eco-logical resource, or habitat type) that is to be protected. Often, these are difficult to measure directly; therefore, in the case of contaminated sedi-ments, a sediment toxicity test (a measurement endpoint) is used as a sur-rogate. This test is conducted in step 6 (site investigation and analysis). The ERA guide for Superfund does not dictate what species should be used in toxicity testing but states that the "selection of the test organism is critical in designing a study using toxicity testing. The species selected should be representative relative to the assessment endpoint, typically found within the exposure pathway expected in the field."

Sediment sampling

Sample design

Accurate assessment of environmental hazards posed by contaminated sed-iment depends greatly on the accuracy and the representativeness of the sediment sample collected for sediment chemistry, benthic community struc-ture, and sediment toxicity tests. It is widely accepted that the methods used in sample collection, transport, handling, storage, and manipulation of sed-iments and ITWs can influence the physicochemical properties and the results of chemical, toxicological, and bioaccumulation analyses (ASTM 2002e; Environment Canada 1994; USEPA 2001b). Addressing these variables in an appropriate and systematic manner helps to ensure more accurate sediment quality data and to facilitate comparison among sediment studies. In 2001, the USEPA Office of Water released a document on the collection, storage, and manipulation of sediments for toxicity and chemical testing (USEPA 2001b). This document builds on guidance from ASTM (2002e) and Environment Canada (1994) and rarely dictates methods that *must* be fol-lowed, but rather makes recommendations for those that *should* be followed. Since every study site and project is unique, sediment monitoring and assess-ment study plans should be carefully prepared to best meet the project objectives (MacDonald et al. 1991; USEPA 2001b; Burton et al. 2002).

    The USEPA (2001b) states that before collecting any environmental data, it is important to determine the type, quantity, and quality of data needed to meet the project objectives (such as the parameters being measured) and to support a decision based on the results of data collection and observation. Generally, sampling designs fall into two major categories: random (or prob-abilistic) and targeted (USEPA 2000c). Random or probability-based sam-pling designs avoid bias in the results by randomly assigning and selecting sampling locations; a requirement is that all sampling units have a known probability of being selected. In targeted sampling, stations are selected based on prior knowledge of other factors, such as contaminant loading, depth, salinity, and substrate type. This type of design is useful if the objec-tive of the study is to screen areas for the presence or absence of unacceptable contamination that can be based on risk-based screening levels, toxicity, or comparisons to a reference or background condition (USEPA 2000c).

Information that should be addressed in the sampling design before collecting the sample includes sample volume (how much material to collect), number of samples, and replication versus composite sampling (USEPA 2001b).

Biological and chemical analyses require specific amounts of sediment (for example, the recommended sediment volume for a 42-day sediment toxicity test with *Hyalella azteca* is 100 ml per replicate [USEPA 2000b]). The required sediment volume per sample location should take into consideration the type and number of analyses as well as the tests that are conducted. The typical amount of sediment needed for a standard acute and chronic whole-sediment toxicity test (assuming one species and eight replicates per sample) is 1 to 2 liters (hereafter, liter is abbreviated as L; milliliter is expressed as ml) of sediment per sample (USEPA 2001b); however, the amount of required sediment may vary considerably depending upon the types of analyses performed. For example, a *Vibrio fischeri* (Microtox™) test requires grams of sediment compared to an ITW assay that requires liters of sediment.

When considering the number of samples to be collected, a better analysis of the areal extent of toxicity generally results when a greater number of sites are sampled. Many programs (such as Superfund) specify the number of samples that must be collected in an area. This must be balanced between the desire to obtain the highest quality data to fully address the project objectives and the constraints imposed by analytical costs, sampling effort, and study logistics (USEPA 2001b). Two approaches that address this issue are the use of replication and compositing. Replication is used to assess the precision of a particular measure (such as separate laboratory analyses on subsamples from the same field sample), and compositing is used to reduce the number of replicates needed for analysis (USEPA 2001b). Compositing refers to combining aliquots (or portions) from two or more samples and analyzing the resulting pooled sample (Keith 1993). Compositing may be a practical, cost-effective way to obtain average sediment characteristics for a particular site. Compositing, however, may dilute the sample if noncontaminated material is combined with contaminated material. If the objective of a study is to define or model physicochemical characteristics of the sediment, it may be important not to composite samples due to model input requirements (EPRI 1999).

Sample collection, processing, transport, and storage

Maintaining the integrity of the collected sediment sample is a major concern in most studies. Disruption of the sediment can change the physical, chemical, and biological characteristics, which may alter contaminant bioavailability and the corresponding toxicity of that sediment. Unfortunately, maintaining the integrity of field-collected sediment during removal, transport, storage, mixing, and testing is extremely difficult. It is virtually impossible to collect sediment samples and remove them from samplers without somewhat altering conditions that control contaminant bioavailability (USEPA 2001b), although some sampling devices are less disruptive than others. It is important to select a sampling technique and apparatus that not only achieve the goals of the study but also minimize changes to the toxicological fraction of the sediment. In sampling efforts, there is a need to balance the sample integrity with the need for efficient collection, processing, transportation, and storage.

There are three main types of sediment sampling devices: core samplers, grab samplers, and dredge samplers. Generally, core and grab samplers are less disruptive than dredge samplers. Core samplers (such as Kajak-Brinkhurst and Phleger) are generally used if (1) deeper sediment characterization is important; (2) one of the goals is to compare deeper, historical sediments to recent surficial sediments; (3) a reduced sediment gradient disruption is required; (4) a reduced oxygen exposure is needed; or (5) sediments are soft and fine grained. Grab samplers (such as Van Veen, Ponar, or Petersen) are typically used if (1) large sediment volumes are needed, (2) larger-grained sediments are common in the study area, or (3) a larger surface area of surficial sediment is needed. Dredge samplers are used primarily to collect benthos, and cause disruption of sediment and ITW integrity, as well as loss of fine-grained sediments. Therefore, only grab and core samplers are recommended for sediment chemistry and toxicity evaluations (USEPA 2001b). Additional information on various samplers, including advantages and disadvantages, has been summarized in USEPA 2001b.

In addition to manipulations that occur during sediment collection, the processing, transportation, and storage of a sample can also affect the bioavailability by introducing contaminants to the sample or by changing its physical, chemical, or biological characteristics. Manipulation processes (such as composite or subsampling) often change the availability of organic compounds by disrupting the equilibrium with organic carbon in an ITW/ sediment system (USEPA 2001b). Similarly, oxidation of anaerobic sediments increases the availability of certain metals (Di Toro et al. 1990; Ankley et al. 1996).Transport and storage methods should be designed, as much as possible, to maintain structural and chemical qualities of sediment and ITW samples. In general, sediments and ITWs contaminated with multiple unknown chemical types should be stored in containers made from high-density polyethylene plastic or polytetrafluoroethylene (PTFE or Teflon®) because these materials are unlikely to add chemical artifacts or interferences and they are much less fragile than glass (USEPA 2001b). All containers should be cleaned prior to filling with samples. Guidance on cleaning new or used sampling containers can be found in Environment Canada (1994), ASTM (2002e), and USEPA (2000b, 2001b). Proper storage conditions should be achieved as quickly as possible after sampling. For example, sediments suspected to be contaminated with organics should be held in brown borosilicate glass containers with PTFE lid liners. The storage condition for most samples is generally either in the dark at 4°C for sediment toxicity analyses or freezing for some chemical analyses of metals and organics (ASTM 2002e). Freezing is not recommended for toxicological analyses. Preferred sample storage times reported for toxicity tests have varied widely (Dillon et al. 1994; Becker and Ginn 1990; Carr and Chapman 1992; Moore et al. 1996; Sarda and Burton 1995; Sijm et al. 1997; Defoe and Ankley 1998), and differences appear to depend primarily on the type or class of contaminants present, similar to storage times for sediment chemical analyses (USEPA 2001b; Ho and Quinn 1993). Considered collectively, these studies suggest that sediment be tested as soon as possible between the time of collection and after eight weeks of storage is appropriate (ASTM 2002b, 2002e; USEPA 2000b, 2001a). Longer storage of sediments that contain high concentrations of labile contaminants (such as ammonia or volatile organics) might lead to loss of these contaminants and a corresponding reduction in toxicity.

Sample manipulation

Manipulation of sediments in the laboratory is often required to achieve certain desired characteristics or forms of material for toxicity and chemical analysis. This can include ITW extraction and sieving (highlighted below), as well as spiking, organic carbon modification, sediment dilution, and elutriate preparation (USEPA 2001b). Generally, manipulation procedures should be designed to maintain sample representativeness for sediment toxicity and sediment chemistry assessment as much as possible. Certain regulatory programs (such as those discussed above) have protocols requiring specific manipulations. For example, the OTM and the ITM specify that if effluent toxicity tests are required, seawater or solvent extractions would be necessary prior to testing. Sieving of sediments is not generally recommended because it can substantially change the physicochemical characteristics of the sediment sample (ASTM 2002e; USEPA 2001b). Day et al. (1995) reported that wet sieving of sediment through fine mesh (with openings of 500 μm or smaller) resulted in a decreased percentage of total organic carbon and a subsequent decrease in concentration of PCBs. This loss may have been due to the PCBs being associated with the fine suspended organic matter that was lost during the sieving process. Sieving can also disrupt the natural chemical equilibrium by homogenizing or otherwise changing the biological activity within the sediment (Environment Canada 1994). In some cases, however, sieving might be necessary to (1) remove foreign materials such as shells, stones, trash, and twigs; (2) increase homogeneity and replicability of samples; (3) remove indigenous organisms prior to toxicity testing; (4) facilitate organism counting, sediment handling, and subsampling; or (5) examine the effects of particle size on toxicity, bioavailability, or contaminant partitioning (ASTM 2002e). If sieving is performed, and the objective of the study is to compare results among stations, it should be done for all samples that will be tested including control and reference sediments (ASTM 2002e).

Also, samples to be used for both chemical analysis and toxicity tests (whole sediment or ITW) should be sieved together, homogenized, and then split for their respective analyses. Additionally, if there is a concern that sieving may affect the outcome of the tests, documenting the effect of sieving by conducting comparative sediment-toxicity tests using sieved and unsieved sediment may be warranted (Environment Canada 1994). Sieving is generally performed by press sieving, where sediment particles are hand-pressed through a sieve using chemically inert paddles, or by wet sieving, which involves swirling sediment particles within a sieve using water to facilitate the mechanical separation of smaller from larger particles. Press sieving is preferable over wet sieving because the use of water during wet sieving dilutes the ITW of the sediment and its chemical constituents (USEPA 2001b).

Extraction of ITW (or porewater) is a common manipulation of sediment. Sediment ITW is defined as the water occupying the spaces between sediment particles. ITW may occupy 50% or more of the volume of a silty sediment, and a general rule of thumb is that 25% to 50% of the sediment volume is extractable as ITW. ITW has relatively high contaminant concentrations due to its intimate contact with contaminated sediment particles and is also the medium by which organisms are exposed to contaminants (along with sediment ingestion). The potential toxicity of sediment-associated non-ionic organic chemicals and divalent metals is often indicated by the amount of the contaminant that is freely available (not bound) in the ITW (Di Toro et al. 1991, 1992; Howard and Evans 1993). Diffusion, bioturbation, and resuspension processes can transport contaminants from ITWs to overlying water (Van Rees et al. 1991). Some investigators have shown that ITW toxicity tests provide increased sensitivity to some toxicants relative to solid-phase tests (Carr et al. 1996, 2000). ITW toxicity tests have also been proven to be useful in sediment toxicity identification evaluation (TIE) studies (Burgess et al. 2000; Carr 1998; Burton et al. 2003), as test procedures and sample manipulations are more established and diverse than solid-phase TIE manipulations (Nipper et al. 2001).

There is no one superior method for the isolation of ITW used for toxicity testing and associated chemical analyses (USEPA 2000b, 2001b). The commonly used methods include filtration, suction, centrifugation, and *in situ* sampling with "peepers" consisting of membrane bags or chambers (Adams 1991; Skalski et al. 1990). Factors to consider in the selection of an isolation procedure may include (1) volume of ITW needed, (2) ease of isolation (materials preparation time and time required for isolation), and (3) artifacts in the ITW caused by the isolation procedure. Each approach has unique strengths and limitations (Bufflap and Allen 1995a, 1995b; Winger et al. 1998) that vary with sediment characteristics, chemicals of concern, toxicity test methods, and the data quality objectives (DQOs) (USEPA 2000b, 2001b). Of the various laboratory methods available, the most commonly recommended is centrifugation (Environment Canada 1994), as this method has been shown to alter the ITW chemistry the least (Ditsworth et al. 1990). Additionally, USEPA recommends that any removal method should be performed without filtration, as filtration removes toxicity in a nonspecific manner (USEPA 1992b). With centrifugation, the force used is important, as small-sized clays and colloids that bind toxicants may not be easily removed. Ho (1997) used double centrifugation to remove finer particles. In this procedure the whole sediment is spun at 5 *G* for 30 minutes. The ITW is then removed and separately spun between 8 to 10 *G* for an additional 30 minutes.

As it is important to maintain the integrity of the sample (minimizing the changes to the *in situ* condition of the water and thereby minimizing the potential alteration of the contaminant bioavailability and toxicity of the sample), *in situ* methods may be superior to *ex situ* methods for collecting ITW. This is due to the fact that *in situ* methods are less subject to sampling-related and extraction-related artifacts and therefore may be more likely to maintain the chemical integrity of the sample (Sarda and Burton 1995; ASTM 2002e; Nipper et al. 2001). However, *in situ* methods have generally produced relatively small volumes of ITW, and often are limited to wadeable or diver-accessible depths (USEPA 2001b). More information on isolation of ITW through both *in situ* and *ex situ* methods can be found in USEPA (2001b).

Recommended procedures for both freshwater and marine test organisms

Currently, a wide range of toxicity tests and test organisms exist, ranging from phytoplankton to worms to bacteria. One should consider the organism's selectivity, sensitivity, appropriateness, preferred test matrix, acceptance levels, and the objectives of the test program when determining an appropriate suite of test organisms.

It is generally accepted that a battery of assays or organisms is appropriate for screening purposes (Adams et al. 2003; Ho 1997; Luoma and Ho 1993), although one needs to balance a large number of test organisms with resource constraints. Because of inherent differences among organisms, a situation rarely exists in which a single test organism can give all the necessary information. Exceptions may include endangered species, specific surrogates for endangered species, specialized environments that have a limited number or type of species (such as brine ponds), or if a specific organism has proven to be an appropriate surrogate for an organism of interest from the test environment.

The objective of developing a suite of assays is to detect possible risks to all organisms. Therefore, a reasonable approach may be to choose organisms from different phyla that may be representative of differences among the phyla, different niches, or different feeding habits that may result in different exposures. However, while choosing organisms representative of different phyla or niches may be the goal, the reality is that there is a limited number of standardized tests among these different phyla.

There has been some discussion of organism appropriateness in terms of habitat and in terms of local species (Nipper et al. 2001). It is not usually the intention of a regulator to protect a specific test amphipod or crustacean; it is assumed that the test organism acts as a surrogate for other organisms or groups of organisms that the regulator is interested in protecting. Given that test organisms are surrogates; it is not uncommon to substitute a pelagic species for a benthic species or vice versa. The theoretical basis for the use of surrogates is that all organisms have similar cellular enzymatic and response systems, and all organisms share similar DNA.

Despite the accepted use of surrogates, there are several reasonable justifications for the use of local species. Local species may have specific sensitivity or selectivity to certain test substances. They are often more easily field collected (as opposed to being purchased and shipped from other areas) and their choice may be more easily justified to the public. Conversely, local species generally lack the standardization of nationally accepted and used test organisms, so results may not be easily compared among programs. Local species may not be as sensitive as other test organisms, and may be present in the sample collected from the field, which can confound results if the sediment sample is not carefully screened.

Often the choice of test organisms is limited by the test matrix. For sediments, the test matrix may be whole sediment, ITW, or elutriates. Although some organisms can be tested in either whole-sediment or aqueous phase (such as certain amphipods including *Ampelisca abdita, Leptocheirus plumulosus, and H. azteca*), requirements of the organism usually include a specific type of matrix (some organisms need a substrate). Typically, the test organism and matrix chosen are dependent on the question being asked. For example, if the concern is about organisms that have more ITW exposure, then an appropriate test could be an ITW toxicity test with a free-burrowing amphipod. When considering ITW tests, a limitation is often the volume of water needed per test replicate. While it is theoretically possible to test a fish that requires 150 ml of test water per replicate, it is often operationally impossible to obtain enough ITW for a replicated dose-response curve. In addition, once ITW is removed from the whole-sediment matrix it becomes relatively unstable (Adams et al. 2001). Changes in the oxidation state can affect the physical-chemical properties of the ITW. Most notably, Fe^{+2}, which is relatively stable and soluble in anoxic ITW, may result in $Fe(OH)_3$ precipitates (a reddish-orange solid) in oxidized ITW. The precipitate will decrease the pH of the ITW (pH of less than 3) by binding hydroxides. Toxicity will occur due to the low pH regardless of the presence of toxic compounds. In addition, $Fe(OH)_3$ may also cause coprecipitation of other compounds. In addition to the change that may occur in metal toxicity due to oxidation of the ITW, sorption of organic toxicants to the sides of test containers may cause a researcher to underestimate the amount of toxicity due to organics. In whole-sediment exposures, the concentration of organics in the ITW remains relatively constant through equilibrium processes; however in an ITW exposure that equilibrium is changed due to the absence of the whole-sediment matrix. Finally, the ITW tests remove any direct or dietary contact the organism may have in a whole-sediment test. While there may be specific reasons to perform ITW testing (see above), whole-sediment exposures are more realistic than ITW tests. Regardless of the question, the organism should have some contact with the sediment or ITW; for example, a fish assay with bedded sediment would not be a sensitive test, simply because there is a limited route of exposure for the organism. Elutriates are normally used in specialized circumstances, such as determining potential exposure during a dredged-material disposal operation.

Researchers can also choose between static and flow-through tests. Both static and flow-through exposures have advantages and disadvantages. Static tests are generally easier to initiate and maintain; a source of clean, running water is not needed, and they simulate field conditions where both sediment and water column are contaminated. Conversely, static tests may overestimate the toxicity of sediments to epibenthic and water-column organisms if overlying water toxicant concentrations begin to resemble the generally higher ITW concentrations. Flow-through toxicity tests are generally more difficult to perform because of the need for a source of clean, running water. These tests also may flush soluble toxicants such as ammonia, metals, and hydrogen sulfide out of the test system (Ankley et al. 1993). However, flow-through tests may provide a better simulation of a field condition where sediment is contaminated but the water column is clean (for example if the sediments are contaminated but the source has been removed or remediated). Finally, flow-through tests may underestimate the toxicity of sediments to organisms that have water-column exposure by removing the overlying water source of exposure.

The number of replicates also depends upon the question asked. Duplicates or even single replicates may be enough for a screening test, whereas a definitive test may require four or more replicates (ASTM 2002b; Environment Canada 1994). For the most part, screening-level toxicity tests are designed to avoid classifying samples as nontoxic when in fact they are toxic (minimizing false negatives).

The level of test acceptance ranges from experimental to highly proscribed. For large, highly visible field programs, one might want to choose test assays with a high level of acceptance, such as those outlined in ASTM, USEPA, or Environment Canada guidelines. This level of acceptance means that the assay has undergone a reasonable level of testing and that data exist on test sensitivity. There may have been round-robin tests so that the data is reproducible in a number of laboratories, and the test is not likely to be useful only in limited situations.

Finally, the choice of test(s) should be appropriate for the objectives of the assessment program. If the objective is to determine if the sediments are genotoxic, assays with genotoxic endpoints should be chosen. If the objective is to assess the chronic toxicity of a sediment, tests with an acute toxicity endpoint or short duration should not be used.

Rather than list the details of each assay here, Table 1.1 summarizes some of the more widely used sediment toxicity tests, including how the organisms can be obtained, the test endpoint, volume of sediment or interstitial water needed, test duration, references for standard methods, and examples of how the test is used in the literature. This table is not inclusive of all assays that can be performed in sediments, nor is every test listed appropriate for every situation. Again, a suite of assays should be chosen that will answer the objectives of the assessment, and that has the appropriate sensitivity, compatibility with the chosen matrix, and level of acceptance.

Interpretation

Laboratory versus field exposures: what is the ecological relevance?

As sediment toxicity tests are essentially surrogates for assessing sediment conditions in the field, it is important to evaluate the ability of laboratory toxicity tests to estimate the effects on benthic populations in the field. For the sake of discussion in this chapter, we will consider field results to be benthic community indices. While the reliability, accuracy, and precision of these indices as a barometer for benthic community effects from anthropogenic sources is in itself a topic for discussion (Canfield et al. 1996; Johnson et al. submitted), that discussion is beyond the scope of this chapter.

The comparison between laboratory results and effects in the field is not always straightforward, as there are a variety of reasons that test results may differ between the laboratory and the field. Factors that may decrease the toxicity of a laboratory test relative to field effects include consumption of contaminated food in the field that would increase body burdens of a contaminant, and delayed or impaired organismal defense or escape mechanisms. Factors that may increase the toxicity of a laboratory test relative to field effects include organism stress during testing, starvation during the test period, and behavioral mechanisms that may occur in the field but not in the lab, such as movement away from a contaminated site. Further, laboratory test design may overestimate or underestimate the response seen in the field. For example, static toxicity tests (in which waters are not renewed) may increase the exposure of organisms to toxicants because of the equilibrium established between the overlying water and sediments (Ankley et al. 1993). Conversely, flow-through tests may flush toxicants from a system (Ferretti et al. 2000).

In addition, changes in the exposure routes of contaminants may result in differences between laboratory and field results. For example, if a tube-dwelling organism has a field exposure to overlying water and ITW, and is then placed in a 100% ITW test, the exposure of the organism to contaminants may be much higher. Also, changes in geochemical factors during sample collection, transport, and storage (see above) that affect bioavailability (such as pH or redox) may result in a change in toxicity relative to the field. While the myriad of results from these changes are unpredictable,

Table 1.1 Characteristics of Some Common Sediment Toxicity Tests

Species (marine)	Collection	Endpoint	Sediment volume or water tested/replicate	Phase tested/duration (ITW = interstitial water)	Standardization	References
Rhepoxynius abronius (free-burrowing amphipod)	Field collected	Acute toxicity	Sediment: 2 cm depth or 175 ml	Whole sediment 10 d	Yes (ASTM 2002c; Environment Canada 1998b; USEPA 1994a)	Swartz et al. 1985, 1989
Ampelisca abdita (tube-dwelling amphipod)	Field collected	Acute toxicity	Sediment: 4 cm depth or 175 ml; ITW: 10–20 ml/5 organisms	Whole sediment 10 d; ITW 96 h	Yes (ASTM 2002c; USEPA 1994a, 1996)	Scott and Redmond 1989; Ho et al. 2000
Eohaustorius estuaries (free-burrowing amphipod)	Field collected	Acute toxicity	Sediment: 2 cm depth or 175 ml/1-l beaker; ITW: 10–20 ml/5 organisms	Whole sediment 10 d; ITW 4–10 d	Yes for whole sediment (ASTM 2002c; Environment Canada 1998b; USEPA 1994a); no for ITW	DeWitt et al. 1989; Thompson et al. 1999
Grandidierella japonica (free-burrowing amphipod)	Field collected	Acute toxicity	Sediment: 2 cm depth or 175 ml/1-l beaker	Whole sediment 10 d	Yes (ASTM 2002c)	Nipper et al. 1989; Hong and Reish 1987
Leptocheirus plumulosus (burrow-dwelling amphipod)	Cultured or field collected	Acute toxicity, partial life cycle tests	Sediment: 4 cm depth or 175 ml sediment/1-l beaker; ITW: 10–20 ml/5 organisms	Whole sediment 10 and 28 d; ITW 4–10 d	Yes for whole sediment (ASTM 2002c; USEPA 1994a, 2001a); no for ITW	McGee et al. 1993; Schlekat et al. 1992

Species (marine)	Collection	Endpoint	Sediment volume or water tested/replicate	Phase tested/duration (ITW = interstitial water)	Standardization	References
Corophium volutator (tube-dwelling amphipod)	Field collected	Acute toxicity	Sediment: 4 cm depth or 175 ml/1-l beaker; ITW: 20 ml/10 organisms	Whole sediment 10 d; ITW 72 h	Yes for whole sediment (Schipper and Stronkhorst 1999 [Dutch Government]); No for ITW	Stronkhorst et al. 2003; Hyne and Everett 1998
Haliotis rufescens (abalone)	Adult field collected; spawned in laboratory	Embryo-larval development	5-10 ml	ITW 48 h	Yes (USEPA 1995, 1996)	Hunt and Anderson 1989, 1993
Merceneria merceneria (hard-shell clam)	Purchased or cultured	Juvenile growth	Sediment: 20-50 ml	Whole sediment	No	Ringwood and Charles 1998, 2002
Crassostrea sp. (oyster)	Adult field collected or purchased	Embryo-larval development	Sediment: 20g; ITW 10-20 ml	Whole sediment[a]; ITW 48 h	Yes (ASTM 2002f; USEPA 1995, 1996)	Chapman and Morgan 1983; Martin et al. 1981
Mytilus sp. (mussel)	Adult field collected or purchased	Embryo-larval development	Sediment: 20g; ITW: 10-20 ml	Whole sediment[a]; ITW 48 h	Yes (ASTM 2002f; USEPA 1995, 1996)	Martin et al. 1981; Hunt and Anderson 1993
Strongylocentrotus purpuratus (sea urchin)	Adult field collected	Fertilization; embryo-larval development	Sediment: 20g; ITW:10-20 ml	Particle-free ITW for fertilization test 1 h; whole sediment[a] or ITW for embryo-larval development 48 h	Yes (ASTM 2002g; USEPA 1995, 1996; Environment Canada 1992a)	Hunt et al. 2001; Anderson et al. 2001

-- continued

Table 1.1 (continued) Characteristics of Some Common Sediment Toxicity Tests.

Species (marine)	Collection	Endpoint	Sediment volume or water tested/replicate	Phase tested/duration (ITW = interstitial water)	Standardization	References
Dendraster excentricus (sand dollar)	Adult field collected	Fertilization; embryo-larval development	Sediment: 20g; ITW: 10–20 ml	Particle-free ITW for fertilization test 1 h; whole sediment[a] or ITW for embryo-larval development 48 h	Yes (ASTM 2002g; USEPA 1995, 1996; Environment Canada 1992a)	Bay et al. 1993; Bailey et al. 1995b
Arbacia punctulata (sea urchin)	Adult field collected	Fertilization; embryo-larval development	Sediment: 20g; ITW: 10–20 ml	Particle-free ITW for fertilization test 1 h; whole sediment[a] or ITW for embryo-larval development 48 h	Yes (ASTM 2002g; USEPA 1994c, 1996; Environment Canada 1992a)	Nacci et al. 1986; Bay et al. 1993
Dinophilus gyrociliatus (polychaete worm)	Field collected or cultured	Acute toxicity; reproductive impairment	10 ml	ITW 4–10 d	Yes (ASTM 2002h)	Carr et al. 1986, 1989
Neanthes arenaceodentata (polychaete worm)	Field collected or cultured	Acute toxicity; growth	Sediment: 2–3 cm depth	Whole sediment 10–28 d	Yes (ASTM 2002d)	Guilherme et al. 2001; Pesch et al. 1995
Americamysis bahia (mysid)	Cultured or purchased	Acute toxicity; life-cycle tests	Whole sediment: 20–50 ml; ITW: 10–20 ml	Whole sediment, ITW 4- to 10-d survival	Yes (ASTM 2002i; USEPA 1996)	Ho et al. 2000; Cripe et al. 2000
Vibrio fischeri (bacterial luminescence [Microtox])	Purchased	Change in bioluminescence	Whole sediment: 7 g; ITW: 2 ml	Whole sediment; ITW 15 min	Yes (Environment Canada 1992b)	Svenson et al. 1996; Tay et al. 1992

Species (marine or freshwater)	Collection	Endpoint	Sediment volume or water tested	Phase tested/duration	Standardization	References
Daphnia magna (water flea)	Cultured or purchased	Acute toxicity; partial life cycle tests	ITW: 10–25 ml	ITW acute toxicity 48–96 h	Yes (USEPA 1992b, 1993b; Environment Canada 2000)	Prater and Anderson 1977; Giesy et al. 1988
Ceriodaphnia dubia (water flea)	Cultured or purchased	Acute toxicity; partial life cycle tests	ITW: 10–20 ml	ITW acute toxicity 48–96 h	Yes (USEPA 1992b, 1993b; Environment Canada 1992c)	Burton et al. 1996; Boucher and Watzin 1999
Hyalella azteca (amphipod)	Cultured or purchased	Acute toxicity; growth, reproduction partial life cycle tests	Sediment: 100 ml; ITW: 10–20 ml	Whole sediment 10 and 42 d; ITW 48–96 h	Yes (ASTM 2002b; Environment Canada 1997a; USEPA 1994b, 2000b)	Schubauer-Berigan, et al. 1993; Call et al. 2001a, 2001b
Chironomus tentans (midge)	Cultured or purchased	Acute toxicity; partial life cycle tests	Sediment: 100 ml; ITW: 10–20 ml	Whole sediment 10 and 65 d; ITW with screen 48–96 h	Yes (ASTM 2002b; Environment Canada 1997b; USEPA 1994b, 2000b)	Call et al. 2001a, 2001b
Chironomus riparius (midge)	Cultured or purchased	Acute toxicity; partial life cycle tests	Sediment: 100 ml; ITW: 10–20 ml.	Whole sediment acute toxicity 10 and 30 d; ITW 48–96 h	Yes (ASTM 2002b; Environment Canada 1997b)	Pittinger et al. 1989; Kemble et al. 1994
Hexagenia spp. (mayfly nymph)	Field collected or purchased	Acute toxicity; partial life cycle tests	Sediment: 200–325 ml; ITW: 50 ml	Whole sediment 10 and 21 d; ITW 7 d with artificial burrows	Yes (ASTM 2002b)	Prater and Anderson 1977; Giesy et al. 1990

[a]Test procedure requires sediment to be mixed with overlying water and then allowed to settle before organisms are added. An alternate method of testing uses a sediment-water interface holding chamber (Anderson et al. 1996, 2001). In general these organisms are small, and if they are allowed to have direct contact with sediment particles, they may be difficult to recover.

Adapted from whole sediment methods outlined in USEPA 2000b and ASTM 2002b.

the likelihood is that some change will occur. Finally, factors such as lag time between the toxic and field effects, insensitivity of relatively short-term (acute toxicity) tests to measure field situations that may be dominated by chronic toxicity, benthic patchiness, ability of researchers to detect changes in benthic community composition, and the sensitivity of field measures to factors other than toxicants (such as grain size or flow) may all contribute to differences between laboratory and field measures.

Despite these issues, there is often a reasonable correlation of effects between laboratory toxicity tests and field (benthic index) results. Swartz et al. (1982, 1994) has shown good concordance between amphipod toxicity test results of *Eohaustorius estuarius, Rhepoxynius abronius,* and *H. azteca* and field populations of amphipods in contaminated sediment. Schlekat et al. (1994) compared benthic macroinvertebrate community structure with the 28-day *H. azteca* toxicity test results and concluded that the two measures were generally in good agreement. Canfield et al. (1994, 1996, 1998) compared results of benthic community assessments to sediment chemistry and toxicity (28-d sediment exposures with *H. azteca* that monitored effects on survival, growth, and sexual maturation) and reported that good concordance occurred in benthic community structure, toxicity tests, and sediment chemistry measures in Clark Ford River, MT, the Great Lakes Areas of Concern (known as AOC, in Buffalo River, NY, Indiana Harbor, IN, and Saginaw River, MI), and the upper Mississippi River. Good concurrence of benthic community structure, toxicity tests, and sediment chemistry measures was evident in very contaminated sediments, although less concurrence was observed in moderately contaminated sediments. Canfield et al. (1994, 1996, 1998) also concluded that laboratory sediment toxicity tests better identified chemical contamination in sediments when compared with many commonly used measures of benthic invertebrate community composition. This was largely because benthic measures reflect other field factors such as habitat alteration and excessive nutrients in addition to responding to contaminants. They also stressed the importance of evaluating noncontaminant factors (total organic carbon [TOC], grain size, water depth, and habitat alteration) in order to better interpret the response of benthic invertebrates to sediment contamination.

Bailey et al. (1995a) concluded that a strong relationship existed between community structure (as measured by 15 major benthic taxa) and sediment toxicity (as measured by eight bioassays) in the Great Lakes. Long and Chapman (1985) concluded that the concordance between sediment chemistry, toxicity, and macroinfaunal condition was good in Puget Sound, WA.

In a synoptic study of sediment chemistry, toxicity, and macroinfaunal condition in estuaries in the Southeastern U.S., Hyland et al. (1998) reported data that indicated a 66% to 76% concurrence (agreement between degraded benthos and sediment chemistry or toxicity tests, or agreement between healthy benthos without contamination or toxicity). Based upon four years of Environmental Monitoring and Assessment (EMAP) data (http://www.epa.gov/emap/), USEPA calculated that 66% of all stations had agreement between degraded benthos and toxicity, or healthy benthos and no toxicity.

An evaluation of the distribution of *L. plumulosus* in Chesapeake Bay, MD, indicates that its distribution is negatively correlated with the degree of sediment contamination (Pfitzenmeyer 1975; Reinharz 1981). A field validation study of the 10-day and 28-day *L. plumulosus* tests by McGee and Fisher (1999) in Baltimore Harbor, MD, indicated good agreement between acute toxicity, sediment-associated contaminants, and responses of the *in situ* benthic community.

Nevertheless, other studies show less concordance and more of a complementary role for toxicity tests and benthic measures (Burton 1989; Chapman et al. 1987). The complementary role of the two measures may be more realistic because they respond to different factors. When differences between toxicity testing and field benthic community analyses occur, toxicity tests are most likely a better indicator of anthropogenic toxic inputs, as the benthic analyzes may respond to field factors other than contaminants (such as flow and grain size) (Canfield et al. 1996).

Future research recommendations

Toxicity assays are a critical and useful tool within clearly defined sediment-assessment objectives. Assays measure biological responses under controlled conditions, and the information gained from them cannot be replaced by either chemical or field measures. In order to be relevant to the assessment, the toxicity endpoints and the sediment-assessment objectives must agree. A clearly defined approach for assessment of assay results within a weight-of-evidence approach is laid out in Grapentine et al. (2002). In addition, a single assay should not be used, but should instead be incorporated within a suite of assays that includes different phyla, feeding, or exposure regimes. The USEPA and USACE recommend a minimum of three species that encompass three different life-history strategies to evaluate sediment toxicity (USACE/USEPA 1991). Therefore, protocols using new test species should be developed to provide sensitive tests (with both lethal and sublethal endpoints), representing a greater range of species and habitat types for both freshwater and marine organisms.

Recommendations to improve the utility of assays in sediment assessments include further demonstration of the relevance of assays to field effects, as was done with amphipod testing by Swartz et al. (1982, 1994). In addition, development and standardization of *in situ* toxicity test methods (Chappie and Burton 2000) may increase their utility in deep and high-energy aquatic systems. Further development of the diagnostic aspects of assays, either as increased understanding of the sensitivity and selectivity of the assays themselves, or within a structured approach such as a TIE framework, would increase their utility (USEPA 1992b). Assays with diagnostic utility would bridge the gap to the next step of sediment assessment, which is source or stressor identification and remediation of impaired sediments.

Summary

Sediment toxicity assays are an integral part of sediment assessments and are often performed for regulatory purposes. Numerous regulations for sediment assessments exist throughout the world and this chapter highlights a few of these regulations and frameworks (sediment remediation and dredged material disposal in the U.S., and dredged material disposal in Canada and Australia). Generally, a suite of assays is recommended and testing proceeds in a tiered fashion. Before sampling occurs, the type, quantity, and quality of data needed as well as the sample design should be chosen to ensure that sediment-assessment objectives are achieved.

Recommendations for selecting appropriate sampling designs given specific objectives are included, as well as procedures for sample collection. Developing the appropriate suite of toxicity test assays depends largely upon the objectives of the sediment assessment. A number of sediment toxicity tests exist for both freshwater and marine sediments, and a brief description, appropriate matrix (ITW or whole sediment), and references to methodological details of widely used sediment assays are included. The ecological significance of sediment toxicity assays is discussed and a number of factors that explain the differences in results between benthic index measures and toxicity tests are outlined. Despite these differences, several studies have demonstrated good concordance between laboratory toxicity tests and benthic measures.

Recommendations to improve the use of toxicity assays within a sediment-assessment program include (1) having clearly defined assessment objectives and matching toxicity test endpoints to those objectives within a weight-of-evidence approach, (2) further demonstrating the relevance of assays to field effects, (3) developing more *in situ* test assays, and (4) further developing the diagnostic aspect of assays, either within the assay itself or within a toxicity identification and evaluation framework.

Acknowledgments

The authors would like to acknowledge the contributions of Dr. Chris Ingersoll, Dr. Warren Boothman, Peg Pelletier, Denise Champlin, and an anonymous reviewer for conducting a thorough peer review of this manuscript.

This paper has been technically reviewed by the Atlantic Ecology Division at the U.S. Environmental Protection Agency (AED 03-036); however, it has not been subjected to an USEPA-wide peer review and therefore does not necessarily reflect USEPA policy. Any mention of trade names does not constitute endorsement by the U.S. Environmental Protection Agency or the federal government.

References

Adams, D.D. 1991. Sampling sediment pore water, in CRC *Handbook of Techniques for Sediment Sampling*. Mudroch, A and MacKnight, S.D. (eds.). CRC Press, Inc., Boca Raton, FL.

Adams, W., Burgess, R.M., Gold-Bouchot, G., LeBlanc, L., Liber, K. and Williamson B. 2001. Porewater chemistry: effects of sampling, storage, handling and toxicity testing. R.S. Carr and M. Nipper eds. Summary of a SETAC technical workshop: Porewater Toxicity Testing: Biological, Chemical and Ecological Considerations with a Review of Methods and Applications, and Recommendations for Future Areas of Research. 18–22 March, 2000; Pensacola, FL. Society of Environmental Toxicology and Chemistry (SETAC). Pensacola, FL.

Adams, W., Green, A., Ahlf, W. Burton, A., Brown, B., Chadwick, B., Crane, M., Gouget, R., Ho, K.T., Reynoldson, T., Savitz, J.D., Sibley, P.K., eds. 2003. *Utilizing Sediment Assessment Tools and a Weight of Evidence Approach*. SETAC Press. Pensacola, FL

ASTM (American Society for Testing and Materials). 2002a. Standard guide for determination of the bioaccumulation of sediment-associated contaminants by benthic invertebrates, E1688-00a., in *Annual Book of ASTM Standards*, Vol. 11, West Conshohocken, PA.

ASTM. 2002b. Standard test methods for measuring the toxicity of sediment-associated contaminants with freshwater invertebrates, E1706-00, in *Annual Book of ASTM Standards*, Vol. 11, West Conshohocken, PA.

ASTM. 2002c. Standard guide for conducting 10-d static sediment toxicity tests with marine and estuarine amphipods, E1367-99, in *Annual Book of ASTM Standards*, Vol. 11, West Conshohocken, PA.

ASTM. 2002d. Standard guide for conducting sediment toxicity tests with polychaetous annelids, E1611-00, in *Annual Book of ASTM Standards*, Vol. 11, West Conshohocken, PA.

ASTM. 2002e. Standard guide for collection, storage, characterization, and manipulation of sediment for toxicological testing, E1391-02, in *Annual Book of ASTM Standards*, Vol. 11, West Conshohocken, PA.

ASTM. 2002f. Standard guide for conducting acute toxicity tests on test materials with fishes, macroinvertebrates and amphipods: E729-96, in *Annual Book of ASTM Standards*. Vol. 11.05. West Conshohocken, PA.

ASTM. 2002g. Standard guide for conducting static acute toxicity tests with echinoid embryos: E1563-98, in *Annual Book of ASTM Standards*. Vol. 11.05. West Conshohocken, PA.

ASTM. 2002h. Standard guide for conducting acute, chronic and life-cycle aquatic toxicity tests with polychaetous annelids: E1562-00, in *Annual Book of ASTM Standards*. Vol. 11.05. West Conshohocken, PA.

ASTM. 2002i. Standard guide for conducting life-cycle toxicity tests with saltwater mysids: E1191-97, in *Annual Book of ASTM Standards*. Vol. 11.05. West Conshohocken, PA.

Anderson, B.S., Hunt, J.W., Hester, M. Phillips, B.M.. 1996. Assessment of sediment toxicity at the sediment-water interface. G K. Ostrander, ed., *Techniques in Aquatic Toxicology*. Lewis Publishers. Ann Arbor, MI. 609–624.

Anderson, B.S., Hunt, J.W., Phillips, B.M., Fairey, R., Puckett, H.M., Stephenson, M., Taberski, K., Newman, J., Tjeerdema, R.S.. 2001. Influence of sample manipulation on contaminant flux and toxicity at the sediment-water interface. *Marine Environmental Research* 51, 191–211.

Ankley, G.T., Benoit, D.A., Hoke, R.A., Leonard, E.N., West, C.W., Phillips, G.L., Mattson, V.R., Anderson, L.A., 1993. Development and evaluation of test methods for benthic invertebrates and sediments: effects of flow rate and feeding on water quality and exposure conditions. *Arch. Environ. Contamination Toxicol.* 25, 12–19.

Ankley, G.T., Di Toro, D.M., Hansen, D.J., Berry, W.J., 1996. Technical basis and proposal for deriving sediment quality criteria for metals. *Environ. Toxicol. Chem.* 15, 2056–2066.

ANZECC/ARMCANZ (Australian and New Zealand Environment and Conservation Council/Agriculture and Resource Management Council of Australia and New Zealand). 2000. *Australian and New Zealand Guidelines for Fresh Water and Marine Water Quality*, National Water Quality Management Strategy.

API (American Petroleum Institute). 1994. *User's guide and technical resource document: evaluation of sediment toxicity tests for biomonitoring programs.* API pub. no. 4607. Prepared for American Petroleum Institute, Health and Environmental Sciences Department, Washington, D.C., by PTI Environmental Services, Bellevue, WA.

Bailey, H.C., Miller, J.L., Miller, M.J., Dhaliwal, B.S.. 1995b. Application of toxicity identification procedures to the echinoderm fertilization assay to identify toxicity in a municipal effluent. *Environ. Toxicol. Chem.* 14: 2181–2186.

Bailey, R.C., Day, K.E., Norris, R.H., Reynoldson, T.B. 1995a. Macroinvertebrate community structure and sediment bioassay results from nearshore areas of North American Great Lakes. *J. Great Lakes Res.* 21, 42–52.

Bay, S.O., Burgess, R., Nacci, D.. 1993. Status and applications of echinoid (Phylum Echinodermata) toxicity test methods. W.G. Landis, J.S. Hughes and M.A. Lewis, eds. *Environmental Toxicology and Risk Assessment*, ASTM STP 1179. American Society for Testing and Materials. Philadelphia. 281–302.

Becker, D.S. and Ginn, T.C. 1990. *Effects of sediment holding time on sediment toxicity*, EOA 910/9-90-009. Prepared by PTI Environmental Services for the USEPA, Region 10 Office of Puget Sound, Seattle, WA.

Bishop, C.A., Brooks, R.J., Carey, J.H., Ng, P., Norstrom, R.J., Lean, D.R.S. 1991. The case for a cause-effect linkage between environmental contamination and development in eggs of the common snapping turtle (*Chelydra s. serpentina*) from Ontario, Canada. *J. Toxicol. Environ. Health* 33, 521–547.

Bishop, C.A., Koster, M.D., Chek, A.A., Hussell, D.J.T. Jock, K. 1995. Chlorinated hydrocarbons and mercury in sediments, red-winged blackbirds (*Agelaius Phoneniceus*), and tree swallows (*Tachycineta bicolor*) from wetlands in the Great Lakes–St. Lawrence River Basin. *Environ. Toxicol. Chem.* 14, 491–501.

Bishop, C.A., Ng, P., Pettit, K.E., Kennedy, S.W., Stegeman, J.J., Norstrom, R.J., Brooks, R.J. 1998. Environmental contamination and development abnormalities in eggs and hatchlings of the common snapping turtle (*Chelydra s. serpentina*) from the Great Lakes–St. Lawrence River Basin (1989-91). *Environ. Pollut.* 101, 143–156.

Bishop, C.A., Mahony, N.A., Trudeau, S., Pettit, K.E. 1999. Reproductive success and biochemical effects in tree swallows (*Tachycineta bicolor*) exposed to chlorinated hydrocarbon contaminants in wetlands of the Great Lakes and St. Lawrence River Basin, USA and Canada. *Environ. Toxicol. Chem.* 18, 263–271.

Bonin, J., Des Granges, J.L., Bishop, C.A., Rodrigue, J., Gendron, A. 1995. Comparative study of contaminants in the mudpuppy (Amphibia) and the common snapping turtle (Reptilia), St. Lawrence River, *Canada. Arch. Environ. Contamination Toxicol.* 28, 184–194.

Boucher, A.M. and Watzin, M.C. 1999. Toxicity identification evaluation of metal-contaminated sediments using an artificial pore water containing dissolved organic carbons. Environ. Toxicol. Chem. 18: 509–518.

Bufflap, S.T. and Allen, H.E. 1995a. Sediment pore water collection methods: A review. *Water Res.* 29, 165–177.

Bufflap, S.T. and Allen, H.E. 1995b. Comparison of pore water sampling techniques for trace metals. *Water Res.* 29, 2051–2054.

Burgess, R.M., Cantwell, M.G., Pelletier, M.C., Ho, K.T., Serbst J.R., Cook, H.F., Kuhn, A. 2000. Development of a toxicity identification evaluation procedure for characterizing metal toxicity in marine sediments. *Environ. Toxicol. Chem.* 19, 982–991.

Burton, G.A., Jr. 1989. Evaluation of seven sediment toxicity tests and their relationship to stream parameters. *Toxicity Assessment* 4, 149–159.

Burton, G.A., Jr., Nelson, J.K., Ingersoll, C.G. 1992. Freshwater benthic toxicity tests, in G.A. Burton Jr., (Ed.), *Sediment toxicity assessment*, 213-240, Lewis Publishers, Chelsea, MI.

Burton, G.A., Jr.,. Ingersoll, C.G, Burnett, L.C., Henry, M., Hinman, M.L., Klaine, S.J., Landrum, P.F., Ross, P., Tuchman, M. 1996. A comparison of sediment toxicity test methods at three Great Lakes areas of concern. *J. Great Lakes Res.* 22:495–511.

Burton, G.A., Jr., Denton, D., Ho, K.T., Ireland, D.S. 2002. Sediment toxicity testing, issues and methods in quantifying and measuring ecotoxicological effects, in D. Hoffman, D. Rattner, G.A. Burton, Jr., J.J. Cairns, (Eds.), *Handbook of ecotoxicology*, CRC/Lewis Publishers, Boca Raton, FL.

Burton, G.A., Jr., Rowland, C.D., Greenberg, M.S., Lavoie, D.R., Nordstrom, J.F., Eggert, L.M. 2003. A tiered, weight-of-evidence approach for evaluating aquatic ecosystems, in M. Kumangai and M. Munawar, (Eds.), *Ecovision Monograph* series. In press.

Call, D.J., Markee, T.P., Geiger, D.L., Brooke, L.T., VandeVenter F.A., Cox, D.A., Genisot, K.I., Robillard, K.A., Gorsuch, J.W., Parkerton, T.F., Reiley, M.C., Ankley, G.T., Mount, D.R. 2001a. An assessment of the toxicity of phthalate esters to freshwater benthos. 1. Aqueous exposures. *Env. Toxicol. Chem.* 20: 1798–1804.

Call, D. J., Cox, D.A., Geiger, D.L., Genisot, K.I., Markee T.P., Brooke, L.T., Polkinghorne, C.N., VandeVenter, F.A., Gorsuch, J.W., Robillard, K.A., Parkerton, T.F., Reiley, M.C., Ankley, G.T., Mount, D.R. 2001b. An assessment of the toxicity of phthalate esters to freshwater benthos. 2. Sediment exposures. *Env. Toxicol. Chem.* 20: 1805–1815.

Canfield, T. J., Kemble, N.E., Brumbaugh, W.G., Dwyer, F.J., Ingersoll, C.G., Fairchild, J.F. 1994. Use of benthic invertebrate communities and the sediment quality triad to evaluate metal-contaminated sediment in the upper Clark Fork River, Montana. *Environ. Toxicol. Chem.* 13, 1999–2012.

Canfield, T.J., Dwyer, F.J., Fairchild, J.F., Haverland, P.S., Ingersoll, C.G. Kemble, N.E., Mount, D.R., La Point, T.W., Burton, G.A., Jr., Swift, M.C. 1996. Assessing contamination in the Great Lakes sediments using benthic invertebrate communities and the sediment quality triad approach. *J. Great Lakes Res.* 22, 565–583.

Canfield, T.J., Brunson, E.L., Dwyer, F.J., Ingersoll, C.G., Kemble, N.E. 1998. Assessing sediments from the Upper Mississippi river navigational pools using a benthic community invertebrate evaluation and the sediment quality triad approach. *Arch. Environ. Contamination Toxicol.* 35, 202–212.

Carr, R.S., Curran, M.D., Mazurkiewicz, M. 1986. Evaluation of the archiannelid Dinophiltis gyrociliatits for use in short term, life-cycle toxicity tests. *Environ. Toxicol. Chem.* 5, 703.

Carr, R.S., Williams, J.W. Fragata, C.T.B. 1989. Development and Evaluation of a Novel Marine Sediment Pore Water Toxicity Test with the Polychaete Dinophilus Gyrociliatus. Environ. Toxicol. Chem. 8, 533–543.

Carr, R.S. and Chapman, D.C. 1992. Comparison of solid-phase and pore-water approaches for assessing the quality of marine and estuarine sediments. *Chemical Ecology* 7, 19–30.

Carr, R.S., Long, E.R., Windom, H.L., Chapman, D.C., Thursby, G., Sloane, G.M., Wolfe, D.A. 1996. Sediment quality assessment studies of Tampa Bay, Florida. *Environ. Toxicol. Chem.* 15, 1218–1231.

Carr, R.S. 1998. Marine and estuarine porewater toxicity testing. In: Wells, P.G., K. Lee, C. Blaise, eds. *Microscale Testing in Aquatic Toxicology: Advances, Techniques, and Practice.* CRC Press, Boca Raton, FL., p. 523–538.

Carr, R.S., Montagna, P.A., Biedenbach, J.M., Kalke, R., Kennicutt, M.C., Hooten, R., Cripe, G.M. 2000. Impact of storm water outfalls on sediment quality in Corpus Christi Bay, Texas. *Environ. Toxicol. Chem.* 19, 561–574.

Chapman, P.M. and Morgan, J.D. 1983. Sediment bioassays with oyster larvae. *Bull. Environ. Contam. Toxicol.* 31: 438–444.

Chapman, P.M., Dexter, R.N., Long, E.R. 1987. Synoptic measures of sediment contamination, toxicity and infaunal community composition (the sediment quality triad) in San Francisco Bay. *Mar. Ecology Progress Ser.* 37, 75–96.

Chapman, P.M. 1989. Current approaches to developing sediment quality criteria. *Environ. Toxicol. Chem.* 8, 589–599.

Chappie, D.J. and Burton, G.A., Jr. 2000. Applications of aquatic and sediment toxicity testing *in situ. J. Soil Sediment Contamination.* 9, 219–246.

Cripe, G.M., Carr, R.S., Foss, S.S., Harris, P.S., Stanley, R.S. 2000. Effects of whole sediments from Corpus Cristi Bay on survival, growth and reproduction of the mysid, *Americamysis bahia* (formerly *Mysidopsis bahia*). *Bull. Environ. Contam. Toxicol.* 64: 426–433.

Day, K.E., Kirby, R.S., Reynoldson, T.B. 1995. The effect of manipulations on freshwater sediments on responses of benthic invertebrates in whole-sediment toxicity tests. *Environ. Toxicol. Chem.* 14, 1333–1343.

Defoe, D.L. and Ankley, G.T. 1998. Influence of storage time on toxicity of freshwater sediments to benthic macroinvertebrates. *Environ. Pollut.* 99, 123–131.

DeWitt, T.H., Swartz, R.C., and Lamberson, J.O. 1989. Measuring the acute toxicity of estuarine sediments. *Environ. Toxicol. Chem.* 8: 1035–1048.

Dillon, T.M., Moore, D.W., Jarvis, A.S. 1994. The effects of storage temperature and time on sediment toxicity. *Arch. Environ. Contamination Toxicol.* 27, 51–53.

Ditsworth G.R., Schults, D.W., Jones, J.K.P. 1990. Preparation of benthic substrates for sediment toxicity testing. *Environ. Toxicol. Chem.* 9, 1523–1529.

Di Toro, D.M., Mahoney, J.H., Hansen, D.J., Scott, K.J., Hicks, M.B., Mayr, S.B., Redmond, M. 1990. Toxicity of cadmium in sediments: the role of acid volatile sulfides. *Environ. Toxicol. Chem.* 9, 1487–1502.

Di Toro, D.M., Zarba, C.S., Hansen, D.J., Berry, W.J., Swartz, R.C., Cowen, C.E., Pavlou, S.P., Allen, H.E., Thomas, N.A., Paquin, P.R. 1991. Technical basis for establishing sediment quality criteria for nonionic organic chemicals using equilibrium partitioning. *Environ. Toxicol. Chem.* 10, 1541–1583.

Di Toro, D.M., Mahoney, J.D., Hansen, D.J., Scott, K.J., Carlson, A.R., Ankley, G.T. 1992. Acid-volatile sulfide predicts the acute toxicity of cadmium and nickel in sediments. *Environ. Sci. Technol.* 26, 96–101.

Environment Australia. 2002. *National ocean disposal guidelines for dredged material.* Commonwealth of Australia, Canberra.

Environment Canada. 1992a. Biological test method: fertilization assay using echinoderms (sea urchins and sand dollars) Environmental Protection Series. Environment Canada, Method Development and Application Section, Environmental Technology Series. Ottawa, Ontario EPS 1/RM/27. December. 97 p.

Environment Canada. 1992b. Biological test method: toxicity test using luminescent bacteria (*Photobacterium phosphoreum*). Environmental Protection Series. Environment Canada, Method Development and Application Section, Environmental Technology Series. Ottawa, Ontario Final. EPS 1/RM/24. pp. 83.

Environment Canada. 1992c. Biological test method: test of reproduction and survival using the cladoceran *Ceriodaphnia dubia*. Environment Canada, Method Development and Application Section, Environmental Technology Series. Ottawa, Canada EPS 1/RM/21. pp. 95.

Environment Canada. 1994. *Guidance document on collection and preparation of sediments for physicochemical characterization and biological testing.* Environmental Protection Series, Environment Canada, Method Development and Application Section, Environmental Technology Series, EPS 1/RM/29, Ottawa, Ontario, 132.

Environment Canada. 1997a. Biological test method: test for survival and growth in sediment using the freshwater amphipod *Hyalella azteca*. Environment Canada, Method Development and Application Section, Environmental Technology Series. Ottawa, Canada EPS 1/RM/33. pp. 123.

Environment Canada. 1997b. Biological test method: test for survival and growth in sediment using the larvae of freshwater midges (*Chironomus tentans* or *Chironomus riparius*). Environment Canada, Method Development and Application Section, Environmental Technology Series. Ottawa, Canada EPS 1/RM/32. pp. 155.

Environment Canada. 1998a. *Biological test method: reference method for determining acute lethality of sediment to marine or estuarine amphipods.* Environmental Protection Series. Environment Canada, Method Development and Application Section, Environmental Technology Series, EPS 1/RM/35, Ottawa, Ontario, 57.

Environment Canada. 1998b. Biological test method: acute test for sediment toxicity using marine or estuarine amphipods. Environment Canada, Method Development and Application Section, Environmental Technology Series. Ottawa, Canada EPS 1/RM/35. pp. 74.

Environment Canada. 2000. Biological test method: reference method for determining acute lethality of effluents to *Daphnia magna*. Second edition. Environment Canada, Method Development and Application Section, Environmental Technology Series. Ottawa, Canada EPS 1/RM/14. pp. 35.

Environment Canada. 2001a. *Biological test method: test for survival and growth in sediment using spionid polychaete worms* (Polydora cornuta). Environmental Protection Series. Environment Canada, Method Development and Application Section, Environmental Technology Series, EPS 1/RM/41, Ottawa, Ontario.

Environment Canada. 2001b. *Disposal at sea regulations*. SOR/DORS/2001-275. Department of the Environment, Ottawa, Ontario.

EPRI (Electrical Power Research Institute). 1999. *Review of sediment removal and remediation technologies at mgp and other contaminated sites*, EPRI, Palo Alto, CA, and Northern Utilities, Berlin, CT, Tr-113106.

Ferretti, J.A., Calesso, D.F., Hermon, T.R. 2000. Evaluation of methods to remove ammonia interference in marine sediment toxicity tests. *Environ. Toxicol. Chem.* 19, 1935–1941.

Gendron, A.D., Bishop, C.A., Fortin, R., Hontela, A. 1997. *In vitro* testing of the functional integrity of the corticosterone-producing axis in mudpuppy (Amphibia) exposed to chlorinated hydrocarbons in the wild. *Environ. Toxicol. Chem.* 16, 1694–1706.

Giesy, J.P., Graney, R.L., Newsted, J.L., Rosiu, C.J., Benda, A., Kreis, R.G., Horvath, F.J. 1988. Comparison of three sediment bioassay methods using Detroit river sediments. *Environ. Toxicol. Chem.* 7:483–498.

Giesy, J.P., Rosiu, C.J., Graney, R.L. 1990. Benthic invertebrate bioassays with toxic sediment and pore water. *Environ. Toxicol. Chem.* 9: 233–248.

Grapentine, L., Boyd, D., Anderson, J., Burton, G.A, Jr., DeBarros, C., Johnson, G., Marvin, C., Milani, D., Painter, S., Pascoe, T., Reynoldson, T., Richman, L., Solomon, K., Chapman, P.M. 2002. A decisions making framework for sediment assessment developed for the Great Lakes. *Human Ecol. Risk Assessment* 8, 1641–1655.

Griffiths, R.P. 1983. The importance of measuring microbial enzymatic functions while assessing and predicting long-term anthropogenic perturbations. *Mar. Pollut. Bull.* 14:162.

Guilherme, L.R., Farrar, J.D., Inouye, L.S., Bridges, T.S., Ringelberg, D.B. 2001. Toxicity of sediment-associated nitroaromatic and cyclonitramine compounds to benthic invertebrates. *Environ. Toxicol. Chem.* 20: 1762–1771.

Ho, K.T.Y. and Quinn, J.G. 1993. Physical and chemical parameters of sediment extraction and fractionation that influence toxicity, as evaluated by Microtox. *Environ. Toxicol. Chem.* 12, 615–625.

Ho, K. 1997. Toxicity-based approach to environmental protection. *European Water Pollution Control 7*, 49–52.

Ho, K.T., Kuhn, A., Pelletier, M., Mc Gee, F., Burgess, R.M., Serbst, J. 2000. Sediment toxicity assessment: comparison of standard and new testing designs. *Arch. Environ. Contam. Toxicol.* 39: 462–468.

Hong, J.S. and Reish, D.J. 1987. Acute toxicity of cadmium to eight species of marine amphipod and isopod crustaceans from southern California. *Bull. Environ. Contam. Toxicol.* 39: 884–888.

Howard, D.E. and Evans, R.D. 1993. Acid-volatile sulfide (AVS) in a seasonally anoxic mesotrophic lake: seasonal and spatial changes in sediment AVS. *Environ. Toxicol. Chem.* 12, 1051–1957.

Hunt, J.W. and Anderson, B.S. 1989. Sublethal effects of zinc and municipal effluents on the larvae of the red abalone. *Mar. Biol.* 101: 545–552.

Hunt, J.W. and Anderson, B.S. 1993. From research to routine: A review of toxicity testing with marine molluscs. W. G. Landis, J. S. Hughes and M. A. Lewis eds. Environmental Toxicology and Risk Assessment. ASTM STP 1179. American Society for Testing and Materials. Philadelphia. 320–339.

Hunt, J.W., Anderson, B.S., Phillips, B.M., Tjeerdema, R.S., Taberski, K.M., Wilson, C.J., Puckett, H.M., Stephenson, M., Fairley, R., Oakden, J. 2001. A large-scale categorization of sites in San Francisco Bay, USA, based on the sediment quality triad, toxicity identification evaluations, and gradient studies. *Environ. Toxicol. Chem.* 20: 1252–1265.

Hyland, J.L., Snoots, T.R., Balthis, L. 1998. Sediment quality of estuaries in the southeastern U.S. *Environmental Monitoring and Assessment* 51, 331–334.

Hyne, R.V. and Everett, D.A. 1998. Application of a benthic euryhaline amphipod, Corophium sp., as a sediment toxicity testing organism for both freshwater and estuarine systems. *Arch. Environ. Contam. Toxicol.* 34: 26–33.

Ingersoll, C.G., MacDonald, D.D., Wang, N., Crane, J.L., Field, L.J., Haverland, P.S., Kemble, N.E., Lindskoog, R.A., Severn, C., Smorong, D.E. 2001. Prediction of sediment toxicity using consensus-based freshwater sediment quality guidelines. *Arch. Environ. Contamination. Toxicol.* 41, 8–21.

Johnson, R.L., Perez, K.T., Davey, E.W., Cardin, J.A., Rocha, K.J., Dettman, E.H., Hetshe., J.F. Discriminating the benthic effects of anthropogenic point sources from salinity and nitrogen loading. *Ecol. Appl.* (submitted).

Keith, L.H. 1993. *Principles of environmental sampling.* ACS Professional Reference Book, American Chemical Society, Washington, D.C.

Kemble, N.E., Brumbaugh, W.G., Brunson, E.L., Dwyer, F.J., Ingersoll, C.G., Monda, D.P., Woodward, D.F. 1994. Toxicity of metal contaminated sediments from the Upper Clark Fork River, MT to aquatic invertebrates in laboratory exposures. *Environ. Toxicol. Chem.* 13: 1895–1997.

Lamberson, J.O., DeWitt, T.H., Swartz, R.C. 1992. Assessment of sediment toxicity to marine benthos, in G.A. Burton, Jr., (Ed.), *Sediment toxicity assessment*, 183-211, Lewis Publishers, Chelsea, MI.

Long, E.R., Chapman, P.M. 1985. A sediment quality triad: measures of sediment contamination, toxicity and infaunal community composition in Puget Sound. *Mar. Pollut. Bull.* 16, 405–415.

Luoma, S.N. and Ho, K.T. 1993. Appropriate uses of sediment bioassays, in P. Calow, (Ed.), *Handbook of ecotoxicology 1*, 193–226, Blackwell Scientific Publications, Oxford.

MacDonald, L.H., Smart, A.W., Wissmar, R.C. 1991. *Monitoring guidelines to evaluate effects of forestry activities on streams in the Pacific Northwest and Alaska,* EPA 910/9-01-001, U.S. Environmental Protection Agency, Region 10, Seattle, WA.

MacDonald, D.D., Ingersoll, C.G., Smorong, D.E., Lindskoog, R.A., Sparks, D.W., Smith, J.R., Simon, T.P., Hanacek, M. 2002a. Assessment of injury to fish and wildlife resources in the Grand Calumet River and Indiana Harbor Area of Concern, USA. *Arch. Environ. Contamination. Toxicol.* 43, 130–140.

MacDonald, D.D., Ingersoll, C.G., Smorong, D.E., Lindskoog, R.A., Sparks, D.W., Smith, J.R., Simon, T.P., Hanacek, M. 2002b. Assessment of injury to sediments and sediment-dwelling organisms in the Grand Calumet River and Indiana Harbor Area of Concern, USA. *Arch. Environ. Contamination Toxicol.* 43, 141–155.

Martin, M., Osborn, K.E., Billig, P., Glickstein, N. 1981. Toxicities of ten metals to *Crassostrea gigas* and *Mytilus edulis* embryos and Cancer magister larvae. *Mar. Poll. Bull.* 12: 305–308.

McGee, B.L., Schlekat, C.E., Reinharz, E. 1993. Assessing sublethal levels of sediment contamination using the estuarine amphipod *Leptocheirus Plumulosus*. *Environ. Toxicol. Chem.* 12: 577–587.

McGee, B.L. and Fisher, D.J. 1999. *Field validation of the chronic sediment bioassay with marine and the estuarine amphipod* Leptocheirus plumulosus. Final Report. Prepared for the U.S. Environmental Protection Agency, Office of Science and Technology, by the University of Maryland, Wye Research and Education Center, Queenstown, MD.

Moore, D.W., Dillon, T.M., Gamble, E.W., 1996. Long-term storage of sediments: implications for sediment toxicity testing. *Environ. Pollut.* 89, 341–342.

Nacci, D., Jackim, E., Walsh, R. 1986. Comparative evaluation of three rapid marine toxicity tests: sea urchin early embryo growth test, sea urchin sperm cell toxicity test and Microtox. *Environ. Toxicol. Chem.* 5: 521–526.

Nipper, M.G., Greenstein, D.J., Bay, S.M. 1989. Short and long-term sediment toxicity test methods with the amphipod Grandidierella japonica. *Environ. Toxicol. Chem.* 8: 1191–1200.

Nipper, M., Burton, G.A., Jr., Chapman, D., Doe, K.G., Hamer, M.A., Ho, K.T. 2001. Issues and recommendations for porewater toxicity testing: methodological uncertainties, confounding factors and toxicity identification evaluation procedures, in *Porewater toxicity testing: biological, chemical, and ecological considerations with a review of methods and applications, and recommendations for future areas of research.* Summary of a SETAC technical workshop, SETAC Technical Publication, SETAC Press, Pensacola, FL.

Pesch, C.E., Hansen, D.J., Boothman, W.S., Berry, W.J., Mahony, W.J. 1995. The role of acid-volatile sulfide and interstitial water metal concentrations in determining bioavailability of cadmium and nickel from contaminated sediments to the marine polychaete Neanthes arenaceodentata. *Environ. Toxicol. Chem.* 14: 129–141.

Pfitzenmeyer, H. 1975. Benthos, in A.T. Koo, (Ed.), *Biological study of Baltimore Harbor.* Contribution No. 621, Center for Environmental and Estuarine Studies, University of Maryland, Solomons, MD.

Pittinger, C.A., Woltering, D.M., Masters, J.A. 1989. Bioavailability of sediment sorbed and soluble surfactants to Chironomus riparius (midge). *Environ. Toxicol. and Chem.* 18: 765–772.

Prater, B.L. and Anderson, M.A. 1977. A 96-hour sediment bioassay of Duluth and Superior Harbor basins (Minnesota) using *Hexagenia limbata, Asellus communis, Daphnia magna,* and *Pimephales promelas* as test organisms. *Bull. Environ. Contam. Toxicol.* 18: 159.

Reinharz, E. 1981. *Animal sediment relationships: a case study of the Patapsco River.* Open file No. 6, Maryland Geological Survey, Baltimore, MD.

Ringwood, A.H.K. and Charles J. 1998. Seed clam growth: an alternative sediment bioassay developed during emap in the carolinian province. *Environmental Monitoring and Assessment* 51, 247–257.

Ringwood, A.H.K. and Charles J. 2002. Comparative in situ and laboratory sediment bioassays with juvenile Mercenaria mercenaria. Environ. Toxicol. Chem. 21, 1651–1657.

Sarda, N. and Burton, G.A., Jr., 1995. Ammonia variation in sediments: spatial, temporal and method-related effects. *Environ. Toxicol. Chem.* 14, 1499–1506.

Schipper, C.A. and Stronkhorst, J. 1999. RIKZ Handboek Toxiciteitstesten voor Zoute Baggerspecie. Rijksinstituut voor Kust en Zee/RIKZ. Den Haag Final. RIKZ/99.012. 1 June 1999.

Schlekat, C.E., McGee, B.L. and Reinharz, E. 1992. Testing sediment toxicity in Chesapeake Bay with the amphipod *Leptocheirus Plumulosus*: an evaluation. *Environ. Toxicol. Chem.* 11, 225–236.

Schlekat, C.E., Velinsky, D.J., Wade, T.L., 1994. Tidal river sediments in the Washington, D.C. area. III. biological effects associated with sediment contamination. *Estuaries* 17, 334–344.

Schubauer-Berigan, M.K., Dierkes, J.R., Monson, P.D., Ankley, G.T. 1993. pH-dependent toxicity of Cd, Cu, Ni, Pb and Zn to *Ceriodaphnia dubia*, *Pimephales promelas*, *Hyalella azteca* and *Lumbriculus variegatus*. *Environ. Toxicol. Chem.* 12: 1261–1266.

Scott, K. J. and Redmond, M.S. 1989. The effects of a contaminated dredged material on laboratory populations of the tubicolous amphipod *Ampelisca abdita*. U. M. Cowgill and L. R. Williams eds. Aquatic Toxicology and Hazard Assessment: ASTM 1027. 12. American Society for Testing Materials. Philadelphia. 289–303.

Sijm, R.T.H., Haller, M., Schrap, S.M. 1997. Influence of storage on sediment characteristics and drying sediment sorption coefficients of organic contaminants. *Bull. Environ. Contamination Toxicol.* 58, 961–968.

Skalski, C., Fisher, R., Burton, G.A., Jr. 1990. *An in situ interstitial water toxicity test chamber.* Abstr. Annual Meeting, Society of Environmental Toxicology and Chemistry 132, 58.

Stronkhorst, J., Schot, M.E., Dubbeldam, M.C., and Ho, K.T. 2003. A toxicity identification evaluation of silty marine harbor sediments to characterize persistent and non-persistent constituents. *Mar. Poll. Bull.* 46: 56–64.

Svenson, A., Edsholt, E., Ricking, M., Remberger, M., and Rottorp, J. 1996. Sediment Contaminants and Microtox Toxicity Tested in a Direct Contact Exposure Test. *Environ. Toxicol. Water Quality* 11: 293–300.

Swain, W.R. 1988. Human health consequences of consumption of fish with organochlorine compounds. *Aquatic Toxicol.* 11, 357–377.

Swartz, R.C., DeBen, W.A., Sercu, K.A., Lamberson, J.O. 1982. Sediment toxicity and the distribution of amphipods in Commencement Bay, Washington, USA. *Mar. Pollut. Bull.* 13, 359–364.

Swartz, R.C., DeBen, W.A., Jones, J.K.P., Lamberson, J.O., and Cole, F.A. 1985. Phoxocephalid amphipod bioassay for marine sediment toxicity. R.D. Cardwell, R. Purdy and R.C. Bahner, eds. Aquatic Toxicology and Hazard Assessment:Seventh Symposium. ASTM STP 854. 854. American Society for Testing Materials. Philadelphia. 284–307.

Swartz, R.C., Kemp, P.F., Schults, D.W., Ditsworth, G.R., Ozretich, R.J. 1989. Acute toxicity of sediments from Eagle Harbor, Washington, to the infaunal amphipod *Rhepoxynius Abronius*. *Environ. Toxicol. Chem.* 8: 215–222.

Swartz, R.C., Cole, F.A., Lamberson, J.O., Ferraro, S.P., Schults, D.W., DeBen, W.A., Lee, H.I., Ozretich, R.J. 1994. Sediment toxicity, contamination and amphipod abundance at a DDT and dieldrin contaminated site in San Francisco Bay. *Environ. Toxicol. Chem* 13, 949–962.

Tay, K.-L., Doe, K., Wade, S.J., Vaughan, D.A., Berrigan, R.E., and Moore, M. 1992. Sediment bioassessment in Halifax Harbour. *Environ. Toxicol. Chem.* 11: 1567–1581.

Thompson, B., Anderson, B., Hunt, J., Taberski, K., and Phillips, B. 1999. Relationships between sediment contamination and toxicity in San Francisco Bay. *Marine Environmental Research* 48: 285–309.

USACE. 1995. *Draft environmental impact statement (EIS): Indiana Harbor and Canal dredging and confined disposal facility, construction and operation, comprehensive management plan, East Chicago, Lake County, IN.* EIS No. 950489, U.S. Army Corps of Engineers.

USACE/USEPA. 1991. *Evaluation of dredged material proposed for ocean disposal—testing manual.* EPA-503-8-91-001, U.S. Environmental Protection Agency, Office of Water, and U.S. Army Corps of Engineers, Washington, D.C.

USACE/USEPA. 1998. *Evaluation of dredged material proposed for discharge in waters of the U.S.—testing manual.* EPA-823-B-98-004, U.S. Environmental Protection Agency, Office of Water, and U.S. Army Corps of Engineers, Washington, D.C.

USEPA. 1992a. *Proceedings of EPA's contaminated sediment management forums, Chicago, IL, April 21-22; Washington, D.C., May 27-28 and June 16, 1992.* EPA-823-R-92-007, U.S. Environmental Protection Agency, Washington, D.C.

USEPA. 1992b. *Sediment toxicity identification evaluation: phase I (characterization), phase II (identification) and phase III (confirmation) modifications of effluent procedures.* Draft technical report, EPA 08-91, U.S. Environmental Protection Agency, Environmental Research Laboratory, Duluth, MN.

USEPA. 1993a. *Guidance manual: bedded sediment bioaccumulation tests.* EPA/600/R-93/183, U.S. Environmental Protection Agency, Office of Research and Development, Washington, D.C.

USEPA. 1993b. *Methods for measuring the acute toxicity of effluents and receiving waters to freshwater and marine organisms,* fourth edition, EPA-600/4-90/027F. U.S. Environmental Protection Agency, Office of Research and Development. Washington, D.C.

USEPA. 1994a. *Methods for measuring the toxicity and bioaccumulation of sediment-associated contaminants with estuarine and marine amphipods.* EPA 600/R-94/025, U.S. Environmental Protection Agency, Office of Research and Development, Washington, DC.

USEPA. 1994b. *Methods for measuring the toxicity and bioaccumulation of sediment-associated contaminants with freshwater invertebrates.* EPA 600/R-94/024, U.S. Environmental Protection Agency, Office of Research and Development, Duluth, MN.

USEPA. 1994c. *Short-term methods for estimating the chronic toxicity of effluents and receiving waters to marine and estuarine organisms.* Second edition. Office of Research and Development. Washington, D.C. EPA-600-4-91-003.

USEPA. 1995. *Short-term methods for estimating the chronic toxicity of effluents and receiving waters to west coast marine and estuarine organisms.* National Exposure Research Laboratory. Cincinati, OH. EPA/600/R-95-136. EPA-600-4-91-003.

USEPA. 1996. Marine toxicity identification evaluation (TIE) procedures manual: Phase I Guidance Document. USEPA/ Office of Research and Development. Washington D.C. 600/R-96/054. Sept. 1996.

USEPA. 1997. *Ecological risk assessment guidance for Superfund: process for designing and conducting ecological risk assessments.* EPA540-R-97-006, U.S. Environmental Protection Agency, Office of Solid Waste and Emergency Response, Washington, D.C.

USEPA. 1998. *EPA's contaminated sediment management strategy.* 823-R-98-001, U.S. Environmental Protection Agency, Office of Water, Washington, D.C.

USEPA. 2000a. *Bioaccumulation testing and interpretation for the purpose of sediment quality assessments.* EPA 823/R-00-001, U.S. Environmental Protection Agency, Office of Water, Office of Solid Waste and Emergency Response, Washington, D.C.

USEPA. 2000b. *Methods for measuring the toxicity and bioaccumulation of sediment-associated contaminants with freshwater invertebrates.* EPA-600-R-99-064, U.S. Environmental Protection Agency, Office of Water, Office of Research and Development, Washington, D.C.

USEPA. 2000c. *Guidance for choosing a sampling design for environmental data collection.* EPA QA/G-5S, U.S. Environmental Protection Agency, Office of Environmental Information, Washington, D.C.

USEPA. 2001a. *Methods for assessing the chronic toxicity of marine and estuarine sediment-associated contaminants with the amphipod* Leptocheirus plumulosus. EPA/600/R- 01/020, U.S. Environmental Protection Agency, Office of Water, Office of Research and Development, and U.S. Army Corps of Engineers, Washington, D.C.

USEPA. 2001b. *Methods for the collection, storage and manipulation of sediments for chemical and toxicological analysis.* EPA/823/B/01/002, U.S. Environmental Protection Agency, Office of Water, Office of Research and Development, Washington, D.C.

USEPA. 2002. *A guidance manual to support the assessment of contaminated sediments in freshwater ecosystems: volume I, an ecosystem-based framework for assessing and managing contaminated sediments.* EPA-905-B-02-001-A, U.S. Environmental Protection Agency, Great Lakes National Program Office, Chicago, IL.

USEPA. 2004. *The incidence and severity of sediment contamination in surface waters of the United States: national sediment quality survey, second edition.* EPA-823-R-04-007, U.S. Environmental Protection Agency, Office of Science and Technology, Washington, D.C.

Van Rees, K.C.J., Sudlicky, E.A., Suresh, P., Rao, C., Reddy, K.R. 1991. Evaluation of laboratory techniques for measuring diffusion coefficients in sediments. *Environ. Sci. Technol.* 25, 1605–1611.

Van Veld, P.A., Westbrook, D.J., Woodin, B.R., Hale, R.C., Smith, C.L., Hugget, R.J., Stegman, J.J. 1990. Induced cytochrome P-450 in intestine and liver of spot (*Leiostomus xanthurus*) from a polycyclic aromatic contaminated environment. *Aquatic Toxicol.* 17, 119–132.

Winger, P.V., Lasier, P.J., Jackson, B.P. 1998. The influence of extraction procedure on ion concentrations in sediment pore water. *Arch. Environ. Contamination Toxicol.* 35, 8–13.

Zarull, M.A., Hartug, J.H., Maynard, L. 1999. *Ecological benefits of contaminated sediment remediation in the Great Lakes Basin.* Sediment Priority Action Committee, Great Lakes Water Quality Board, International Joint Commission.

Appendix

Toxicity tests for sediment quality assessments

The following are examples of select toxicity test species and their associated methods for both freshwater and marine species. This text has been taken from Burton et al. (2002) and ASTM (2002) for *Chironomus riparius,* that goes into great detail on the culturing and testing for a variety of test organisms. The purpose of this appendix is not to be all inclusive, but to provide some examples.

Freshwater test organisms

Hyalella azteca

Hyalella azteca is a small freshwater amphipod that has been shown to be a sensitive indicator of the presence of contaminants in freshwater sediments. *H. azteca* is an epibenthic detritivore and herbivore and will burrow in the surface sediments in search of food. Its short life cycle, widespread and abundant distribution, ease of culture, and wide tolerance of sediment grain size and salinity make it a very suitable test species. Methods for culture and testing are summarized by the ASTM and USEPA (ASTM 2002; USEPA 1994c, 2000). They have been used extensively for whole sediment toxicity testing in North America.

   H. azteca can be obtained from a commercial supplier or laboratory culture. The amphipods can be held in 80-L glass aquaria filled with about 50 L of moderately hard reconstituted water, 80 to 100 mg/L as $CaCO_3$ (Ingersoll and Nelson 1990). A flaked food (such as Tetrafin®) is added to each culture chamber receiving daily water renewals to provide about 20 g of dry solids per 50 L of water twice weekly in an 80-L culture chamber (USEPA 2000). Each culture chamber has a substrate of maple leaves and artificial substrates (six 20-cm diameter sections per 80-L aquaria of nylon "coiled-web material"; 3-M, St. Paul, MN). Before use, leaves are soaked in 30‰ salt water for about 30 d to reduce the occurrence of planaria, snails, or other organisms on the substrate. The leaves are then flushed to remove the salt water and residuals of naturally occurring tannic acid before placement in the cultures (USEPA 2000). Cultures should be maintained at 23°C under a 16:8 h light-dark cycle at an illuminance of 100 to 1000 lx. Gentle aeration is provided. Water in culture chambers is changed weekly. Survival of adults and juveniles and production of young should be measured at this time. Mixed-age amphipods may be separated by sieving the amphipods through 250-μm, 425-μm, and 600-μm sieves. Sieves should be held under water to isolate the amphipods. Artificial substrates or leaves are placed in the 600-μm sieve. Culture water is rinsed through the sieves and small amphipods stopped by the 250-μm sieve are washed into a collecting pan. Larger amphipods in the other two sieves are returned to the culture chamber. The smaller amphipods are then placed in 1-L beakers containing culture water and food (about 200 amphipods per beaker) with gentle aeration. Newborn amphipods should be held for 6 to 13 d to provide 7- to 14-day-old organisms to start a 10-d test or should be held for 7 d to provide 7- to 8-day-old organisms to start a long-term test (USEPA 2000).

Assessment of whole-sediment toxicity involves a 10- to 42-d exposure of juvenile amphipods to sediments, using procedures described by EPA and ASTM (ASTM 2002; USEPA 2000). Both the short-term (10-d) and long-term (42-d) sediment toxicity tests are conducted in 300-ml high-form lipless beakers. The sediment volume for both test is 100 ml with 175 ml of overlying water. The recommended number of replicate chambers for routine testing for the 10-d sediment test is 8 and the recommended number of replicate chambers for the 42-d sediment test is 12 (USEPA 2000). Sediments are prepared the day before test initiation and allowed to equilibrate overnight. The following day (day 0), test organisms are added (10 organisms per chamber for both the short- and long-term sediment tests) and the experiment begins. Prior to distribution to the test containers, the overlying water is renewed, with two volume replacements per day thereafter, continuous or intermittent (for example, one volume addition every 12 hours) (USEPA 2000) thereafter. Other successful test methods have consisted of a 1:4 sediment to water ratio in 30-ml beakers (1 organism per beaker, 10 replicates) to 250-ml beakers (10 organisms per beaker, 3 replicates) with daily water renewal; however, these methods do not follow the USEPA methods (Burton et al. 1989). A negative (clean) control, consisting of fine silica sand (culture material) or mesh is tested concurrently. Juvenile amphipods (as outlined above) are sieved or picked from their holding containers and 10 amphipods are randomly distributed to each test container. Sieved organisms should be held for 1 to 3 d prior to testing to check for sieve-related mortality.

The short-term sediment test is allowed to proceed for 10 d, and the long-term sediment test is allowed to proceed for 42 d. Both tests are conducted at 23 ±1°C under a 16:8 h light-dark cycle. Gentle aeration is provided if needed to keep the dissolved oxygen (DO) greater than 2.5 mg/L throughout the test. Hardness, alkalinity, conductivity, and ammonia are monitored at the beginning and the end of both sediment tests. For the 10-d sediment test, pH is also monitored at the beginning and end of the test. Also for the 10-d sediment test with *H. azteca*, temperature and DO are monitored daily. For the 42-d sediment test, temperature is monitored daily, conductivity is monitored weekly, and DO and pH are monitored three times per week. Amphipods are fed 1 ml of yeast-cerophyll-trout chow (YCT) daily to each test chamber. At test termination of the 10-d sediment test, the sediments are sieved and the number of live, dead, and missing amphipods in each test container is recorded. The test endpoints from the short-term sediment test with *H. azteca* are survival and growth measured on day 10. For the long-term sediment test, on day 28 the amphipods are isolated from the sediment and placed in water-only chambers where reproduction is measured on day 35 and day 42. Endpoints measured in the long-term amphipod test include survival (on day 28, day 35, and day 42), growth (on day 28 and day 42), and reproduction (number of young per female from day 28 to day 42). For best recovery of the live organisms, the test beakers should be gently swirled several times to resuspend the upper layer of sediment, and then quickly poured into the sieve. Since *H. azteca* does not burrow into the deeper layers, this method allows recovery with minimal sieving. Test acceptability for the 10-d sediment test is a minimum mean control survival of 80% and measurable growth of test organisms in the control sediment. For the 42d sediment test, acceptability is defined as a minimum mean control survival of 80% on day 28. For both tests, additional performance-based criteria specifications are outlined by the USEPA (USEPA 2000).

Chironomus riparius

Chironomus riparius is a fairly large freshwater midge that has a short generation time, is easily cultured in the laboratory, and (like *H. azteca*) has been shown to be sensitive to many contaminants associated with sediments (Pittinger et al. 1989; Ingersoll and Nelson 1990; USEPA 1991; Kemble et al. 1994; Burton et al. 1995). This species has often been used in Canada and western Europe (Burton et al. 2002).

C. *riparius* can be reared in aquaria in either static or flowing water with a 16:8-h light-dark photoperiod at 20 to 23°C at about 500 lx. For static cultures, the water should be gently aerated and about 25% to 30% of the water volume should be replaced weekly. The water should be replaced more often if organisms appear stressed or if the water is cloudy (ASTM 2002). Ingersoll and Nelson (1990) reared *C. riparius* in 30 ∞ 30 ∞ 30-cm polyethylene containers covered with nylon screen. Each culture chamber contains 3 L of culture water. To start a culture, 200 to 300 mg of cereal leaves (ASTM 2002) is added to the culture chamber; additionally, green algae (*Selenastrum capricornutum*) is added as desired to maintain a growth of algae in the water column and on the bottom of the culture chamber. Cultures are fed about 3 ml of a suspension of commercial dog treats (Biever 1965) daily. Additional procedures for culturing *C. riparius* can be seen in ASTM (2002).

Adult emergence will begin about two to three weeks after hatching at 23°C. Once adults begin to emerge, they can be gently siphoned into a dry aspirator flask on a daily basis. Sex ratios of the adults should be checked to ensure that a sufficient number of males are available for mating and fertilization. One male may fertilize more than one female, and a ratio of one male to three females improves fertilization (ASTM 2002).

About two to three weeks before the start of a test, three to five egg cases should be isolated for hatching. Tests with *C. riparius* have been started with larvae less than 24 h old in 1-L test chambers with 200 ml of sediment and 800 ml of overlying water (Ingersoll and Nelson 1990), with 3-d-old larvae in 13-L test chambers with 2 L of sediment and 11 L of overlying water (USEPA 1991), and with 5- to 7-d old larvae (second instar) in 300-ml test chambers with 100 ml of sediment and 175 ml of overlying water (Burton et al. 1995).

Decisions concerning the various aspects of experimental design, such as the number of treatments, number of test chambers and midges per treatment, and water-quality characteristics should be based on the purpose of the test and type of procedure that is used to calculate results (ASTM 2002). The recommended requirements for test acceptability are (1) the age of the *C. riparius* at the start of the test must be within the required range; (2) the average survival of *C. riparius* in the control sediment must be greater than or equal to 70% at the end of the test; and (3) hardness, alkalinity, and ammonia of overlying water typically should not vary more than 50% during the test, and dissolved oxygen should be maintained above 2.5 mg/L in the overlying water.

Duration of tests with *C. riparius* ranges from less than 10 d to tests continuing up to 30 d. Larval survival, growth, or adult emergence can be monitored as biological endpoints (ASTM 2002). Larval survival and growth can be assessed by ending tests on day 10 to day 14 when larvae have reached the third or fourth instar (Ingersoll and Nelson 1995; Burton et al. 1995). A consistent amount of time should be taken to examine sieved material for recovery of test organisms (such as 5 min/replicate).

Ingersoll and Nelson (1995), Pittinger et al (1989), and USEPA (1991) describe procedures for conducting *C. riparius* sediment toxicity tests until the larvae pupate and emerge as adults. Cast pupal skins left by emerging adult *C. riparius* should be removed and recorded daily. These pupal skins remain on the water surface for over 24 h after emergence of the adult. The test should be ended after the test organisms have been exposed for up to 30 d, when about 70% to 95% of the control larvae should have completed metamorphosis into the adult life stage (ASTM 2002). Endpoints calculated in these adult emergence tests include percentage of emergence, mean emergence time, or number of days to first emergence.

Marine test organisms

Ampelisca abdita

Ampelisca abdita is a commonly used test organism in marine and estuarine systems. This organism has been found on the east coast of the U.S. from

Maine to northern Florida and in the eastern Gulf of Mexico. There has been an introduced population in San Francisco Bay (Scott and Redmond 1989).

A. abdita is a tube dweller, and builds soft 2- to 3-cm long tubes in fine silty surface sediments. It has a shorter life cycle than other amphipods (six weeks at 20°C), which makes it a candidate for tests using reproduction as an endpoint (Scott and Redmond 1989). Test organisms can be purchased or obtained from field collections. For field collections *A. abdita* is collected using a small dredge or grab, or by skimming surficial sediments with a handheld long-handled net. The amphipod tubes are gently sieved in the field to separate them from surrounding sediments. Amphipods and tubes are then immediately transported back to the laboratory in clean buckets with overlying seawater, and maintained at a temperature at or below collection temperature. In the laboratory the amphipods and tubes are placed on a sieve series consisting of a 2- 1- and 0.5-mm sieve. Collection-temperature seawater is sprayed over the tubes to separate the amphipods from the tubes. The amphipods fall through the sieve series and are sorted according to size. In the laboratory, *A. abdita* are held in clean sediment, under flow-through conditions at collection. If need be, they are acclimated 2 to 3°C per day until they reach test conditions (20°C). During holding and acclimation they are fed a diatom algae daily *(Phaeodactylum tricornutum* or *Skeletonema)* (USEPA 1994b). Organisms should be used within 2 to 10 days of collection.

Test containers for 10-d whole-sediment toxicity tests are 1-L glass jars with a screened hole drilled near the top to allow water overflow in a flow-through arrangement. Each container has a 2-cm layer of sediment and 800 ml of filtered seawater (adjusted to the appropriate salinity as necessary). Negative controls consist of sediment from the amphipod collection site or "clean" sediments proven to be nontoxic, combined with filtered seawater. The sediments are prepared the day before test initiation and allowed to equilibrate overnight, and the amphipods are added the following day (day 0). Replication of treatments is dependant upon the objectives of the experiment, although a minimum of five replicates is recommended for each treatment and each replicate should contain 20 amphipods. Daily monitoring of water-quality parameters (temperature, pH, DO, and salinity) should be conducted and may be done in one of the replicates or in an additional separate monitoring replicate. The test is conducted at 20 ±1°C under continuous illumination for 10 d, with gentle aeration provided. Water quality is measured daily. Organisms are not fed during the test. After 10 d, the sediments are sieved and the number of living, dead, and missing amphipods is determined for each replicate. The level of effort required to recover surviving amphipods varies with the species. Some effort is required to recover *A. abdita* from the tubes. Response criteria include mortality, emergence from sediment, and ability to rebury in clean sediment after a 10-d exposure. Test acceptability in all the amphipod 10-d exposures is 90% survival.

Microtox

The Microtox test uses the photoluminescent marine bacterium *Vibrio fischeri* and may be used in freshwater, estuarine, or marine studies (USEPA 1994a; Environment Canada 1992). Changes in bioluminescence in response to exposure to test solutions are detected using a Microtox analyzer. The solid-phase protocol is similar to the liquid phase; although there is a sediment exposure and filtration step added.

Freeze-dried bacteria are rehydrated with 1 ml of reconstitution solution and stored in the Microtox analyzer at 4°C. Serial dilutions of 0%, 12.5%, 25%, 50%, and 100% are prepared with the elutriate or pore and Microtox diluent (2% NaCl in sterile water). The 0% dilution is a control blank containing only diluent. Each of 10 cuvettes receives 10 µl of the bacterial suspension and 350 µl of diluent. The dilutions are allowed to incubate at 15°C for 5 min or 15 min, and an initial light reading is taken. Aliquots of 500 µl are added to two of the cuvettes from each extract dilution (two replicates per dilution). At 5 min or 15 min after addition of the extract, a final light reading is taken.

Measured values are relative to standard light levels set at the beginning of the test. The light given off by the bacteria diminishes naturally over time. Light loss in the blank controls is used as a ratio to normalize results from the cuvettes containing sample supernatant. The blank response is the ratio of light levels read at time 0 and after 15 min. Light-level results of the test chambers are normalized against the blank response to yield gamma values. The gamma value is a percentage decrease (represented as a positive value) in light that has been adjusted to take into account the natural loss in luminescence over time of the bacteria. Negative gamma values represent an increase in light output in the system. The 15-min EC_{50} values (effective concentration yielding a 50% response in the test system) for each sample can be calculated.

With some samples, it is necessary to alter the method to optimize sensitivity. Some samples have turbidity or color associated with them that can be determined and removed from the effect determination. For samples containing ionic compounds, such as ammonia, the response can be optimized by replacing the NaCl osmotic adjusting solution with 20.4% sucrose (Hinwood and McCormick 1987). Because turbidity and color play a large role in the interpretation of the solid-phase test, factoring in these effects should be considered in the assessment of toxicity.

References

ASTM (American Society for Testing and Materials). 2002. Standard test methods for measuring the toxicity of sediment-associated contaminants with freshwater invertebrates, E1706-00, in *Annual book of ASTM standards*, Vol. 11.05, West Conshohocken, PA.

Biever, K.D. 1965. A rearing technique for the colonization of chironomid midges. *Annals Entomol. Soc. America*, 58, 135–136.

Burton, G.A., Jr., Stemmer, B.L. Winks, K.L. Ross, P.E. and Burnett, L.C. 1989. A multitrophic level evaluation of sediment toxicity in Waukegan and Indiana harbors. *Environ. Toxicol. Chem.*, 8, 1057.

Burton, G.A., Jr., Ankley, G.T. Ingersoll, C.G. Norberg-King, T.J. and Winger, P.V. 1995. *Evaluation of sediment toxicity test methods: round-robin testing design.* Presented at the annual meeting of the Society of Environmental Toxicology and Chemistry (SETAC), Houston, Texas.

Burton, G.A., Jr., Denton, D., Ho K.T., and Ireland, D.S. 2002. *Sediment toxicity testing, issues and methods in quantifying and measuring ecotoxicological effects*, in D. Hoffman, D. Rattner, G.A. Burton Jr., and J.J. Cairns, (Eds.), *Handbook of ecotoxicology.* CRC/Lewis Publishers, Boca Raton, FL.

Environment Canada. 1992. *Biological test method: toxicity test using luminescent bacteria* (Photobacterium phosphoreum). Final, EPS 1/RM/24, Environmental Protection Series, Environment Canada, Method Development and Application Section, Environmental Technology Series, Ottawa, Ontario, 83.

Hinwood, A.L. and McCormick, M.J. 1987. The effect of ionic solutes on EC 50 values measured using the Microtox test. *Toxicol. Assessment* 2, 499.

Ingersoll, C.G. and Nelson, M.K. 1990. Testing sediment toxicity with *Hyalella azteca* (amphipoda) and *Chironomus riparius* (diptera). In W.G. Landis and W.H. van der Schalie, (Eds.), *Aquatic toxicology and risk assessment: 13th volume*, 93–109. ASTM STP 1096, ASTM, Philadelphia, PA.

Kemble, N.E., Brumbaugh, W.G. Brunson, E.L. Dwyer, F.J. Ingersoll, C.G. Monda, D.P. and Woodard, D.F. 1994. Toxicity of metal-contaminated sediments from the Upper Clark Fork River, MT to aquatic invertebrates in laboratory exposures. *Environ. Toxicol. Chem.* 13, 1895–1997.

Pittinger, C.A., Woltering, D.M., Masters, J.A. 1989. Bioavailability of sediment sorbed and soluble surfactants to *Chironomus riparius* (midge). *Environ. Toxicol. Chem.* 18, 765–772.

Scott, K.J. and Redmond, M.S. 1989. The effects of a contaminated dredged material on laboratory populations of the tubicolous amphipod *Ampelisca abdita*, in U.M. Cowgill and L.R. Williams, (Eds.), *Aquatic toxicology and hazard assessment*, 12th volume, 289. STP 1027; ASTM, Philadelphia, PA.

USEPA. 1991. *Technical support document for water-quality based toxic control.* EPA 505/ 2-90/001, USEPA, Washington, D.C.

USEPA. 1994a. *Assessment and remediation of contaminated sediments (ARCS) program.* Assessment Guidance Document, EPA-905-B94-002, USEPA Great Lakes National Program Office, Chicago, IL.

USEPA. 1994b. *Methods for measuring the toxicity and bioaccumulation of sediment-associated contaminants with estuarine and marine amphipods.* EPA 600/R-94/025, USEPA Office of Research and Development, Washington, D.C.

USEPA. 1994c. *Methods for measuring the toxicity and bioaccumulation of sediment-associated contaminants with freshwater invertebrates.* EPA 600/R-94/024, USEPA Office of Research and Development, Duluth, MN.

USEPA. 2000. *Methods for measuring the toxicity and bioaccumulation of sediment-associated contaminants with freshwater invertebrates.* EPA-600-R-99-064, USEPA Office of Water, Office of Research and Development, Washington, D.C.

chapter two

Bioassays and tiered approaches for monitoring surface water quality and effluents

M. Tonkes, P.J. den Besten, and D. Leverett

Contents

Summary 45
Introduction 45
    Limitations of the chemical-oriented approach 46
    Bioassays 46
    Assessment of surface water quality 48
    Assessment of effluents 48
Bioassays for the assessment of surface water quality 48
Bioassay types for effluent monitoring and assessment 49
    Genotoxicity or mutagenicity 51
    Bioaccumulation 51
    Toxicity 51
        Standardized tests 51
        Nonstandardized tests 52
        Validity criteria 53
    Pretreatment of effluents 54
        Turbidity 54
        Aeration 54
        Adjustment of pH 54
    Effluent sampling 55
Tiered approaches for the assessment of effluent toxicity 55
    The Netherlands...56

Germany 58
United Kingdom 59
United States 62
Conclusions 64
Surface water 65
Effluents 65
References...66
Appendix 69
Regulatory test batteries 69
Freshwater acute tests using fish 69
Freshwater acute tests using invertebrates 70
Daphnia immobilization test 70
Gammarid toxicity test 70
Toxicity tests with rotifers 70
Toxicity tests with protozoans 71
Freshwater acute tests using bacteria 71
Activated sludge respiration inhibition test............................ 71
Nitrification inhibition test 71
Vibrio fischeri toxicity test 71
Freshwater short-term chronic tests 72
Early life stage (ELS) fish toxicity test 72
Ceriodaphnia dubia survival and reproduction test 72
Chronic rotifer toxicity test 73
Pseudomonas putida growth inhibition test 73
Vibrio fischeri growth inhibition test 73
Anaerobic bacteria inhibition test 73
Growth inhibition of activated sludge microorganisms 73
Algal growth inhibition test 74
Lemna toxicity test 74
Freshwater long-term chronic tests 75
Chronic fish toxicity test 75
Daphnia magna reproduction test 75
Renewal toxicity test with ceriodaphnia dubia 76
Chronic toxicity test with higher plants..................................76
Marine acute tests using fish 76
Marine acute tests using invertebrates 76
Marine copepod toxicity test 76
Mysid shrimp toxicity test 77
Oyster toxicity test (shell deposition) 77
Toxicity tests with rotifers 77
Toxicity tests with protozoans 77
Marine acute tests using bacteria 77
Vibrio fischeri assay 77
Marine short-term chronic tests 77
Bivalve embryo-larval development toxicity test 77
Marine algae growth inhibition test 78

        Early life stage fish toxicity test 78
    Marine long-term chronic tests 79
        Mysid shrimp toxicity test 79
        Tisbe battagiai population level test 79
Genotoxicity tests 79
    Ames assay 80
    UmuC assay 80
    Chromosomal aberration..81
Biodegradation and sorption tests 81
    Sorption to activated sludge 82
    Sorption to solids and sediments 82
    Removal by evaporation 82
    Zahn-Wellens test 83
    Treatment plant simulation model..83
    Elimination of biological effects 84
References 84

Summary

Surface waters, wastewater discharges and industrial effluents are all complex mixtures with many constituents, both known and unknown. For many decades, a solely chemical-oriented approach was used to assess the quality of water and effluent samples. Being confronted with an ever-increasing number of constituent substances, however, has led to the need for the development of new approaches. An effect-oriented approach, using bioassays, makes possible a more complete quality assessment. A large number of bioassays are available, and can be selected depending on factors such as the chemical mode of action on test organisms, sample type, trophic level, cost, and other technical requirements. Tiered approaches are suggested to enable a cost-effective assessment of both water and wastewater quality.

Introduction

This chapter deals with the use of bioassays for the monitoring and toxicity assessment of surface waters and effluents. In many cases similar bioassay types and organisms are used for both surface water and effluent assessment. Both compartments have their own characteristics, and may differ considerably; therefore the application of bioassays requires that specific criteria be met in each case. Bioassays are often used as part of a tiered approach to save resources and support a step-by-step process of increasing weight of evidence.

This chapter gives an overview of the type of bioassays that are used for both compartments. The focus, however, is on the use of bioassays for the assessment of effluent toxicity.

Limitations of the chemical-oriented approach

The chemical-oriented approach plays a major role in the water-quality policies of many countries. When considering complex mixtures such as surface water, sediments, or effluents, however, the potential of a chemical assessment is limited because of several aspects (Tonkes and Baltus 1997):

- Many substances cannot be identified or detected through analysis.
- The number of substances can be so large that a chemical-specific approach is unattainable.
- There are missing or incomplete data on the environmental characteristics for many substances.
- Micropollutants and degradation products are undefined and therefore not accounted for.
- Combined effects are not being considered — a mixture can have very different environmental characteristics when compared to the characteristics of the separate substances.

Because of these limitations, environmental samples can only be partly characterized or assessed. This is a problem for industry, government authorities and regulators, and the environmental movement.

Some of the limitations of the substance-oriented approach can be avoided by using chemical group parameters (such as chemical oxygen demand [COD], total organic carbon [TOC], and adsorbable organic halides [AOX]) that give a better impression of the constituents of an effluent, since all substances are considered regardless of their chemical specification (UBA 1999). In general, only a small proportion of the concentrations measured by group parameters can be attributed to specific chemicals. Additionally, to date, no direct relationship has been found between chemical group parameters and ecotoxicological effects in effluents.

Bioassays

A bioassay is a tool that enables us to investigate the effects of an environmental or waste sample on an organism. An example is exposing water fleas (daphnia) to river water to determine the effects on survival, growth, or reproduction. Bioassays are most commonly carried out on discrete water samples in a laboratory, but they can also be conducted *in situ* in order to integrate the effects of varying exposures to pollutants in the environment (such as the assessment of effects on the feeding rate of freshwater shrimps *in situ*). They can also be set up to operate online (for example, fish and invertebrate activity monitors such as those used to assess water quality on the Rhine). In the aquatic environment, bioassays can be conducted on both water samples and sediments.

Bioassays have the advantage of directly measuring toxic effects of bio-available substances on aquatic organisms. Bioassays consider both known and unknown hazardous substances, including degradation products.

In the early 1970s, the first acute ecotoxicological testing guidelines were developed. In 1980 the U.S. Environmental Protection Agency (USEPA) began developing short-term toxicity tests for estimating chronic toxicity in an effort to obtain data on the chronic effects of effluents in a cost-effective manner.

Bioassays present an opportunity for a more holistic (and therefore more meaningful) way of assessing effects on ecosystems than is possible using chemical-based monitoring alone. They can:

- Integrate the effects of all the substances present in a complex mixture, including breakdown products
- Take into account the effects of interactions among the substances present
- Provide predictions and an early warning of environmental impacts, whereas ecological community measures can only determine impacts after they have occurred
- Enable the cause of poor ecological quality to be determined and traced back to the source (serving as diagnostic tools)

The introduction of microscale/high-throughput laboratory-based methodologies in recent years has enabled large numbers of samples to be tested at minimal cost, while still ensuring the data generated are of high quality and "fit for purpose." Bioassays need not be any more difficult or costly to perform than either chemical or ecological community measures. Overall, bioassays should be viewed as an important tool, adding complementary information to that provided by chemical and ecological community measures (such as the Triad Approach [van de Guchte 1992]). These features enable bioassays to be used to:

- Prioritize receiving-water sites and effluents as a first tier of investigation, thus focusing subsequent resources where they are needed most
- Aid decision-making in a weight-of-evidence approach as part of a triad of surface water monitoring techniques, alongside chemical analysis and ecological survey methods (though not necessarily all three together), or in support of the chemical analysis of complex effluents
- Inform relationships between chemical and biological quality, including the identification of cause and effect

Assessment of surface water quality

Three different approaches can be followed for the assessment of surface water quality. First, a water sample may be analyzed for well-known substances and the contaminant levels compared to environmental quality standards. Second, biological monitoring (used in many countries) may be used to assess the ecological quality of the water system. Even within countries, many different techniques are used to perform this biological assessment. Third, bioassays may be used for surface water-quality assessment, but this is less common. This approach, however, is being used more frequently.

Assessment of effluents

The most common way of assessing effluents is using an emission-based approach in combination with a water-quality–based system (Tonkes et al. 1995). The emission-based approach plays a key role in reducing water pollution in many countries. It is based on the intrinsic (toxic) properties of chemicals in effluents and requires data on chemical, ecotoxicological, and technological characteristics. Discharges into a water body must then be treated to bring them within certain defined limiting values. The water-quality–based approach is focused on criteria for preventing toxic effects in the receiving water, and thus has its foundation in the actual or desirable state of the receiving-water body.

Bioassays for the assessment of surface water quality

There are numerous documents describing the use of bioassays for water-quality monitoring. For instance, the United Nations Economic Commission for Europe (UN/ECE) guideline on water-quality monitoring and assessment of transboundary rivers (Niederländer et al. 1996) describes how pollution of surface water with toxic substances can be monitored by ecotoxicological indicators and by bioassays. The Environment Agency for England and Wales UKEA (in collaboration with others) has recently completed an extensive literature review of the role, application, and guidance for the use of bioassays in the monitoring and management of the water environment (UKEA 2001a, 2001b).

The selection of ecotoxicological test methods in the quality assessment of environmental samples requires careful consideration and should account for the following:

- Random short-term testing is less sensitive than regular long-term testing. The discriminatory power needed to distinguish temporal or spatial differences is essential.
- Species having different physiologies and feeding strategies have different sensitivities to different pollutants. In general, representatives of algae, crustaceans, and fish, if used in combination, can cover

a wide variety of chemicals, assuming concentrations are high enough to elicit responses.
- As a substitute for regular long-term testing, environmental samples can be preconcentrated to improve detection levels and subsequently tested over short timescales. The extraction techniques currently available, however, cause the loss of some of the chemicals present.

The appendix to this chapter summarizes a number of bioassays that are recommended for use in different monitoring strategies. These biotesting methods are well described in test protocols (see the Organisation for Economic Co-operation and Development (OECD), the American Society for Testing and Materials (ASTM), the Society of Environmental Toxicology and Chemistry (SETAC), and the International Organization for Standardization (ISO).

Recently, *in situ* bioassays have been developed that can be used for the assessment of water quality over longer periods of time. These bioassay techniques require that caged test organisms be deployed at sites of interest in the field. After a fixed exposure time, the organisms can be taken back to the laboratory for measuring endpoints, which can be similar to the laboratory bioassays (survival, growth, and reproduction). Additionally, the application of biomarker techniques to *in situ* bioassays is also possible (see Chapter 3, "Biomarkers in Environmental Assessments" and Chapter 5, "Bioassays and Biosensors: Capturing Biology in a Nutshell" in this book). The UKEA is also developing more sensitive sublethal methodologies for the assessment of receiving waters (Simpson and Grist 2003).

Bioassay types for effluent monitoring and assessment

This section gives a current state-of-the-art overview of suitable bioassays for effluent monitoring and assessment. This overview is based on the Federal Environment Agency in Germany, known as UBA (UBA 1999).

The most important objective of aquatic toxicity tests is to estimate the "safe" or "no adverse effect" concentration for separate chemicals or environmental samples. This is defined as the concentration that will permit normal propagation and development of fish and other aquatic life in the receiving water (Klemm et al. 1994).

Since the early 1970s, the number of ecotoxicological test types, and the experience in performing tests, has grown rapidly. The ability to detect acute and chronic toxicity plays an increasing role in identifying and controlling the toxicity of discharges to surface water.

Early experience in effluent testing indicated that even discharges that had passed the chemical quality criteria of regulators could still show acutely toxic effects on aquatic life (Heber et al. 1996). Limitations on the specific compounds present in complex effluents do not necessarily provide adequate protection for aquatic life. The toxicity of effluent components may often be unknown; furthermore, it is not possible to examine additive, synergistic, or antagonistic effects or to evaluate the toxicity of an effluent that has not been chemically characterized (USEPA 1995).

A first review of the environmental hazard assessment of effluents was published by Bergmann et al. (1986). In 1995 a workshop in whole-effluent toxicity at the University of Michigan provided a detailed overview (Grothe et al. 1996). SETAC held a conference at the Univeristy of Luton (England) in July 1996, and a major symposium and workshop was hosted by Zeneca (Brixham Environmental Laboratory) in Torquay, England in October 1996. In 1997, a Convention for the Protection of the Marine Environment of the North-East Atlantic (OSPAR) workshop on the ecotoxicological evaluation of wastewater was organized by the Federal Environment Agency in Berlin. In the recent workshop "Effluent Ecotoxicology: A European Perspective," held in Edinburgh in March 1999, experience with numerous test methods was presented from different European countries.

For monitoring wastewater discharges, attention was paid to bioassays that were:

- Performed to an internationally accepted standard with clearly defined endpoints
- Able to provide reproducible, repeatable, and comparable results
- Sensitive to many chemicals
- Able to measure biologically relevant toxic effects to representative organisms of the aquatic environment (juridical reliability)
- Able to clearly demonstrate the success of wastewater treatment
- Practicable for routine measurements (available through the year and suitable for laboratory cultivation)
- Of moderate resource burden
- Able to provide rapid and unambiguous test results

There are both acute and chronic international standardized methods available that fit all of these requirements. The main test principles are described in the appendix to this chapter.

While direct discharges of industrial wastewater into the receiving environment may cause direct effects upon the aquatic community, indirect discharges are treated together with household water in municipal biological treatment plants. Municipal wastewater treatment plants usually consist of a mechanical treatment (grit removal or primary clarification), a biological treatment (TOC removal, nitrification, denitrification, or phosphate precipitation) and a final clarification tank (sedimentation of activated sludge or effluent). In this context ecotoxicity tests are applied to assess possible adverse effects of effluents on the biological process. The respiration and nitrification inhibition tests with activated sludge are widely accepted as good tools for predicting impacts on purification efficiency. Additionally, biodegradation tests are used to assess the behavior of effluents within the treatment plant.

Genotoxicity or mutagenicity

Until recently, the number of available tests to assess genotoxic effects appeared to be limited. However, work by de Maagd (2000) has shown that many tests (more than 200) have been or are being developed. De Maagd has also shown why this particular parameter is of concern for effluent assessment, and that it is useful to use at least one primary DNA damage test for effluent testing. De Maagd also draws some conclusions regarding genotoxicity protocols:

- Data evaluation should preferably be based on dose-response curves
- A sample should be tested in a dilution series to prevent artifacts due to cytotoxicity
- Genotoxicity data derived with the S9-addition should only be used in a qualitative way
- Although the use of filtration or a concentration procedure can be necessary for both effluent and surface water samples, care should be taken to avoid the loss of genotoxic compounds

Bioaccumulation

De Maagd (2000) has presented a review on the use of different tests or techniques in order to estimate or assess possible bioaccumulation owing to discharges. De Maagd concludes that an assessment of potentially bioaccu-mulating substances (PBS) leads to a more comprehensive hazard assessment of effluents. He also concludes that this parameter should therefore be included in whole-effluent assessments. The preference lies with validated solid-phase microextraction (SPME) techniques in combination with high-performance liquid chromatography (HPLC) or gas chromatogra-phy–mass spectrometry (GC-MS) analysis.

Toxicity

Standardized tests

The principle of acute toxicity tests is that test organisms are exposed to a sample under standard, well-defined conditions. The aim is to estimate the toxicity of the sample. Acute toxicity deals with short-term endpoints, a maximum of 96 hours.

The tests are relatively simple and cheap to perform, and internationally standardized methodologies are available for different trophic levels (Beck-ers-Maessen 1994; Tonkes and Botterweg 1994; de Graaf et al. 1996; Tonkes and Baltus 1997).

Traditional base-set type approaches comprise tests with organisms over four trophic levels, namely bacteria, algae, crustaceans, and fish. More recently, such tests have been developed into ecotoxicity testing kits, called toxkits. These are fast and simple to perform and are significantly cheaper than standard tests. They have only recently become operational for application within water management or for regulatory purposes, so issues regarding quality assurance (QA) may remain.

For all tests, internationally accepted protocols (ISO, OECD, Beckers-Maessen 1994; de Graaf et al. 1996) or standard operational procedures are utilized (including toxkits) (Creasel 1990a, 1990b, 1990c, 1990d). Base-set organisms may be bacteria (*Vibrio fischeri*); algae (*Pseudokirschneriella subcapitata* [previously *Selenastrum capricornutum, Raphidocelis subcapitata*] or *Skeletonema costatum* [marine]); crustacean (*Daphnia* sp. [freshwater] or *Acartia tonsa, Tisbe battagliai, Crassostrea gigas* [marine]); fish (*Brachydanio rerio* [*Danio rerio*], *Poecilia reticulata, Oncorhynchus mykiss* [freshwater], or *Scopthalmus maximus* [marine]); rotifer (toxkit, *Brachionus calyciflorus* [freshwater], or *B. plicatilis* [marine]); crustacean (toxkit, *Thamnocephalus platyurus* [freshwater], or *Artemia salina* [marine]).

Very important for all tests are the validity criteria (see the discussion later in this chapter). These criteria are essential because if they are not met, the results of the test cannot be interpreted as intended. Important parameters include water-quality measurements such as pH, dissolved oxygen, ammonia, salinity, and conductivity, as well as the effect on test organisms of a reference substance (of known toxicity).

In a recent paper on aquatic toxicity testing methods for pesticides and industrial chemicals, about 450 pelagic and 260 benthic test methods from national and international test standards and the scientific literature were reviewed (OECD 1998a). In addition, about 20 test methods for determining biodegradation and elimination are listed in the current ISO work program on water quality. Only a few of the described test methods have been applied in effluent assessment. The principles of most test standards are based on OECD or ISO guidelines, as well as national standards. Test species and test methods, and (where possible) their ISO, OECD, and national standards are summarized in the appendix to this chapter.

Nonstandardized tests

The criteria recommended for selecting alternative test species or test design include the following topics (Weber 1993; Klemm et al. 1994; Chapman et al. 1995; OECD 1998b):

- Proposed species should have an ecological, commercial, or recreational importance in the receiving water.
- Species should be at least as sensitive to toxic substances as the current test species representing that phylogenetic category.
- An early life stage (ELS) should be used because it is usually the most sensitive stage.
- The ELS of the species should be readily available throughout the year.
- The species must be easy to handle in the laboratory.

- The species must give consistent and reproducible responses to toxicants.
- The toxicological endpoints should be easily quantifiable and suited for statistical analysis.
- Interlaboratory and intralaboratory validation of the test procedures should be performed.

The OECD (1998a) recommends the following tests with a high priority for OECD guideline development, although some are already standardized within specific countries or organizations (such as Environment Canada/ICES):

- Pelagic tests
- Saltwater crustacean — acute and reproduction tests
- Higher plant (cormophyta [lemna]) — growth test
- Fish — full or partial life cycle test
- Microalgae (freshwater and saltwater spp.) — growth test
- Mollusca saltwater sp. — acute on ELS and shell deposition tests
- Bacteria, sludge bacteria, and nitrification tests

Validity criteria

In protocols for toxicity tests, criteria are usually specified for checking validity. These criteria are meant, among other things, to limit deviations between replicate analyses, such as variations in oxygen content or acidity (pH) during the test, or mortality rate in the blank analysis. Other physico-chemical components may also act as modifying factors, such as nitrite, ammonia, chloride, salinity, conductivity, and temperature. If validity criteria are exceeded, the test result is unreliable and the test must be repeated. There is no obligation to report the measured validity parameters, although certain critical validity criteria should be reported (see the discussion later in this chapter). For details, the reader should refer to the specific protocols (see the appendix to this chapter).

It is essential that all validity criteria that are part of the specific tests are measured, both prior to and after the test. These validity criteria should be determined in the undiluted effluent, and if exceedance is observed there, in all concentrations of the dilution series. Only if all set validity criteria have been met can it be concluded that detected toxic effects are caused by toxic components in the investigated effluent. If there is no insight into potential exceedance of validity criteria, detected toxic effects may be erroneously attributed to toxic components present in the effluent. If the validity criteria are exceeded, it is permitted, in some cases, to apply a correction. This is possible for pH, oxygen, chloride concentration, salinity, and conductivity. It is necessary to report how and to what level corrections are made. If one of the validity criteria has been exceeded, there are three options:

- To adapt the effluent to be tested for the relevant parameter
- To test the effluent, despite exceedance of the precondition
- To abandon the tests

Adaptations in the effluent, such as pH correction, salinity increase, or dilution, may have effects on the composition of the effluent, either by changing chemical balances (pH adaptation), or by eliminating volatile components (aeration). Furthermore, the bioavailability of toxic components may be increased or decreased. The influence of these changes in the effluent on the test results is difficult to assess, and can complicate the interpretation of the test results.

Pretreatment of effluents

Turbidity

Tests using algae and crustaceans with some detection devices (such as photometry) may be disturbed by particulate matter present in the sample. In practice, such effluents may be filtered, but this can remove potentially toxic substances that may be bound to or integrated within the particles. This can lead to an underestimation of acute toxicity. The USEPA and the UKEA recommend determining the toxicity of effluents without pretreatment, if at all possible. Only when suspended matter or turbidity of the sample can affect the test result do they recommend pretreating, and in such cases, concurrent tests with and without pretreatment should be performed.

Centrifugation or settlement (30 min to 2 h) is generally preferred over filtration, and is routinely included in some testing guidelines (Deutsch Einheitsverfahren zur Wasser-, Abwasser-, und Schlammuntersuchung, 1989). Other tests methods (such as the *V. fischeri* assay) offer the possibility of determining a correction factor for parameters such as turbidity or color.

Aeration

Low oxygen content in effluents may be caused by high temperature, biodegradation (biological oxygen demand [BOD]), or chemical oxidation (COD). If the oxygen pressure is too low for organisms, aeration is necessary. This may affect the availability of some compounds, and volatile compounds may be removed from the effluent. Furthermore, oxidation may cause specific compounds to be released from complexes (such as metals from sulfides). It is therefore advisable to aerate only in those cases in which the test organisms are threatened with actual damage.

Adjustment of pH

Samples with extreme pH values (exceeding the tolerance limits of the test organisms) are generally neutralizd prior to testing. Neutralization should be omitted if the effect of pH will be reflected in the result or if physical or chemical reactions (such as precipitation) are observed owing to pH adjustment.

Effluent sampling

Sampling procedures, as well as procedures for the preservation and pre-treatment of samples, are described in detail in ISO 5667-16. The choice of representative sampling points, frequency of sampling, and so on is highly dependent on the objective of the study. The material of sample vessels should be chemically inert, easily cleaned, and resistant to heating and freezing. Glassware, polythene, or polytetrafluoroethene (PTFE) vessels are recommended. When cooled to between 0°C and 5°C and stored in the dark, most samples are normally stable for up to 24 hours. Deep-freezing below -18°C may allow a general increase in preservation but will be highly dependant on the chemical composition of the effluent in question. In general, biotests are carried out with the sample as received.

Sampling should take place at a point appropriate to the objectives of the testing. It is proposed that routine regulatory testing take place at the end of pipe, but the way in which the result is interpreted and used should take into account the dilution available in the receiving water, as well as other receiving-water characteristics. During the characterization of the effluent, sampling may take place at many different places, such as at the end of pipe, at a point in the receiving water, or upstream and downstream of the discharge outlet, in order to see how the toxicity in the water changes (UKWIR 2001b, 2001c). If unacceptable toxicity is found in the effluent, sampling may take place further up in the process to determine the sources of the toxicity. (UKEA 1996a, UKWIR 2001a).

Tiered approaches for the assessment of effluent toxicity

A combined chemical and effect-oriented assessment of effluents is now generally regarded as the most effective approach. For example, at the level of the European Union this is established in the IPPC Directive and in the Water Framework Directive. The combined approach makes use of two elements: the application of the best available technology (BAT) to reduce emissions (an emission-based approach), and the use of monitoring to check whether water-quality objectives are met.

As already mentioned, a chemical-specific approach has limitations, and it is not possible to assess the true environmental hazard of a complex effluent based on the levels of specific substances alone. Whole-effluent assessment (WEA) or direct toxicity assessment (DTA) can offer solutions to this problem (Tonkes et al. 1998). The aim of whole-effluent assessment is to gather data on the combined effects of all known and unknown hazardous substances in effluents, and of the interactions between them, by making use of measurements of biological effects using bioassays.

The same persistence, bioaccumulation, and toxicity characteristics (PBT) that are used for the chemical-oriented approach are all incorporated into WEA. They are assessed by means of persistence, bioaccumulation, toxicity (acute and chronic), and genotoxicity parameters.

At this moment specific research in the field of WEA is being performed in various countries such as the Netherlands and the U.K. A number of countries use WEA (or parts of it) within regulatory practice, including the U.S. and Germany. The U.S. and Germany already have extensive experience in determining acute toxicity that dates back 10 to 15 years. The results are used to start or enforce discharge-quality improvements at production plants. The acute toxicity parameter has been included in legislation in both countries. Another similarity between the U.S. and Germany is that there are interstate differences in the way in which WEA is applied.

Other countries with experience in acute toxicity for the regulation of effluents are the U.K., the Netherlands, Belgium, Sweden, Denmark, Ireland, France, Portugal, and Canada (Tonkes and Botterweg 1994; Tonkes et al. 1995; UBA 1999).

In the following country-specific examples, the potential use of tiered approaches to assess effluents is shown in more detail. More information can be gathered from extensive overviews by Tonkes and Botterweg (1994), Tonkes et al. (1995), and UBA (1999).

The Netherlands

Within the Dutch emission policy, the assessment of wastewater discharges or effluents is focused on the precautionary principle: the reduction of specific pollutants or substances. Depending on the characteristics and the environmental hazard of a substance, the discharger must remediate a discharge that is known to contain the substance.

This emission approach has three phases:

- Prevention of pollution
- Reuse of water and substances where possible
- End-of-pipe treatment

The substance-oriented approach focuses on BAT and further demands are based on certain national criteria (such as maximum permissible risk). In addition, the Netherlands uses a water-quality approach, which is based on environmental quality criteria. Finally, a stand-still approach is used for new discharges and for the extension of existing discharges.

Many effluents in the Netherlands are nevertheless of a complex nature. In the last few decades, numerous measures have been taken to limit surface-water emissions. This has led to an improvement in surface-water quality, but not all the water-quality targets have been reached. In addition to certain substance-specific standards being exceeded, biological effects have also been observed in numerous places in the surface water (Hendriks 1994).

Only a limited number of these effects can be explained by the presence of known substances. Clearly, there is a need for methods that fully define the potential effects or identify the relevant substances or sources.

The Institute for Inland Water Management and Waste Water Treatment (RIZA) started work on the development of effect-oriented methods or techniques in the early 1990s. This resulted in a first report on the use of acute toxicity tests for the assessment of complex effluents (Beckers-Maessen 1994). RIZA is currently developing a method for whole-effluent assessment that considers the following five parameters (see Figure 2.1):

- Acute toxicity: specific short-term, lethal, or potentially lethal effects that occur as a result of exposure to a substance or medium
- Chronic toxicity: specific longer-term, nonlethal effects that occur as a result of exposure to a substance or medium
- Bioaccumulation: the net accumulation of a substance in an organism as a result of combined exposure via direct surroundings and food
- Genotoxicity: the ability to cause damage to genetic material or cause an adverse effect in the genome, such as mutation, chromosomal damage, and so on
- Persistence: a substance property indicating how long a substance remains in a certain environment before being converted physically, chemically, or biologically

For WEA the same assessment parameters are used as for the assessment of specific substances. The WEA method is not meant to predict the effects

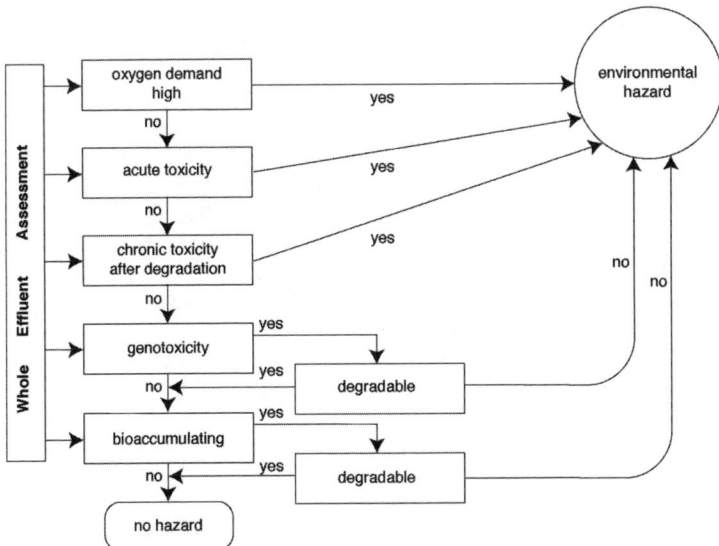

Figure 2.1 Whole effluent assessment in the Netherlands.

on the receiving-water body, but to complement the assessment of components that are known to be present in a complex effluent (Tonkes et al. 1995). Figure 2.2 shows a possible stepwise procedure for the hazard and risk assessment of complex mixtures (after Tonkes et al. 1995).

The use of WEA is to be an extension of the Dutch emission policy. The possible effects from effluents are only monitored at the end of pipe, and within the process or sewerage systems. Assessing the biological effects of discharges in the receiving water is not yet practiced in the Netherlands.

Germany

In Germany, WEA has been included in routine regulatory practice since 1976 (UBA 1999). The environmental policy emphasizes the emission-based approach, and the water-quality–based approach has been developed in parallel. According to Section 7a of the German Federal Water Act (WHG), discharge permits are granted if the waste load is kept within the current BAT level. The requirements for BAT were established by the federal gov-

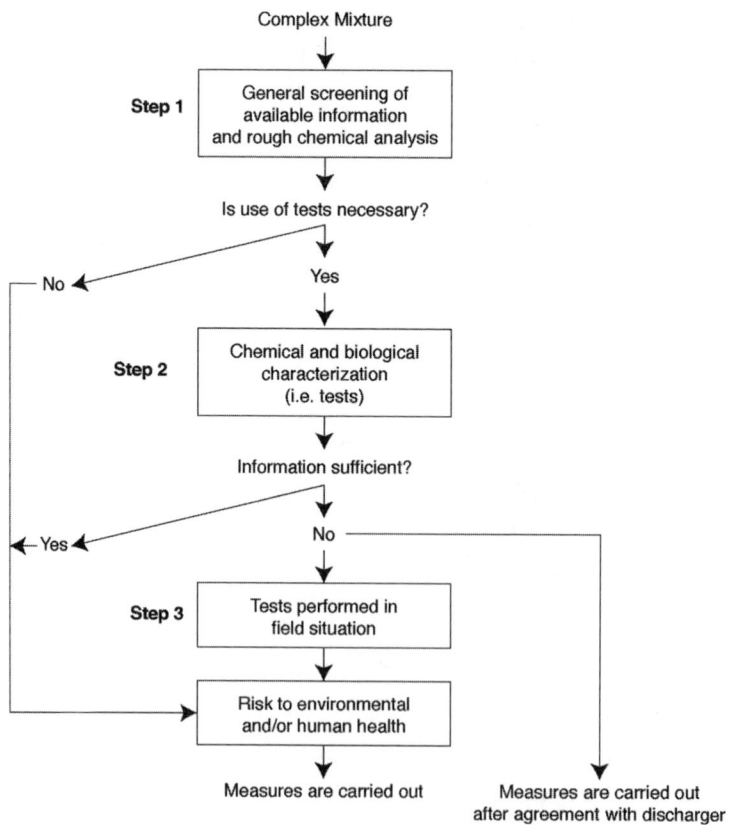

Figure 2.2 Complex mixtures.

ernment in the appendices of the Waste Water Ordinance (AbwV) for the different industrial branches and processes, and updated according to further developing knowledge. Discharge limits to different wastewater sectors are set in about 50 annexes of the AbwV.

The guiding philosophy for implementation of biotests in WEA is the precautionary principle (to do all that can be reasonably expected to prevent unnecessary risks) and the polluter pays principle (PPP — the principle that transfers the financial burden for the prevention and control of pollution onto the party responsible for its generation). The emphasis of the German approach is on emission reduction at the source; so it does not include environmental risk assessment that considers the flow capacity of the receiving body.

German experience over the last 23 years has shown that this approach assists in the further development of BAT. Coupling WET with BAT guarantees equal treatment of the dischargers in the different branches of industry, regardless of the water quality of the receiving waters.

The evaluation of toxicity tests follows the concept of lowest ineffective dilution (LID) (ISO 1998), which is exclusively applied in Germany. LID is the most concentrated effluent dilution at which there is no observed effect on the test organism, or there are only effects that do not exceed the test-specific variability. LID is expressed as the reciprocal value of the volume fraction of wastewater in the test dilution.

Currently biotests for other endpoints such as bioaccumulation, endocrine disruptors, immunotoxicity, and mutagenicity (with eukaryotic cells) are all under development.

Apart from the emission-based approach described here, water-quality surveys using bioindicators are active. Passive monitoring for emission control became routine in Germany in the 1950s. In the 1970s, coastal areas were also included in the monitoring programs. Recently, chemical quality assessment has been implemented in addition to the biological quality assessment, which describes water quality by means of seven categories.

In special cases, ambient toxicity close to the effluent discharge location is also determined, but not on a routine basis. In large rivers (such as the Rhine and the Elbe), continuous biological monitoring devices (daphnids, dreissena) are in operation as early-warning systems.

United Kingdom

U.K. water-quality management policy requires, on the whole, that consideration is taken of the quality of receiving watercourses; this is known as the water-quality approach. Environmental quality standards (EQSs) are used to protect the ecosystem and maintain the quality for specific use, taking into account dilution and dispersion (Tonkes et al. 1995).

Recommendations have been made to include direct toxicity assessment (DTA) in the assessment of effluents. Whole-effluent parameters such as bioaccumulation and persistence are also in development. DTA has been used widely in the context of research, development, and demonstration, and numerous projects have been completed to support the use of DTA to monitor and control effluents. These include projects to:

- Develop and evaluate existing methods specifically for effluent and receiving-water assessment, such as a *Daphnia magna* reproduction test, and a *Tisbe battagliai* population-level test
- Improve and standardize methods, such as producing method guidelines for effluent and receiving-water assessment (UKEA 1999a, 2001a, 2001b)
- Develop quality-control and assurance procedures, such as performance standards for ecotoxicity tests (WRC 1996)
- Improve the way in which ecotoxicity test data are used in risk assessment, such as developing a risk framework for direct toxicity assessment of effluent discharges (UKEA 1999b; UKWIR 2001a)
- Demonstrate the use of the tests in the management of effluents, such as the Direct Toxicity Assessment Demonstration Programme (UKWIR 2001a, 2001b, 2001c)

Research and development has been undertaken to investigate and demonstrate the benefits of using DTA in assessing effluents. DTA offers these benefits:

- DTA provides a synopsis of the effects of all constituents. This includes unknown and unidentified chemicals, and chemicals that may be breakdown products.
- DTA can provide a measure of additivity and other combined effects, and is effective in assessing complex mixtures.
- DTA can help where known chemicals are present in the effluent, but where little or no toxicity data exist.
- DTA measures relate to the monitoring endpoints (receiving-water biological status) better than chemical surrogates, and for some tests this relationship may be modelled.
- DTA is a proactive biological measure, which can be used to predict potential impact, and to provide a measure of hazard.
- DTA can provide a useful summary measure for process control, and is a holistic measure for determining variability in the composition of complex effluents.

Some DTA tests are cost-effective compared to chemical analysis, considering the relevance and holistic nature of the measurements made (Boumphrey et al. 1999).

Nationally standardized (UKEA and the Scottish Environmental Protection Agency) and internationally standardized (OECD) acute-toxicity tests with fish (*Oncorhynchus mykiss and Cyprinus carpio*), acute and chronic tests with *Daphnia magna,* and tests with algae (selenastrum, skeletonema), *Vibrio fischeri,* and various other organisms (oyster embryo-larval, *Tisbe battagliai, Acartia tonsa, Gammarus pulex,* and *Lemna minor*) have all been used in research and development projects (UKEA, 1996a, 1996b, 1996c). The UKEA (2001a) defines those tests to be used within DTA assessments.

The UK has developed a seven-stage protocol for assessing and regulating effluents (UKWIR 2001a; see Figure 2.3). This protocol has been derived as a result of previous research and development (National Rivers Authority 1993) and public consultation, and was tested in the DTA Demonstration Programme, a collaborative among the UK regulators, industry, and water companies.

The protocol enables the regulator to prioritize resources, and investigate and manage complex effluents. The first stage of the protocol directs the investigation toward receiving waters where the biological quality of the aquatic system is already impaired (the existing "worst cases"), and where there is a likelihood that this is due to toxic substances (as opposed to, for example, oxygen depletion). The effluents are then characterized using a range of toxicity tests, a risk assessment is made, and a level of toxicity is derived at which no harm is thought to occur in the receiving water. If unacceptable toxicity is found in the receiving water, a site and process audit and toxicity identification evaluation (TIE) would be undertaken, and a

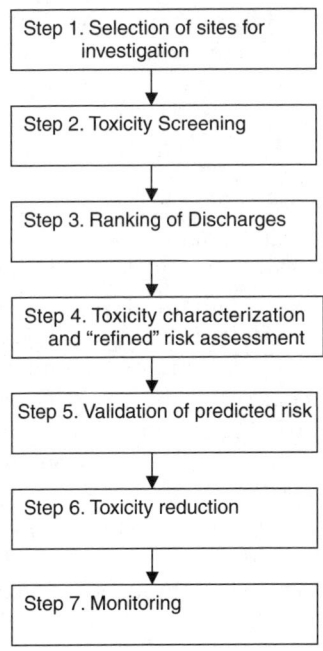

Figure 2.3 Proposed scheme for direct toxicity assessment (DTA) in England.

toxicity-reduction program derived. This would be assessed using BAT criteria; a plan for implementation, with associated timescales, would be put forward to the regulator. The plan would be implemented, and the success of the program, in terms of toxicity reduction and changes in the receiving environment, appraised and fed back into the management process.

The British approach focuses on three levels:

- End of the pipe
- Toxicity close to the outlet
- Changes of the ecosystem related to toxicity and other anthropogenic effects

Development of DTA is ongoing, and toxicity assessment methods that will better predict the effects of continuous low-level exposures of chemical mixtures on populations of organisms, as well as *in situ* receiving-water tests, biomarkers, and biosensors, are being developed and validated. Toxicity limits may not be applied to industry on a sector-by-sector basis, but on a site-specific, case-by-case basis, taking into account the needs of the receiving-water environment.

Most recently (Leverett 2003), the UKEA has prioritized a number of industrial effluents based on intrinsic hazard (measured toxicity). The final ranking of these effluents will eventually also account for the environmental risk (volume of discharge, dilution in the receiving environment, flows, tides, and so on). Once complete, this will allow the focusing of resources on the control and remediation of effluents with the potential to cause most toxicity problems in the environment.

United States

The U.S. is believed to be the most progressive country outside Europe as far as the prescription of toxicity requirements in discharge permits is concerned. Many states have legally based toxicity requirements (Tonkes and Botterweg 1994). WET testing has an important role in the USEPA water-quality program. Most industries are regulated by effluent guidelines based on the best available (economic) technology. Heber et al. (1996) reported over 6500 effluent permits including WET monitoring or WET limits on a case-by-case basis. The WEA guidelines developed by the USEPA were published in detail, and technical documents are available on the Internet (Weber 1993; Lewis et al. 1994).

Since the 1980s, acute and chronic toxicity limits have also been incorporated into the wastewater discharge permits of industrial and municipal treatment facilities, but the test methods vary geographically. There are guidelines for conducting toxicity identification and reduction evaluations of toxic effluents using BAT.

The detailed environmental hazard and risk assessment scheme is shown in Figure 2.4 and Figure 2.5.

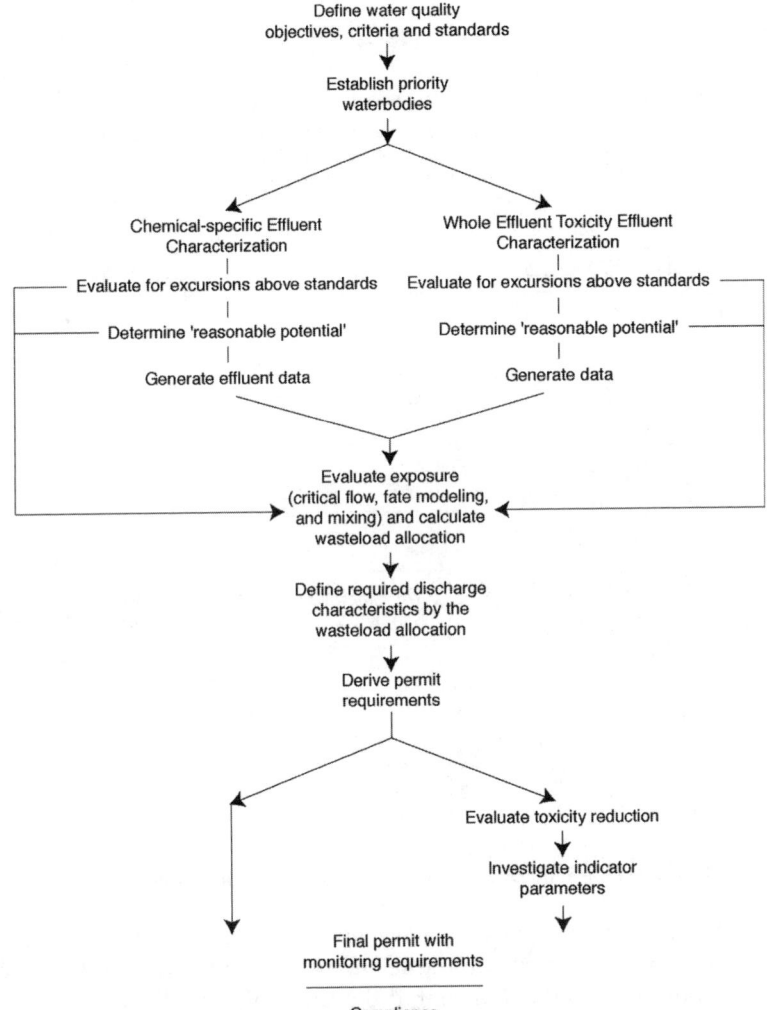

Figure 2.4 Overview of water-quality–based "standards to permits" process for toxics control.

The Clean Water Act and EPA regulations authorize and require the use of an integrated strategy to achieve and maintain water-quality standards, considering chemical-specific analysis, biosurveys in the receiving water, and WET. The WET program gives a characterization of the whole toxicity of an effluent without necessarily knowing all of its components and considering the effects of bioavailable substances. The strategy is completed with toxicity-reduction evaluations (TREs) and toxicity-identification evaluations (TIEs) (Huwer et al. 1999) in order to identify and reduce pollutants at the source (Tonkes et al. 1995).

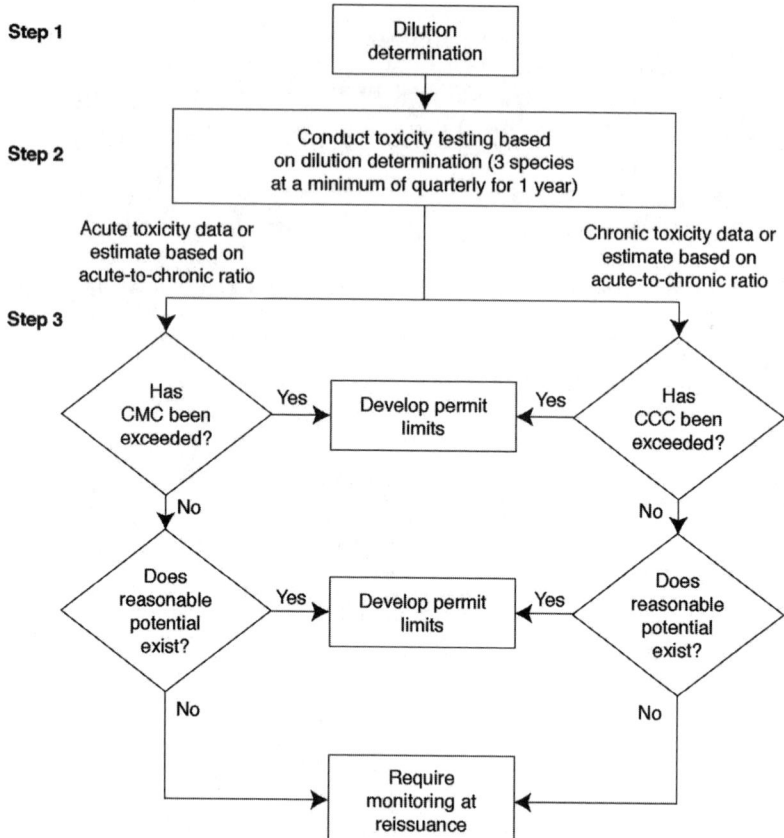

Figure 2.5 Effluent characterization for whole-effluent assessment.

Grothe et al. (1996) gives an overview of a workshop held in Pellston, MI, in 1995 that was focused on the science of WET testing. Grothe provides a state-of-the-art overview (current at the time) of the following topics:

- The appropriateness of the endpoints used in routine WET methods
- The degree and causes of method variability in WET testing
- Biotic and abiotic factors that can influence measured field responses to effluents
- The relationship between effluent toxicity, ambient toxicity, and receiving-system impacts

Conclusions

Based on the preceding information, a number of concise conclusions can be drawn regarding the use of bioassays for the assessment of surface waters and effluents.

Surface water

Recommendations have been made for the use of biotests to monitor chemical pollution in surface waters, and biotests have been implemented to varying degrees in monitoring programs across the world.

The full development of a battery of longer-term tests with sublethal endpoints is required to provide ecotoxicological measurements of sufficient sensitivity to adequately detect biological effects (especially those at a population level) within the environment.

Effluents

Chemical group parameters and biological tests should be employed in combination in order to assess the toxicity of complex effluent mixtures (UBA 1999). There is strong agreement among scientists that the toxicity testing of effluents should be based on a battery of tests covering different trophic levels. The most widespread taxonomic groups used in effluent toxicity testing are bacteria, algae, crustaceans, and fish.

Most countries that make use of an emission-based approach (end of pipe) have started to use acute toxicity tests to assess the quality of effluents. The use of other parameters such as chronic toxicity, genotoxicity, bioaccumulation, and persistence are not currently commonplace. A hazard assessment or even risk assessment of the recipient water is rarely performed.

It is clear that the focus of effluent-quality assessment lies with hazard assessment.

The water-quality–based approach focuses on the evaluation or assessment of ambient toxicity and takes into account the flow capacity of the receiving river.

There are numerous international testing guidelines to determine aquatic toxicity or degradability of single substances that can be modified for use with wastewater evaluation. A limited number of suitable test methods have been developed to address the specific conditions of wastewater.

There is an urgent need to create a wider range of internationally accepted standards for toxicity and degradability tests of wastewater. Test principles should focus on the same type of endpoints as are used for the evaluation of hazardous substances in order to attain a broad acceptance of the methodologies. It is, however, generally accepted that methods developed for single-substance tests cannot be used without some consideration of the inherent differences in testing effluent and wastewater samples.

The need to integrate genotoxicity testing in WEA is widely agreed upon, although possible hazard effects of genotoxins to the environment remain unclear. It is accepted that no individual test represents all possible endpoints. A test battery is therefore recommended. Up to now only bacterial tests (umuC test, Ames test, SOS chromo test) have been applied to a wide range of wastewater samples. The need for other test systems on a higher organism level is recognized, but currently no internationally accepted guidelines exist.

There is also a need for further development, validation, harmonization, and implementation of test systems to measure bioaccumulation (de Maagd 2000), endocrine disruptors, and genotoxicity (de Maagd and Tonkes 2000).

References

Beckers-Maessen, C.M.H. (1994). *Toxicity tests in the WVO law regulatory framework.* In Dutch. RIZA document 94.071X.

Bergmann H.L., Kimerle R.A., and Maki A.W. (1986). *Environmental hazard assessment of effluents.* Elmsford N: Pergamon.

Boumphrey, R., Tinsley D., Forrow, D., and Moxon, R. (1999). *Whole effluent assessment in the UK.* Presentation to OSPAR workshop on whole effluent assessment, Lelystad, the Netherlands, 28-29 October 1999.

Chapman, G.A., Anderson, B.S., Bailer, A.J., Baird, R.B., Berger, R., Burton, D.T., Denton, D.L., Goodfellow, W.L., Heber, M.A., McDonald, L.L., Norberg-King, T.J., and Ruffier, P.J. (1995). Session 3: methods and appropriate endpoints, discussion synopsis, 83-90, in R. Grothe; K.L. Dickson and D.K. Reed-Judkins, *Whole effluent toxicity testing: an evaluation of methods and prediction of receiving system impacts,* SETAC Press, Pensacola, FL.

Creasel (1990a). THAMNOTOXKIT FTM, *Crustacean toxicity screening test for freshwater. Standard operational procedure.* V071090.

Creasel (1990b). ROTOXKIT FTM, *Rotifer toxicity screening test for freshwater. Standard operational procedure.* V071090.

Creasel (1990c). ROTOXKIT MTM, *Rotifer toxicity screening test for estuarine and marine waters. Standard operational procedure.* V071090.

Creasel (1990d). ARTOXKIT FTM, *Artemia toxicity screening test for estuarine and marine waters. Standard operational procedure.* V071090.

de Graaf, P.J.F., Graansma, J., Tonkes, M., Ten Kate, E.V. and Beckers, C.M.H. (1996). *Acute toxicity tests, an addition to the chemical specific approach?* Research of industrial effluents in North-Netherlands and South-Holland. In Dutch. FWVO report 96.03.

Deutch Einheitsverfahren zur Wasser-, Abwasser-, und Schlammuntersuchung (1989). Bestimmung der nicht akut giftigen Wirkung von Abwasser gegenüber Daphnien über Verdünnungstufen DEV. Teil 30, DIN 38412.

de Maagd, P.G.-J. (2000). Bioaccumulation tests applied in whole effluent assessment: a review. *Environ. Toxicol. Chem.* 19, 25–35

de Maagd, P.G.-J and M. Tonkes. (2000). Selection of genotoxicity tests for risk assessment of effluents. *Environ. Toxicol.* 15, 81–90

Falcke, H. and Irmer, H. (1994). Monitoring of wastewater from chemical industry with summative parameters and bioassays. In *Federal environmental agency and Netherlands institute for inland water management and waste water treatment (RIZA) workshop on emission limits inorganic chemical industry,* UBA-Texte 47/94.

Grothe, D.R., Dickson, K.L. and Reed-Judkins, D.K., (Eds.), (1996). *Whole effluent toxicity testing: an evaluation of methods and prediction of receiving systems impacts.* SETAC Pellston workshop on whole effluent toxicity; 16-25 Sept. 1995, Pellston, MI.

Heber, M.A., Reed-Judkins, D.K., and Davies, T.T. (1996). USEPA's whole effluent toxicity testing program: a national regulatory perspective, 9–15, in D.R. Grothe; K.L. Dickson and D.K. Reed-Judkins, *Whole effluent toxicity testing: an evaluation of methods and prediction of receiving system impacts*, SETAC Press, Pensacola, FL.

Hendriks, A.J. (1994). Monitoring and estimating concentrations and effects of microcontaminants in the Rhine-delta: chemical analysis, biological laboratory assays and field observations. *Water Sci. Technol.* 29, 223–232.

Houk, V. S. (1992). The genotoxicity of industrial wastes and effluents: a review. *Mutation Res.* 277, 91–138.

Hubbard, S.A. et al. (1984). The fluctuation test in bacteria, in *Handbook of mutagenicity test procedures*, second ed. , B.J. Kilbey, M. Legator, W. Nichols, and C. Ramel (Eds.). Elsevier Sciences, New York, 141–161.

Huwer, S.L. and Brils, J.M. (1999). *The role and application of toxicity identification evaluation (TIE) in the United States.* TNO report TNO-MEP-R 99/331, Aperdorn, the Netherlands.

International Standards Organisation (ISO) (1998). *Water quality — sampling — part 16: guidance on biotesting of samples*, ISO 5667-16:1998(E).

Klemm, D.J., Morrison, G.E., Norberg-King, T.J., Peltier, W.H., and Heber, M.A. (1994). *Short-term methods for estimating the chronic toxicity of effluents and receiving waters to marine and estuarine organisms.* 2nd ed., EPA/600/4-91/003, U.S. Environmental Protection Agency, Cincinnati, OH.

Leverett, D.H. (2003). *Derivation of a baseline list of point source discharges from industrial effluent plants causing acute toxic effects.* Environment Agency for England and Wales internal report)

Lewis, P.A., Klemm, D.J., Morrison, G.E., Lazorchak, J.M., Norberg-King, T.J., Peltier, W.H., and Heber, M.A. (1994). *Short-term methods for estimating the chronic toxicity of effluents and receiving waters to freshwater organisms*, third ed. EPA/600/4-91/002, U.S. Environmental Protection Agency, Cincinnati, OH.

National Rivers Authority. (1993). *R&D note 210: definition study for the implementation of toxicity based consents.*

Niederländer, H.A.G., Dogterom, J., Buijs, P.H.L., Hupkes, R., & Adriaanse, M. (1996) State of the art on monitoring and assessment of rivers. *UN/ECE task force on monitoring and assessment*, vol. 5. RIZA report 95.068.

OECD. (Organisation for Economic Cooperation and Development) (1998a). *Detailed review paper on aquatic testing methods for pesticides and industrial chemicals.* OECD Environmental Health and Safety Publications, Series on Testing and Assessment, No. 11.

OECD. (1998b). *Report of the OECD workshop on statistical analysis of aquatic toxicity data.* OECD Environmental Health and Safety Publications, Series on Testing and Assessment, No. 10.

Simpson, P. and Grist, E.P.M. (2003). *Predicting the long-term sub-lethal toxicity of complex effluents using* Tisbe battagliai. Environment Agency for England and Wales internal project report.

Tonkes, M. and Botterweg, J. (1994). *Totaal Effluent Milieubezwaarlijkheid. Beoordelings-methodiek milieubezwaarlijkheid van afvalwater.* Literatuur- en gegevensevaluatie. RIZA Notanr. 94.020.

Tonkes, M., van de Guchte, C., Botterweg, J., de Zwart, D., and Hof, M. (1995). *Monitoring strategies for complex mixtures: monitoring water quality in the future,* Vol. 4, Ministry of Housing, Spatial Planning and the Environment, Department of Information and International Relations, the Hague, The Netherlands.

Tonkes, M. and Baltus, C.A.M. (1997). *Praktijkonderzoek aan complexe effluenten met de totaal effluent milieubezwaarlijkheit (TEM)-methodiek – resultaten van 10 complexe effluenten.* Ministerie van verkeer en waterstaat, rijksinstituut voor integraal zoewaterbeheer en afvalwaterbehandling (RIZA) (mei 1997).

Tonkes M., Pols, H., Warmer, H. and Bakker, V. (1998). *Whole effluent assessment.* RIZA report 98.034, Institute for Inland Water Management and Waste Water Treatment (RIZA).

UBA. (1999). *Ecotoxicological evaluation of wastewater within whole effluent assessment* (Draft). OSPAR background document concerning the elaboration of programs and measures relating whole effluent assessment, November 1999, presented by the German Federal Environmental Agency (Umweltbundesamt, UBA).

UKEA (Environment Agency for England and Wales). (1996a). *R&D technical report P23 and project record R&D P2 – 493/11, toxicity based consents – pilot study.*

UKEA. (1996b). *R&D technical report P28, toxicity reduction evaluation: case summary for the pulp and paper industry.*

UKEA. (1996c). *R&D technical report P29, toxicity reduction evaluation: case summary for the chlor-alkali industry.*

UKEA. (1999a). *R&D technical report E83, short-term ecotoxicological method guidelines for effluent and receiving water assessment* (draft).

UKEA. (1999b). *R&D technical report E88, towards a risk framework for direct toxicity assessment of effluent discharges.*

UKEA. (2001a). *Ecotoxicity test methods for effluent and receiving water assessment — comprehensive guidance.*

UKEA. (2001b). *Ecotoxicity test methods for effluent and receiving water assessment — supplementary advice to international test guidelines.*

UKWIR (UK Water Industry Research). (2001a). *UK direct toxicity assessment (DTA) demonstration programme, technical guidance,* Report Ref 00/TX/02/07.

UKWIR. (2001b). *UK direct toxicity assessment (DTA) demonstration programme, Lower Tees estuary project—Part 1,* Report Ref 00/TX/02/03.

UKWIR. (2001c). *UK direct toxicity assessment (DTA) demonstration programme, Lower Tees estuary project—Part 2,* Report Ref 00/TX/02/04.

USEPA (1995). *Whole effluent toxicity: guidelines establishing test procedures for the analysis of pollutants.* (Final rule October 16, 1995, Vol 60, No. 199).

van de Guchte, C. (1992) The sediment quality TRIAD: an integrated approach to assess contaminated sediments. In P.J. Newman, M.A. Piavaux, and R.A. Sweeting (Eds.) *Riverwater quality: ecological assessment and control,* EC, Luxembourg, pp. 425–431

WRC (Water Research Centre). (1996). *Performance standards for ecotoxicity tests.* Report No. SR 4166/1, Medmenham. England.

Appendix

Regulatory test batteries

The following acute tests are most commonly used in Germany (UBA 1999): Fish (*Leuciscus idus, Brachydanio rerio, Cyprinus carpio, Dicenthrarcus palrax, Gasterosteus aculeatus, Salmo trutta* and *Salmo salar*), daphnia (*Ceriodaphnia dubia, Daphnia magna, D. pulex*), algae (*Scenedesmus subspicatus, Pseudokirschneriella subcapitata*), and bacteria (*Vibrio fischeri, Pseudomonas putida*, activated sludge, anaerobic digestor sludge).

In the U.S., 10 freshwater and marine organisms are commonly used: *Ceriodaphnia dubia, D. magna, D. pulex*, fathead minnow (*Pimephales promelas*), rainbow trout (*Oncorhynchus mykiss*), brook trout (*Salvelinus fontinalis*), mysids (*Nysudiosis bahia* and *Holmesimysis costata*), bannerfish shiners (*Notropis leedsi*), sheepshead minnows (*Cyprinodon variegatus*), and silversides (*Menida menidia, M. beryllina*, and *M. peninsulae*) (USEPA 1995).

In the U.K., the following tests are recommended by the Environment Agency for England and Wales (UKEA) and the Scottish Environmental Protection Agency (SEPA) for use in effluent and receiving-water assessments (UKEA 2001a, 2001b): fish (*Oncorhyncus mykiss, Scophthalmus maximus*), crustacea (*D. magna, Tisbe battagliai, Crassostrea gigas*), algae (*Pseudokirschneriella subcapitata, Skeletonema costatum*), and bacteria (*Vibrio fischerei* — SEPA only).

Freshwater acute tests using fish

The acute fish toxicity test is one of the most commonly used biotest methods. It is well established and internationally standardized (ISO, OECD). It is also applied to determine the wastewater fee (Abwasserabgabe) in Germany, for wastewater permits in the U.S., and for environmental hazard and risk assessment of industrial effluents in Sweden.

The fish are exposed to the test substance or wastewater for a period of 48 to 96 hours. Mortalities are recorded and the LC50, LID/LOEC, and NOEC are determined. In general at least seven fish are used for each concentration and in the controls. In order to derive a point estimate value (such as LC50)

at least five concentrations must be examined. In Germany, a national standard method with three fish for each concentration is applied for the determination of toxic effects in wastewater.

In England and Wales, although both freshwater and marine fish tests
are recommended for use in effluent characterization (if they are considered
to be the most sensitive species), they are not used for screening, TIE, monitoring, or receiving-water assessment because of ethical concerns regarding
the use of vertebrate test species for toxicity testing.

Freshwater acute tests using invertebrates

Daphnia immobilization test

As with the toxicity test with fish, the test with daphnia is also widely
recommended and internationally standardized (ISO, EN, OECD). Labora-
tory-bred daphnids less than 24 hours old at the beginning of the test,
apparently healthy, and with a known history (breeding method and pretreatment) are used in the test. At least 20 animals, preferably divided into
four replicate groups of five animals each, should be used for each test
concentration and for the controls. The test duration is 24 or 48 hours. The
proportional immobility at 24 hours and (if required) at 48 hours is used to
calculate a point estimate result (such as EC50) with associated confidence
limits. Further EC values (EC0, 10, 100) may also be reported. In Germany,
a more cost-effective, standard method with 10 animals per concentration is
used to determine the lowest ineffective dilution (LID). Acute toxicity tests
with *Cerodaphnia dubia* are recommended by the USEPA for assessing effluent
toxicity, but no reference guideline is available (Weber 1993).

Gammarid toxicity test

Experience with gammarids is reported by the Netherlands and Denmark
as contracting parties, but the test is so far only standardized by the USEPA.

The USEPA specifies the amphipods *Gammarus fasciatus, G. pseudolim-
naeus,* and *G. lacustris* for the test. The mortalities of the test organisms are
recorded at 24, 48, 72, and 96 hours and the LC50 is determined. A minimum
of 20 gammarids per concentration are exposed to five or more concentrations. The 48-, 72-, and 96-hour LC50 and their corresponding 95% confidence
limits are reported. If sufficient data have been generated, the 24-hour LC50
value is reported.

Toxicity tests with rotifers

Rotifers are used in WET testing (in Denmark and Belgium), but to date no
internationally standardized test procedures exist. The rotifer *Brachionus
calyciflorus* has been applied in toxkit methods, and an ASTM standard is
available.

Toxicity tests with protozoans

Protozoans are used in WET testing (in Denmark and Belgium), but to date no internationally standardized test procedures exist. The ciliate tetrahymena is part of the Danish risk-assessment scheme for effluents, and an OECD standard protocol has been elaborated (Pauli and Berger 1996).

Freshwater acute tests using bacteria

Bacterial toxicity tests have been applied to assess indirect discharges with the objective of protecting biological treatment plants. In this sector toxicity tests with activated sludge (inhibition of respiration or nitrification) are used. Additionally, bacterial toxicity is measured routinely to assess impacts of specific direct discharges, such as cooling water, on the aquatic environment.

Activated sludge respiration inhibition test

This method is used to determine the toxicity of individual substances and effluents. It is reported by different countries (Spain and the UK) and also used for environmental hazard and risk assessment of industrial effluents (in Sweden). The method is internationally standardized (ISO, OECD).

This method assesses the effect on microorganisms by measuring the respiration rate. The purpose of this test is to provide a rapid-screening method whereby substances that may adversely affect aerobic microbial treatment plants can be identified. The test is also used to indicate suitable noninhibitory concentrations of test substances to be used in biodegradability tests. The oxygen consumption of aerobic sludge is measured to determine the respiration rate. At least five concentrations should be used. The test duration is 3 hours. An EC50 value is calculated.

Nitrification inhibition test

This test is useful to identify toxic effects on nitrification processes in aerobic microbial treatment plants. It is internationally standardized (ISO) but is a fairly novel test and therefore not yet widely used in effluent assessment.

For determining nitrification inhibition, the concentration of nitrite and nitrate is examined. Most commonly, five different concentrations of test substance are used. Inhibition is calculated comparing the nitrification rate of test vessels with that of the controls. The test duration is 4 hours.

Vibrio fischeri *toxicity test*

The acute toxicity test with *V. fischeri* (previously known as *Photobacterium phosphoreum*) is one of the most commonly used biotest methods. It is well established and internationally standardized (ISO). It is used by many member states (such as Belgium, Finland, Germany, and Sweden) for environmental hazard and risk assessment of industrial effluents.

The method uses inhibition of light emission by the marine bacterium *V. fischeri*. The test is performed using a specially designed apparatus. Light emission is measured photometrically from a suspension of exposed bacteria. Test duration is most commonly 15 or 30 minutes. In a number of publications, a duration of only 5 minutes is also prescribed. EC50 values or LIDs are determined. The procedure is used to examine aqueous effluents, leachates, surface water, and chemicals.

Freshwater short-term chronic tests

The endpoints generally used in short-term chronic tests are survival, growth, and reproduction. Most test guidelines specially adapted for effluent testing have been developed in recent years in the United States.

Early life stage (ELS) fish toxicity test

Fathead minnow (Pimephales promelas*) survival and growth test.* Larvae (preferably less than 24 hours old) are exposed in a semistatic system to control water and to at least five concentrations of effluent or receiving water for 7 days. Test results are determined from the survival and weight of the exposed larvae as compared to controls. Toxicity endpoints are NOEC, with no adverse effect on survival or growth observed, and IC25 (inhibition concentration for a 25% effect). USEPA recommends the IC25 for regulatory use.

Fathead minnow (Pimephales promelas*) embryo-larval survival and teratogenicity test.* Fathead minnow embryos are exposed in a semistatic system to control water and at least five different concentrations of effluent or receiving water, from shortly after egg fertilization to hatch. The larvae are exposed an additional 4 days posthatch (for a total of 8 days). Test results are determined on the combined frequency of both mortality and gross morphological deformities (terata) in test solutions, compared to the controls. The test is useful for screening for teratogens, which may be most biologically active during embryonic development. Toxicity endpoints are NOEC, with no adverse effect on survival, growth, or reproduction observed, and IC25. USEPA recommends the IC25 for regulatory use.

Ceriodaphnia dubia *survival and reproduction test*

Ceriodaphnia is closely related to daphnia, but is smaller and has a shorter generation time of 3 to 5 days, compared with 6 to 10 days for daphnia (Weber 1993). For that reason ceriodaphnia is increasingly used to determine the reproductive toxicity of test substances. Ceriodaphnia neonates are exposed to control water and at least five different concentrations of effluent or receiving water, in a semistatic system, for a maximum of 8 days. Test results are based on survival and reproduction in test solutions, compared to controls. Toxicity endpoints are NOEC, with no adverse effect on survival, growth, or reproduction observed, and IC25 (inhibition concentration for 25% effect). EPA recommends the IC25 for regulatory use.

The UKEA, in collaboration with others, has also developed a short-term chronic test (7 to 12 days) using *D. magna*.

Chronic rotifer toxicity test

A 48-hour reproduction test method with *Brachionus calyciflorus* has been ring-tested by 12 laboratories in France. The standardized method was published in early 2000.

Pseudomonas putida *growth inhibition test*

The growth inhibition test with *Pseudomonas putida* is internationally standardized (ISO, OECD) and also recommended in Denmark for investigation and assessment of hazard and risk to freshwater environments.

The test is used to determine the growth inhibition of *P. putida* in relation to a control culture. The test duration is 16 hours. The test is performed as a LID procedure with dilution factors 2, 4, 8, 16, 32, and 64. The results are given as EC10 and EC50 values.

The bacterium *P. putida* is used as a representative for heterotrophic microorganisms in freshwater. The test is used to determine the toxicity of water, wastewater, and water-soluble substances. The test procedure is not suitable to examine strongly colored or highly turbid samples.

Vibrio fischeri *growth inhibition test*

In addition to the acute toxicity test with *V. fischeri*, an inhibition growth test has been developed to determine chronic toxicity effects. Bacteria are incubated for 7 hours with the test item and the inhibition of growth is determined. In Germany a national standard guideline is available.

Anaerobic bacteria inhibition test

This standard (ISO draft) prescribes a screening method for assessing the potential toxicity of substances, mixtures, wastewaters, effluents, sludge, or other environmental samples to the production of gas from anaerobic digestion of sewage sludge over periods of up to 3 days. Aliquots of a mixture of undiluted anaerobically digesting sludge and a degradable substrate are incubated alone and simultaneously with a range of concentrations of the test material in sealed vessels. The amounts of gas production by the various concentrations of the test material are calculated from the amounts produced in the respective test and control bottles. The EC50 and other effective concentrations are calculated.

This is an important toxicity test regarding digestion sludge in wastewater treatment plants, but it is rarely used in such applications.

Growth inhibition of activated sludge microorganisms

This method is also internationally standardized (ISO), but only a few test results have been reported up to now.

The test method is applicable to water, wastewater, and chemical substances. Flasks containing organic test medium and test material are inoculated with an overnight culture of activated sludge. The biomass of these cultures, and of controls, is determined. The recommended method is measurement of the turbidity in a spectrophotometer at a wavelength of 530 nm and expression in relative units (OD530). The test gives information on inhibitory effects on the microorganisms over incubation periods up to 6 hours. EC50, EC20, or EC80 values may be calculated.

Algal growth inhibition test

The algal growth inhibition test is one of the most commonly used biotest methods. It is well established and internationally standardized (ISO, OECD). It is also used for environmental hazard and risk assessment of industrial effluents (in Sweden and the UK) and reported by many countries.

Exponentially growing cultures of selected unicellular green algae are exposed to various concentrations of the test substance over several generations. The inhibition of growth in relation to control cultures is determined over a fixed period.

The cell concentration in each concentration of test substance is determined at least at 24, 48, and 72 hours after the start of the test. The measured cell concentrations in the test cultures and controls are tabulated together with the concentration of the test substance and the times of measurements.

The percentage inhibition of the cell growth and the average specific growth rate is calculated. Both endpoints are given as an EC50 value. In addition, NOEC and LOEC values may be calculated.

In Germany, a standardized, cost-effective method with fewer replicates (especially designed for the examination of wastewater samples) is used to determine the LID. A French Association Francaise de Normalisation (AFNOR) standard is also published. Such adapted protocols for wastewater testing will be incorporated as an annex in the next revision of the ISO standard.

The UKEA has developed a microscale method for algal growth inhibition using fluorescence as a surrogate measure of cell density.

Lemna toxicity test

Up to now only USEPA and ASTM standards as well as an AFNOR standard (ref. NFXP T 90-337 "Testing water – Determination of the inhibitory effect on the growth of *Lemna minor*") are available, although this test method is reported by many countries (Sweden and the Netherlands). The OECD recommended the method for inclusion in the OECD Test Guidelines Programme (OECD 1998a).

The recommended procedure is to expose lemna species to a chemical concentration series and to determine the EC5, EC50, EC90, LOEC, and NOEC values for lemna growth based on total frond number, growth rate, and frond mortality. Other optional endpoints include dry weight and chlorophyll and phaeophytin pigment analyses. At least five concentrations are chosen. The test duration is 7 days. Observations should be made on days 0, 3, 5, and 7. The test species are *L. gibba* and *L. minor.*

Freshwater long-term chronic tests

Longer-term ecotoxicity testing methods are rarely used for effluent testing. The Swedish and Danish environmental protection agencies recommend chronic tests with fish and daphnia for environmental hazard and risk assessment of industrial effluents.

Chronic fish toxicity test

Prolonged toxicity test.   The prolonged toxicity test with fish has been standardized by the OECD.

The test is used to measure lethal and other observed effects in fish exposed to test substances (including all effects observed on the appearance, size, and behavior of the fish that make them clearly distinguishable from the control animals). The fish are inspected at least once a day. Threshold levels and NOEC are determined at intervals during the test period (at least 14 days). If necessary, the test period may be extended by one or two weeks.

Early life stage toxicity test.   The early life stage toxicity test with fish is internationally standardized by the OECD.

Tests with the early life stages of fish are intended to define the lethal and sublethal effects of chemicals on the life stages and species tested. The fish are exposed to a range of concentrations of the test substance, preferably under flow-through conditions, or (where appropriate) under semistatic conditions. The test is started by placing fertilized eggs in the test chambers and is continued at least until all the control fish are free-feeding. Lethal and sublethal effects are assessed and compared with control values. The LOEC and NOEC are determined. Fish species recommended include *Oncorhynchus mykiss, Pimephales promelas, Brachydanio rerio, Oryzias latipes,* and *Cyprinodon variegatus.* The test duration is 28 to 32 days, depending on the species.

Daphnia magna *reproduction test*

The *D. magna* reproduction test is internationally standardized (ISO Draft, OECD). The primary objective of the test is to assess the effect of chemicals on the reproductive output of daphnia. To this end, daphnia less than 24 hours old are exposed to at least five concentrations of the test sample. The test duration can be up to 21 days, but reduced duration tests have been developed. The number of living offspring produced per parent animal is assessed and, as far as possible, the data are analyzed using a regression model in order to estimate the EC50, EC20, EC10, LOEC, and NOEC values.

Renewal toxicity test with ceriodaphnia dubia

The test principle is described together with the short-term chronic ecotoxicity methods. The only difference is that test duration is prolonged to reach three broods of young (about 9 to 15 days). ASTM and USEPA standards are available. The test can, with appropriate modifications, also be used with other cladocera.

Chronic toxicity test with higher plants

The objective of this test is to determine effects to plants during critical stages of development. It is performed under natural conditions and in the environment. It is a multiple-dose test (at least 5 concentrations) designed to evaluate the phytotoxicity of substances, especially pesticides. Aquatic plant representatives of the following plant groups are used: dicotyledonae, monocotyledonae, vascular cryptogamae, algae, bryophyta, and hepatophyta. The test duration should be of sufficient length to assess multiple applications and observations should continue for the entire life cycle of test plants, with observations every 2 to 4 weeks.

Marine acute tests using fish

Acute toxicity tests with marine fishes are recommended by several agencies (Danish and Swedish EPAs, USEPA, UKEA) for risk evaluation of industrial effluents to marine environments. The test is also applied by other member states (the Netherlands). Up until now only USEPA and the UKEA guidelines for *Scophthalmus maximus* are available. Currently an ISO working group is preparing a test guideline with *S. maximus*. The following saltwater species are also recommended: Atlantic silverside (*Menidia menidia*), sheepshead minnow (*Cyprinodon variegatus*), and tidewater silverside (*Menidia penisulae*). The LC50 and NOEC are determined in a static, semistatic, or flow-through test. The duration of the test is 96 hours. At least five concentrations are used.

Marine acute tests using invertebrates

Marine copepod toxicity test

The application of this test is often reported (Belgium, the Netherlands) and is also recommended by the Danish Environmental Protection Agency as a procedure for investigation and assessment of hazard or risk to marine environments from industrial effluents. Recently ISO guidelines with the test species *Acartia tonsa, Tisbe battagliai,* and *Nitocra spinipes* were published. The guideline for *T. battagliai* has been ring-tested by the UKEA.

Copepods are exposed to a range of concentrations of seawater or effluent. Mortality is recorded after 24 and 48 hours, and the LC50 and other point estimate values are determined. The NOEC and LOEC values can also be estimated.

Mysid shrimp toxicity test

The application of the acute mysid toxicity test is reported by Belgium; until now, however, only a USEPA guideline has been available.

A minimum of 20 mysids should be exposed to at least five concentrations for up to 96 hours. The LC50 is determined at 48 and 96 hours. Each test chamber should be checked for dead mysids at 24, 48, 72, and 96 hours. In addition to death, any abnormal behavior or appearance of the exposed mysids should also be reported.

Oyster toxicity test (shell deposition)

An USEPA guideline prescribes tests to be used to determine the acute toxicity of chemical substances and mixtures to the Eastern oyster (*Crassostrea virginica*). At least 20 prepared oysters are placed in each of the test chambers and exposed to at least five test concentrations for a period of up to 4 days. The oysters are inspected at least every 24 hours. Shell deposition (the measured length of growth that occurs within the test period) is the primary criterion. At the end of the test the EC50 is determined.

Toxicity tests with rotifers

In addition to the acute toxicity test with the freshwater rotifer *Brachionus calycilforus*, the ASTM standard describes another toxicity test with the rotifer *B. plicatilis* for estuarine and marine waters. The procedure is applicable to most chemicals and also aqueous effluents, leachates, oils, particulate matter, sediments, and surface water.

Toxicity tests with protozoans

This test system using the marine ciliate *Uronema marinum* is proposed by the Danish EPA, but no test guideline is available.

Marine acute tests using bacteria

Vibrio fischeri *assay*

The acute toxicity test with the marine bacterium *Vibrio fischeri* is also applied for fresh-water effluent toxicity testing. The test method is described earlier.

Marine short-term chronic tests

Bivalve embryo-larval development toxicity test

A *Crassostrea gigas* embryo-larval guideline has been ring-tested and widely applied in England and Wales. This method has undergone further development and is now miniaturized and analyzed using imaging technology. Existing USEPA and ASTM guidelines prescribe methods for the evaluation of the acute toxicity of chemicals and mixtures to different bivalves: eastern oysters (*C. virginica*), pacific oysters (*C. gigas*), quahogs (*Mercenaria mercenaria*) or bay mussels (*Mytilus edulis*). The ASTM guideline also recommends the test with appropriate modifications for aqueous effluents, leachates, oils, particulate matter, sediments, and surface water. The test is started about 4 hours after fertilization while embryos are in the 16- to 32-cell stage. At least five concentrations are tested in a static system. The endpoint is shell abnormality and the 24- or 48-hour EC50 (and other point estimates) can be calculated, along with LOEC and NOEC values. The high sensitivity of this test allows identical methodologies to be used for both effluent and receiving-water assessment.

Marine algae growth inhibition test

Algal cells or chains are cultured for several generations in a medium containing a range of concentrations of the test substance. The method is available as an ISO draft. The minimum test duration is 72 hours, during which the cell density in each sample is measured at least every 24 hours. Inhibition is measured as a reduction in growth and growth rate. The EC10, EC50, LOEC, and NOEC can be determined. Recommended algal species are *Skeletonema costatum*, *Phaeodactylum tricornutum*, and red macroalgae. The UKEA has developed a microscale method for algal growth inhibition using fluorescence as a surrogate measure of cell density.

Early life stage fish toxicity test

Larval survival and growth test. This method is recommended by the USEPA for evaluating chronic toxicity of effluents and receiving waters to sheepshead minnow (*C. variagatus*), using newly hatched larvae in a 7-day semistatic test. The effects include the synergistic, antagonistic, and additive effects of all the chemical, physical, and biological components that adversely affect the physiological and biochemical functions of the test species. This method is commonly used in one of two forms: (1) a definitive test, consisting of a minimum of five effluent concentrations and a control; and (2) a receiving-water test, consisting of one or more receiving-water concentrations and a control. In a similar test the toxicity for inland silverside (*M. beryllina*) is also used in which 7- to 11-day old larvae are exposed. Results are based on the survival and weight of the larvae.

Embryo-larval survival and teratogenicity test. This method is recommended by the USEPA for evaluating the chronic toxicity of effluents and receiving waters to the sheepshead minnow (*C. variegatus*), using embryos and larvae in a 9-day static renewal test. The effects include the synergistic, antagonistic, and additive effects of all the chemical, physical, and biological components that adversely affect the physiological and biochemical functions of the test organisms. The test is useful in screening for teratogens because organisms are exposed during embryonic development. This method is commonly used in one of two forms: (1) a definitive test, consisting of a minimum of five effluent concentrations and a control; and (2) a receiving-water test, consisting of one or more receiving-water concentrations and a control.

Marine long-term chronic tests

Mysid shrimp toxicity test

The long-term chronic toxicity test with mysids is performed as a flow-through test over 28 days. Mysids are exposed to five test concentrations and the LC50 and MATC (maximum acceptable toxicant concentration) values and the effects on growth and reproduction are determined.

Tisbe battagiai *population level test*

The UKEA has recently conducted research into the development and appraisal of a new test protocol, using *Tisbe battagliai*, to elucidate the long-term (chronic) hazard of effluents (measuring effects at both the individual and population level) while incorporating the standard short-term methodology.

Individual cohorts of juvenile (5-day) *T. battagliai* are exposed to a range of test concentrations for up to 21 days. Observations of individual survival and fecundity are performed at 48-hour intervals (allowing the construction of partial life-tables). The following endpoints and point estimates are calculated: survival (48-hour EC50, $NOEC_{acute}$ and $LOEC_{acute}$), inhibition of reproduction (IC with 95% CI by ICP nonlinear interpolation, $LOEC_{chronic}$ and $NOEC_{chronic}$), and the inhibition of the intrinsic rate or population increase (I_rC_x with 95% CI by Leslie population matrix modeling with double bootstrap).

Genotoxicity tests

Genotoxicity generally summarizes all effects that may damage DNA. DNA damage might be repaired enzymatically so that changes are not inherited to daughter cells or may lead to a change in DNA sequence (mutation).

There are three reasons to consider genotoxic effects in effluents (de Maagd et al. 1999):

- Genotoxicity can affect fitness and reproduction of organisms.
- Higher mutation frequencies can increase the instability of ecosystems.
- Genotoxic compounds might be relevant to humans when contaminated surface water is used downstream for other purposes such as agriculture, recreation, or drinking water.

The first two arguments are based on a few studies (Lynch and Bürger 1995) but have not been proved to have a clear cause-and-effect relationship. Until now it is not clear what relevance genotoxic effects have on an ecosystem level (de Maagd 1999; de Maagd and Tonkes 2000; Depledge 1998). The third argument can be extended because genotoxic effluents will always provide an indication of the possible effects of compounds on human health. Effluent and surface-water samples may be highly concentrated on solid phase or extracts in order to enhance their sensitivity to genotoxicity tests. This may lead to unrealistically high and ecologically irrelevant exposure concentrations and there is no agreement on what concentration factor is acceptable (de Maagd 1999). Additionally, each concentration procedure recovers different fractions of the sample, and volatile substances may be lost. Testing crude samples should therefore be favored over such concentration procedures in order to get a realistic estimate of the genotoxicity of an effluent (de Maagd and Tonkes 2000).

There are numerous test procedures for the genotoxicity testing of wastewater, but only a few of these are based on international standardized test guidelines. The most frequently applied test procedures are summarized in de Maagd (2000).

Ames assay

The Ames assay is a bacterial *in vitro* test using mutant *Salmonella typhimurium* strains that have lost their ability to grow (in the absence of histidine). Reverse mutations caused by exposure to mutagenic compounds can reactivate their ability to form colonies in the absence of histidine. The number of colonies at different concentrations of the test compound are compared with that of the negative controls and are a measure for mutagenicity. The most commonly used salmonella strains in wastewater screening are TA 98 and TA 100, designed for detecting frame shift mutations and point mutations respectively. Usually the test is performed in the absence and the presence of S9 liver homogenate in order to activate promutagens. The Ames test has been the most widely used method in wastewater mutagenicity testing (Stahl 1991; Houk 1992), but in the last decade other genotoxicity tests have been established that are faster and easier to handle. Recently a microplate version of the Ames test based on color changes has been developed (Hubbard et al. 1994)

UmuC assay

The umuC assay was originally developed by Oda et al. in 1985. The assay is based on the use of a genetically modified *S. typhimurium* strain TA 1535 that contains plasmids with the umuC gene and the lacZ gene, which encode for β-galactosidase. The activation of the umuC gene by DNA-damaging agents (as part of the SOS pathway) is measured by an increase of β-galactosidase that induces a color reaction at 420 nm. The test is carried out both with and without S9. Bacterial growth is measured as turbidity at 600 nm and growth factors are considered in the test results.

Experience in wastewater testing with the umuC test are reported by Rao et al. (1995) with extracts of bleached kraft mill effluents in Canada. In Switzerland and Germany, hospital and municipal wastewater have been investigated (Hartmann et al. 1998, 1999; Gartiser and Stiene 1999). The test method has also been introduced in Germany as a routine regulatory measurement for chemical and pharmaceutical effluents (Miltenburger 1997).

Chromosomal aberration

There are several standards for determining chromosomal damage in eukaryotic cells. The OECD guidelines contain 13 tests for the genetic toxicology testing of chemicals. For wastewater testing several approaches have been pursued, but no broadly accepted standards or procedures exist.

Biodegradation and sorption tests

The biodegradability of wastewater samples is most commonly estimated by determining the BOD over 5 days. The BOD is compared with the COD and a BOD/COD ratio of approximately 0.5 is assumed to indicate biodegradability of wastewater. As a parameter for readily degradable organic substances, the BOD serves as an important criterion for choosing the dimensions of sewage treatment plants. Nevertheless, the short test time might indicate an excessively weighted criterion for evaluation of total biodegradation. Therefore standard tests for ready biodegradation with a low inoculum concentration (about 30 mg/L suspended solids) are also occasionally applied to complex mixtures over 28 days. Standardized procedures are available from the OECD 301 test series and ISO guidelines (annex I-5). Endpoints of ultimate biodegradation are oxygen consumption and CO_2 evolution. Other endpoints such as dissolved organic carbon (DOC) elimination are also used, but strictly speaking this can be interpreted as biodegradation only when degradation follows a typical curve with lag, degradation, and plateau phases. Nevertheless the test design of the ready biodegradation tests assumes relatively low test concentrations of 10 to 50 mg/L TOC, so that in the future standardized adaptations for wastewater applications should be considered. It is known that longer-term BOD testing is interfered with by the oxygen consumption of nitrification processes, and this reduces accuracy, especially for wastewater samples with high ammonia loads. Other test systems using CO_2 evolution as an endpoint may avoid this problem, but at this time these have seldom been applied due to the greater effort involved in performing the test. In Sweden a modified DOC "die-away" test, performed according to EN ISO 7827, is used to determine degradability of wastewater (the STORK project, Swedish EPA 1997). In the Netherlands Tonkes and Baltus (1997) followed a similar approach using a modified OECD 301 E procedure with surface water as inoculum (Tonkes, personal communication 1999). In both studies the test duration was 28 days.

Sorption to activated sludge

The sorption of wastewater compounds to activated sludge in biological wastewater treatment systems is an important clarification process. The sorbed fraction might be removed from the system with the excess sludge or might be degraded in the adsorbed phase.

There is still no internationally accepted test guideline for determining the adsorbable fraction of wastewater. Most commonly this fraction is estimated by the Zahn-Wellens test method according to EN ISO 9888, where the 3-hour value is used to estimate sorption processes; this test, however, is not designed to distinguish between adsorption and biodegradation. One method, developed by the USEPA for determining the sorption of chemicals onto activated sludge, uses common model kinetics such as the Freundlich or Langmuier isotherm. The activated sludge is washed, settled, and lyophilized into a dry powder prior to being used as a sorbent (USEPA 1996).

Other methods described in the literature use fresh (Pagga and Taeger 1994) or dried activated sludge at different water hardness classes (Kördel and Willme 1996). Adsorption kinetics and isotherms are determined by DOC or chemical analysis from 2 to 48 hours using laboratory shakers or stirrers.

It must be noted that adsorption tests cannot be used to assess the fate of chemicals, as chemicals might be biodegraded in the adsorbed phase. In this context the retention time of chemicals adsorbed to activated sludge is determined by the sludge retention time (usually 20 to 30 days) and not by the hydraulic retention time (usually less than 8 hours) in wastewater treatment plants.

Sorption to solids and sediments

The possibility that hazardous wastewater constituents may be adsorbed onto suspended solids and deposited in rivers has not yet been considered. Initial experimental approaches were discussed by Pardos and Blaise (1999). Developmental degradation tests using suspended sediments under development such as the shake flask batch test, performed according to ISO/CD 14592, consider the adsorption processes but are not practicable for wastewater testing, as 14C labelled substances are added.

Removal by evaporation

There is no accepted standard to determine the removal of wastewater samples by evaporation. In the Zahn-Wellens test an additional abiotic control without inoculum but with a biocide to inhibit biodegradation is tested, where DOC elimination in the abiotic control may be interpreted as stripping or other physico-chemical processes.

Zahn-Wellens test

The behavior of wastewater in municipal treatment plants can be simulated by determining the elimination of organic sum parameters and by combining a biodegradation test with ecotoxicity tests.

The Zahn-Wellens test is the most commonly used test for determining the inherent biodegradability of chemicals. International (ISO, EN, OECD) as well as national standard guidelines (USEPA, ASTM, DIN) are available. The principle consists of an activated sludge static test with a high inoculum concentration (200 to 1000 mg/l suspended solids). The test concentration is relatively high compared with the biodegradation tests discussed earlier (50 to 400 mg/L DOC). DOC/COD elimination is determined for the filtered samples over a period of 28 days. Along with the test vessels containing the test compound, blank vessels are assayed and an abiotic degradation check (abiotic control) is carried out.

In Germany, three different modifications of this test are used within the Wastewater Ordinance. Here the inoculum concentration has been fixed at 1000 mg/L suspended solids and the test duration varies between 3 and 28 days according to the respective requirements of the different wastewater sectors. A DOC/COD elimination of 80% (less the part eliminated in the abiotic control) is considered to indicate that the wastewater is treatable by municipal plants. The test is also used to determine elimination of other group parameters such as AOX. The amount eliminated by biodegradation and that eliminated by adsorption cannot be distinguished, especially for complex mixtures. Results are given as elimination (bioelimination).

Treatment plant simulation model

A laboratory sewage flow-through treatment plant is used to determine degradability of organic compounds. This test is also known as the coupled units test or OECD confirmatory test. The test item is dissolved in a synthetic sewage matrix and continuously dosed into the activated sludge vessel. A control unit is fed synthetic sewage only. Both units may be coupled by swapping a defined volume of activated sludge between them on a daily basis. DOC is measured in the effluent, and the daily DOC eliminations are calculated after correcting for the material transfer due to the transinoculation procedure. ISO, OECD, and USEPA methods are available. In a recent modification, the concentration of synthetic sewage was halved in order to guarantee stable nitrification conditions (DIN 38412 L26, ISO/NP 16821). The test has been used occasionally to assess elimination of effluents in sewage treatment plants (Gartiser and Brinker 1996), but the considerable effort involved prevents its broader application. Further extensions of the test method with an additional anoxic vessel for denitrification processes are under development (Deutsche Einheitsverfahren, DEV L 43).

Elimination of biological effects

Degradability of biological effects may be of special interest if effluent samples indicate ecotoxic or genotoxic effects. Usually this additional information is obtained from coupling degradation tests with the respective effect test. Therefore effects can be classified as degradable or "hard." This very useful approach might be considered as a part of a TIE procedure. Up to now there is no international accepted guideline for coupling degradation tests with effect tests.

In the Netherlands Tonkes and Baltus (1997) combined a DOC die-away test with effect tests and also determined the degradability of potentially bioaccumulating substances (PBSs).

In Germany, hospital and textile effluents have been assessed with a combination of elimination and genotoxicity-ecotoxicity tests. The Zahn-Wellens test and treatment plant simulation model have also been used as a degradation device and the practicability of the Zahn-Wellens test has been confirmed (Jäger and Meyer 1995; Jäger et al. 1996a, 1996b; Gartiser et al. 1996b, 1997).

The combination of a treatment plant simulation model with ecotoxicity tests has been integrated into the German Wastewater Ordinance for the sector relating to landfill leachates. Here the limits regarding effluent toxicity might be achieved after the biological treatment process.

In the U.S., a guideline for assessing microbial detoxification of chemically contaminated water exists using an unspecified degradation test and the *V. fischeri* assay. The OSPAR WEA Demonstration Program (2003) also advocates this approach. De Groot (1999) proposed combining a 28-day biodegradation test with the chronic daphnia reproduction test and the ELS test with fish, but up to now no test results are available. Whale and Battersby (1999a) used a respirometer biodegradation test to assess the recalcitrant (hard) or easily biodegradable (soft) toxicity of three effluents.

References

de Groot, W.A. (1999). Development of a tiered approach to determine the environmental hazard of effluents, in *Effluent ecotoxicology: a European perspective.* Society of Environmental Toxicology and Chemistry, 14-17 March 1999, Edinburgh.

Depledge, M. H. (1998). The ecotoxicological significance of genotoxicity in marine invertebrates. *Mutation Res.* 399, 198–212.

de Maagd, P.G.-J., Tonkes, M., Maas, J.L., and van de Guchte, C. (1999). *Genotoxicity as effect parameter in whole effluent assessment,* RIZA work document no. 99.110X.

de Maagd, P.G.-J. (1999). Determining the concentration of potential bioaccumulatable compounds in effluents, in *Effluent ecotoxicology: a European perspective.* Society of Environmental Toxicology and Chemistry, 14-17 March 1999, Edinburgh.

de Maagd, P.G.-J. (2000). Bioaccumulation tests applied in whole effluent assessment: a review. *Environ. Toxicol. Chem.* 19, 25–35

de Maagd, P.G.-J and M. Tonkes. (2000). Selection of genotoxicity tests for risk assessment of effluents. *Environ. Toxicol.* 15, 81–90

Deutch Einheitsverfahren zur Wasser-, Abwasser-, und Schlammuntersuchung (1989). Bestimmung der nicht akut giftigen Wirkung von Abwasser gegenüber Daphnien über Verdünnungstufen DEV. Teil 30, DIN 38412.

Gartiser, S., Meyer, M., and Jäger, I. (1996). Zur Interpretation des Zahn-Wellens-Tests bei der Untersuchung von Abwasserproben. *gwf-Wasser* 137 (7), 345–352.

Gartiser, S. and Brinker, L. (1996). *Abwasserbelastende Stoffe und Abwassersituation in Kliniken.* Forschungsbericht Nr. 102 06 514 des Bundesministeriums für Umwelt, Naturschutz und Reaktorsicherheit im Auftrag des Umweltbundesamtes, UBA-FB 95-075.

Gartiser, S., Meyer, M. and Jäger, I. (1997). Abbau ökotoxischer und mutagener Abwasserinhaltsstoffe im Zahn-Wellens-Test und Labor- bzw. technischen Kläranlagen. *gwf-Wasser,* 138, 28–34.

Gartiser, S. and Stiene, G. (1999). *Umweltverträgliche Desinfektionsmittel im Krankenhaus-abwasser.* Forschungsbericht Nr. 29727526. Im Auftrag des Umweltbundesamtes (in press).

Hartmann, A., Alder, A.C., Koller, T., and Widmer, R.M. (1998). Identification of fluoroquinolone antibiotics as the Maine source of umuC genotoxicity in native hospital wastewater. *Environ. Toxicol. Chem.* 17, 377–382.

Hartmann, A., Golet, E., Gartiser, S., Alder, A.C., Koller, T. and Widmer, R.M. (1999). Primary DNA damage but not mutagenicity correlates with Ciprofloxacin concentrations in German hospital wastewaters. *Arch. Enviro. Contamination Toxicol.* 36, 115–119.

Houk, V. S. (1992). The genotoxicity of industrial wastes and effluents: a review. *Mutation Res.* 277, 91–138.

Hubbard, S.A. et al. (1984). The fluctuation test in bacteria, in *Handbook of mutagenicity test procedures,* second ed., B.J. Kilbey, M. Legator, W. Nichols, and C. Ramel (Eds.). Elsevier Sciences, New York, 141–161.

Jäger, I. and Meyer, G. (1995). *Toxizität und Mutagenität von Abwässern der Textilproduktion.* Forschungsbericht 102 06 519 des Bundesministeriums für Umwelt, Naturschutz und Reaktorsicherheit im Auftrag des Umweltbundesamtes, UBA-FB 95-045, 7/95.

Jäger, I., Gartiser, S., and Willmund, R. (1996a). Einsatz von Biotestsystemen zum Strommanagement in Textilveredlungsbetrieben. *Melliand Textilberichte* 1-2/96, 72–75.

Jäger, I., Gartiser, S., and Willmund, R. (1996b). Anwendung von biologischen Testverfahren an Abwässern der Textilindustrie. *Acta hydrochim. hydrobiol.* 24/96, 22–30.

Kördel, W. and Willme, M. (1996). *Bestimmung der Sorption organischer Chemikalien an Klärschlämmen.* Forschungsbericht Nr. 106 04 142. Im Auftrag des Umweltbundesamtes (1996).

Miltenburger, H.G. (1997). *Quantitative Beurteilung von Mutagenität in Abwasserströmen der chemischen Industrie.* Studie 250503; Verband der chemischen Industrie; Abschlußbericht.

Oda, Y.S., Nakamura, S., Oki, I., Kato, T., and Shinagawa, H. (1985). Evaluation of the new system (umu-test) for the detection of environmental mutagens and carcinogens. *Mutation Res.* 147, 219–229.

OECD. (1998a). *Detailed review paper on aquatic testing methods for pesticides and industrial chemicals.* OECD Environmental Health and Safety Publications, Series on Testing and Assessment, No. 11.

Pagga, U. and Taeger, K. (1994). Development of a method for adsorption of dye stuffs on activated sludge. *Water Res.* 28, 1051–1057.

Pardos, M., Blaise, C. (1999). Assessment of (geno)toxicity of hydrophobic organic compounds in wastewater, in *Effluent ecotoxicology: a European perspective.* Society of Environmental Toxicology and Chemistry, 14-17 March 1999, Edinburgh.

Pauli, W. and Berger, S. (1996). *Proceedings of the international workshop on a protozoan test protocol with Tetrahyemna in aquatic toxicity testing.* German Federal Environmental Agency, UBA-Texte 96/34.

Rao, S.S., Quinn, B.A., Burnison, B.K., Hayes, M.A. and Metcalfe, C.D. (1995). Assessment of the genotoxic potential of pulp mill effluent using bacterial, fish and mammalian assays. *Chemosphere* 31, 3553–3566.

Stahl, R.G., Jr. (1991). The genetic toxicology of organic compounds in natural waters and wastewaters. *Ecotoxicology Environ. Saf.* 22, 94–125.

Swedish EPA (1997). *Characterization of discharges from chemical industry — the STORK project.* Swedish Environmental Protection Agency, Report No. 4766, Stockholm.

Tonkes, M. and Baltus, C.A.M. (1997). *Praktijkonderzoek aan complexe effluenten met de totaal effluent milieubezwaarlijkheit (TEM)-methodiek – resultaten van 10 complexe effluenten.* Ministerie van verkeer en waterstaat, rijksinstituut voor integraal zoewaterbeheer en afvalwaterbehandling (RIZA) (mei 1997).

UBA. (1999). *Ecotoxicological evaluation of wastewater within whole effluent assessment* (Draft). OSPAR background document concerning the elaboration of programmes and measures relating whole effluent assessment, November 1999.

UKEA (Environment Agency for England and Wales). (2001a). *Ecotoxicity test methods for effluent and receiving water assessment — comprehensive guidance.*

UKEA. (2001b). *Ecotoxicity test methods for effluent and receiving water assessment — supplementary advice to international test guidelines.*

USEPA. (1995). *Whole effluent toxicity: guidelines establishing test procedures for the analysis of pollutants.* (Final rule October 16, 1995, Vol 60, No. 199).

USEPA. (U.S. Envirnmental Protection Agency) (1996). Activated sludge sorption isotherm (Draft guideline OPPTS 835.1110).

Weber, C.I. (1993). *Methods for measuring the acute toxicity of effluents and receiving waters to freshwater and marine organisms, fourth ed.* EPA/600/4-90/027F, U.S. Environmental Protection Agency, Cincinnati.

Whale, G.F. and Battersby, N.S. (1999). A combined biodegradation and toxicity approach to improve effluent risk assessment, in *Effluent ecotoxicity: a European perspective.* Society of Environmental Toxicology and Chemistry, 14-17 March 1999, Edinburgh.

chapter three

Biomarkers in environmental assessment

R. van der Oost, C. Porte-Visa, and N.W. van den Brink

Contents

Abstract 89
Introduction 91
Biomarkers overview 94
    Phase I biotransformation enzymes 95
    Phase II biotransformation enzymes 96
    Oxidative stress parameters 97
    Biotransformation products 97
    Stress proteins, metallothioneins, and multixenobiotic resistance 98
    Hematological parameters 98
    Immunological parameters 99
    Reproductive and endocrine parameters 99
    Neuromuscular parameters 99
    Genotoxic parameters 100
    Physiological and morphological parameters 100
    Proteomics and genomics 101
Invertebrate biomarkers 101
    Phase I enzymes 102
    Oxidative stress 102
    Stress proteins, metallothioneins, and multixenobiotic resistance ...103
    Reproductive and endocrine parameters 104
    Neuromuscular parameters 104
    Genotoxic parameters 104
    Physiological and morphological parameters 104
Fish biomarkers 105
    Phase I enzymes 105
    Phase II enzymes 106

Oxidative stress parameters 106
Biotransformation products 107
Stress proteins, metallothioneins, and MXR 108
Reproductive and endocrine parameters 108
Neuromuscular parameters 109
Genotoxic parameters 109
Physiological and morphological parameters 110
Amphibian biomarkers 110
Phase I enzymes 111
Phase II enzymes 111
Oxidative stress 112
Hematological parameters 112
Stress proteins 112
Reproduction and endocrine parameters 112
Neuromuscular parameters 113
Physiological and morphological parameters 113
Mammalian and avian biomarkers 114
Phase I enzymes 115
Phase II enzymes 115
Hematological parameters 116
Stress proteins and metallothioneins 116
Endocrine parameters 116
Neuromuscular parameters 117
Morphological and histological parameters 117
Summary, discussion, and conclusions 118
Invertebrates 119
Fish 119
Amphibians 120
Birds and mammals 120
Perspectives and recommendations 121
References 123
Appendix 1 138
Overview of all biomarker assays 138
Appendix 2 141
Standard operating procedures of biomarker assays 141
SOP 1: Isolation of microsomes and cytosol from liver tissues 141
Solutions 141
Further requirements 142
Needed per sample 142
Procedure 142
SOP 2: EROD activity in liver microsomes 142
Principle 142
Solutions 143
Further requirements 143
Needed per determination (duplo) 143
Procedure 143

Calculations 144
SOP 3: Cyt P-450 and Cyt b5 Contents of Liver Microsomes 144
Principle 144
Solutions 144
Further requirements 144
Needed per determination (duplo) 144
Procedure for cyt b5 145
Procedure for cyt P-450 145
Calculations 145
SOP 4: GST Activity of Liver Cytosol 146
Principle 146
Solutions 146
Further requirements 146
Needed per determination (duplo) 146
Procedure 146
Calculations 147
SOP 5: SOD Activity in Liver Cytosol 147
Principle 147
Solutions 147
Further requirements 147
Needed per determination (duplo) 148
Procedure 148
Calculations 148
SOP 6: ACHE Activity in Tissue Homogenates 148
Principle 148
Solutions 149
Further requirements 149
Needed per determination (triplicates in microplate method) 149
Procedure 149
Calculations 150
SOP 7: Lysosomal Membrane Stability in Cells 151
Principle 151
Solutions and chemicals 151
Further requirements 151
Procedure 151
Interpretation of results 152

Abstract

It has been demonstrated that an environmental risk assessment based solely on pollutant levels in the environment is generally not considered to be reliable. There is a growing awareness that risk assessors have to focus on the effects of the total mixture of contaminants present in the environment, instead of on the presence of some selected compounds. Biomarkers are defined as changes in biological responses (ranging from molecular through cellular, and from physiological responses to behavioral changes) that can be related to exposure to, or to the toxic effects of, environmental chemicals. The main objective of this chapter is to provide an overview of studies that use biomarkers for the assessment of exposure and toxic impact of environmental contaminants to various organisms. The possibilities for biomarkers to detect, identify, and assess pollutant exposure, primary actions, and effects on invertebrates, fish, amphibians, birds, and mammals are discussed in separate sections and reviewed for site-specific case studies.

The use of biomonitoring methods in the control strategies for chemical pollution has advantages over chemical monitoring. First, these methods measure effects in which the bioavailability of the compounds of interest is integrated with the concentration of the compounds and their intrinsic toxicity. Second, most biological measurements form the only way of integrating the effects on a large number of individual and interactive processes. Although it has been demonstrated that biomarkers are useful monitoring tools, it is clear that more information is needed about the relation between biomarker responses and the health and fitness of organisms, and even more so between biomarker responses and the risks for the ecosystem. A limitation of most of the biological-effect measurements is that biomarker data interpretation must always be carefully controlled for false-negative and false-positive results, since the effects of non–pollution-related confounding factors may interfere with biomarker responses. With respect to future biomarker research, it is important to realize that different concepts are needed for the specific purposes of environmental monitoring programs.

In conclusion, it can be stated that biomarkers are promising tools for environmental risk assessment (ERA). In view of the present chemically oriented pollution-abatement policies and the need to reveal specific chemical problems, it is most probable that biological-effect analysis will never totally replace chemical analyses. The biomarker approach, therefore, should not be considered as a replacement for conventional assessment techniques, but as an important supplementary approach of great ecological relevance. Several guidelines for ecotoxicological research leading to an actual incorporation of biomarkers in ERA monitoring are proposed in this chapter. Much work has to be done in order to test and interpret biomarker responses and to develop acceptable quality assurance (QA) procedures. Only when both scientific and legal credibility of this information is established can the biomarker techniques be fully applied in routine monitoring programs. Since it seems obvious that chemical monitoring alone is insufficient for a reliable classification of water quality, the efforts to incorporate biological compounds to the ERA research will eventually be worthwhile.

Introduction

Environmental risk assessment (ERA) in its classic form was mainly focused on the relationship between partitioning of toxic substances in the environment and the potential hazards of these pollutants if they exceed certain threshold levels. In the last decades, however, it has been demonstrated that a risk assessment based solely on pollutant levels in the environment is not considered to be reliable, partly due to the ability of various pollutants (and their derivatives) to mutually effect their toxic actions. There is a growing awareness that risk assessors have to focus on the effects of the total mixture of contaminants present in the environment, instead of on the fate of some selected compounds. The main objective of this chapter is to provide an overview of studies that use biomarkers for the assessment of exposure and toxic impact of environmental contaminants to various organisms.

The term biomarker is generally used in a broad sense to include almost any measurement that reflects an interaction between a biological system and a potential hazard, whether chemical, physical, or biological (WHO 1993). Many more biomarker definitions can be found in the literature. A biomarker has been defined as a change in a biological response (ranging from molecular through cellular and from physiological responses to behavioral changes) that can be related to exposure to, or the toxic effects of, environmental chemicals (Peakall 1994). Van Gestel and van Brummelen (1996) defined a biomarker as any biological response to an environmental chemical at the subindividual level, measured inside an organism or in its products (urine, feces, hair, feathers, and so on), indicating a deviation from the normal status that cannot be detected in the intact organism. When biological responses are measured using whole organisms, Van Gestel and van Brummelen (1996), as opposed to most others, refer to bioindicators instead of biomarkers. Although most biomarkers discussed in this chapter are measured at molecular to tissue levels, the definition by the World Health Organization (WHO 1993) is preferred since it is not focused on a specific integration level.

A pollutant stress situation normally triggers a cascade of biological responses, each of which may, in theory, serve as a biomarker (McCarthy et al. 1991). Above a certain threshold (in pollutant dose or exposure time) the pollutant-responsive biomarker signals deviate from the normal range in an unstressed situation, finally leading to the manifestation of a multiple-effect situation at higher hierarchical levels of biological organization. When deleterious effects on populations or ecosystems become clear, the destructive process has often gone too far to save the ecosystem by remedial actions or risk reduction. The sequential order of responses to pollutant stress within a biological system, from the molecular to the ecosystem level (Bayne et al. 1985) have triggered the research to establish early-warning signals reflecting the adverse biological responses towards anthropogenic environmental toxins (Bucheli and Fent 1995).

Most biomarkers are measurements in body fluids, cells, or tissues indicating biochemical or cellular modifications due to the presence and magnitude of toxicants or of host responses (NRC 1987). Effects at higher hierarchical levels are always preceded by *earlier* changes in biological processes, allowing the development of early-warning biomarker signals of effects at *later* response levels (Bayne et al. 1985). The most promising feature of biomarker investigation is the early and adjustable indication of potentially toxic effects. This early indication can be seen in the perspective of both time and concentration. In an environmental context, biomarkers offer promise as sensitive indicators demonstrating that toxicants have entered organisms, have been distributed between tissues, and are eliciting a toxic effect at critical targets (McCarthy and Shugart 1990). By screening multiple biomarker responses, important information may be obtained about organism toxicant exposure and stress. Since biomarkers can be applied both in the laboratory and in the field, they may provide an important linkage between laboratory toxicity and field assessment. For field samples, biomarkers provide an important index of the total external load that is biologically available in the real-world exposure.

Improper application or interpretation of biomarker responses, however, may lead to false conclusions as to pollutant stress or environmental quality. Certain responses established for one species are, for instance, not necessarily valid for other species. Moreover, ecotoxicological data obtained by laboratory studies can be difficult to translate into accurate predictions of effects that may occur in the field (ECETOC 1993). Since both overestimation and underestimation of effects may occur, laboratory observations on biomarkers must always be validated when used for site-specific assessments. A successful implementation of biomarkers in environmental monitoring programs requires a good understanding of the mechanisms underlying the responses. Biomarker responses are powerful because they integrate a wide array of environmental, toxicological and ecological factors that control and modulate exposure to, as well as effects of, environmental contaminants. However, these same factors may also complicate interpretation of the significance of the biomarker responses in ways that may not always be anticipated (McCarthy 1990). Many non–pollution-related variables may have an additional impact on the various enzyme systems, and thus may interfere with biomarker responses when experimental conditions are not thoroughly analyzed or controlled. Examples of such confounding factors are the organisms' health, condition, sex, age, nutritional status, metabolic activity, migratory behavior, reproductive and developmental status, and population density, as well as factors like season, ambient temperature, heterogeneity of the environmental pollution, and so on. Unfortunately, most available toxicity data rarely quantify the potency that confounding factors are likely to exhibit in natural environments (De Kruijf 1991). Moreover, estimates of confounding-factor interactions are scarce, as evidenced by the extensive use of uncertainty factors in risk assessment to address unknowns (Power and McCarty 1997).

In order to objectively evaluate the strength and weaknesses of biomarkers that can be broadly adopted for an overall ERA, six criteria have been proposed comprising the most important information that should be available or has to be established for each candidate biomarker (Van der Oost et al., 2003):

- The assay to quantify the biomarker should be reliable (with quality assurance [QA]) and preferably cheap and easy to perform.
- The biomarker response should be sensitive to pollutant exposure or effects in order to serve as an early warning.
- Baseline data of the biomarker should be well defined in order to distinguish between natural variability (noise) and contaminant-induced stress (signal).
- The impacts of confounding factors to the biomarker response should be well established.
- The underlying mechanism of the relationships between biomarker response and pollutant exposure (dosage and time) should be established.
- The toxicological significance of the biomarker, such as the relationship between its response and the (long-term) impact to the organism, should be established.

It is obvious that virtually no biomarker will meet all of the criteria mentioned above, and that the importance of each criterion highly depends upon the aim of the study in which the biomarker is used. However, since all of these criteria have their specific importance, they should be taken into account if the general value of (potential) biomarkers is discussed. In addition to these criteria it has been suggested that biomarkers should preferentially be noninvasive or nondestructive, to allow or facilitate environmental monitoring of pollution effects in protected or endangered species (Fossi and Marsili 1997). The basic biology and physiology of the test organism should be known so that sources of uncontrolled variation (growth and development, reproduction, and food sources) can be minimized (Stegeman et al. 1992).

Biomarker responses may be applied in so-called integrated monitoring programs, which are studies consisting of coordinated monitoring activities comprising both chemical and biological measurements in a variety of environmental media or compartments. An example of an effect-based integrated monitoring program is the Triad approach, consisting of analyses of chemical contamination (chemical monitoring), assessment of toxicity using bioassays (biological-effect monitoring) and determination of the in-faunal community structure (ecosystem monitoring). The Triad approach is described in detail by Chapman (1990). Analyses of Triad data in order to determine the pollution status can involve comparisons of ratio-to-reference (RTR) values, ranking, multivariate analyses, and Mantel's test on disease clustering (Mantel 1967). Van Gestel and van Brummelen (1996) proposed a stepwise integrated ERA of chemicals, consisting of four different biomonitoring levels using biomarkers, bioassays, bioindicators, and ecological indicators.

With respect to the indication of exposure to toxic xenobiotics, it is also interesting to study the development and application of sensitive laboratory bioassays that are based upon the responses of biomarkers, such as the CYP1A and estrogenic response assays with cell lines (Sawyer and Safe, 1982; Murk et al. 1996; Legler et al. 1999). Bioassays offer many advantages for comparing the relative toxicity of specific chemicals or specific effluents. Toxicity tests, however, also have serious limitations for biological monitoring because most do not account for the effect of chemical speciation in the environment, kinetics and sorption of chemicals to sediment, accumulation through food chains, and modes of toxic action that are not readily measured as short-term effects (McCarthy and Shugart 1990). Depledge and Fossi (1994) suggested the use of biomarkers in toxicity tests as an attempt to link biomarker responses to effects on life-history characteristics (such as survival and reproduction), which provides a further foundation for the use of biomarkers in environmental assessment.

In the second section of this chapter, an overview of the potential types of biomarkers is given. These include biochemical and physiological responses, such as phase I and phase II metabolic transformation activities, biliary metabolites, DNA adducts, and plasma hormone levels that may have promise as biomarkers for exposure and effect. Furthermore, parameters are addressed that may serve as indicators of effect propagation to higher levels of biological response, such as tissue damage, the prevalence of tumors, or scope for growth. Finally, the possibilities of biomarkers to detect, identify, and assess pollutant exposure and primary actions and effects on invertebrates, fish, amphibians, birds, and mammals is discussed in separate sections and reviewed for site-specific case studies.

Biomarkers overview

It would be ideal for ERA purposes to have a limited set of specific biomarkers indicating the exposure and assessing the hazards of all major classes of pollutants, as well as nonspecific biomarkers that assess accurately and completely the health condition of the organism and the ecosystem (Peakall and Walker 1994). In the real world, there are some promising biomarkers for assessing exposure and effects of toxic substances, but much more research is required before the ideal situation is achievable. Given the current understanding of the different mechanisms of environmental contaminants, several biomarkers may reflect different aspects of molecular interaction. Most of the biomarkers that are currently being used to monitor the quality of the aquatic environment are discussed in this section.

Phase I biotransformation enzymes

The first phase of metabolism, unmasking or adding reactive functional groups, involves oxidation, reduction, or hydrolysis (Goeptar et al. 1995), which leads to the formation of less-hydrophobic compounds that are more easily excreted (Figure 3.1, route I). For the majority of xenobiotic compounds, the phase I reactions are catalyzed by microsomal monooxygenase (MO) enzymes, also known as the mixed-function oxidase (MFO) system. Generally, alterations in the levels and activities of phase I biotransformation enzymes are among the most sensitive biomarkers known. The activity of these enzymes may be induced or inhibited upon exposure to xenobiotics (Bucheli and Fent 1995). Enzyme induction is an increase in the amount or activity of these enzymes, or in both (Sijm and Opperhuizen 1989). It is generally assumed that de novo protein synthesis is the most important enzyme induction process (Stegeman and Hahn 1994). Different isoforms of CYP450 are known to be inducible by different types of contaminants. CYP1A, related to ethoxyresorufin-O-dealkylase and methoxyresurufin-O-dealkylase (EROD and MROD) activities, is induced by 3-MC types of inducers, such as dioxins and the so-called planar PCBs (Bosveld 1995).

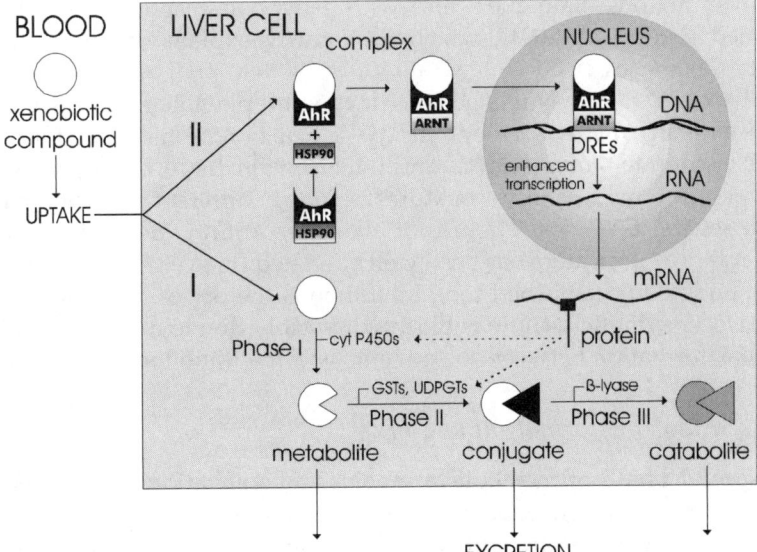

Figure 3.1 Simplified presentation of the fate of certain xenobiotic compounds in the liver cell. Route I: a possible mechanism for detoxification or toxification. Route II: a possible mechanism for enzyme induction. AhR: aryl hydrocarbon receptor; HSP90: 90 kD heat shock protein; ARNT: Ah receptor nuclear translocator; DREs: dioxin responsive elements; CYP 450s: cytochromes P450; GSTs: glutathione S-transferases; UDPGTs: UDP-glucuronyl transferases. Adapted from Van der Oost, R., A. Goksøyr, M. Celander, H. Heida, and N.P.E. Vermeulen, *Aquat. Toxicol.* 36, 189–222, 1996. With permission from Elsevier Science.

The mechanisms and consequences of enzyme induction have been particularly well studied for cytochrome P450-1A (CYP1A). The induction of CYP1A occurs through ligand binding of contaminants to a cytoplasmic receptor, the aryl hydrocarbon receptor (AhR). The resulting ligand-receptor complex leads to gene activation (Figure 3.1, route II). Compounds have to fulfil certain structural requirements, particularly planarity and size, in order to serve as a ligand of the AhR (Mekenyan et al. 1996). The AhR binding and effect chain has been described in detail for 2,3,7,8-tetrachloro dibenzo-dioxin (TCDD). Binding of the ligand activates the receptor, which leads to dissociation of the heat shock proteins (HSPs) and formation of a complex with the aryl hydrocarbon nuclear translocator (ARNT). This complex moves to the nucleus and specifically binds to dioxin (or xenobiotic) responsive elements (DRE or XRE) at the DNA, upstream of promoters of, for example, CYP1A genes (Safe 2001), which results in an up-regulation of gene transcription and subsequent rises in CYP1A mRNA, CYP1A protein, and CYP1A catalytic activity (Stegeman and Hahn 1994). In addition to CYP1A, AhR regulates a number of other enzymes, such as the glutathione-S-transferase (phase II) family (George 1994). AhR-mediated gene expression provides a mechanistic model for induction of metabolic enzymes by xenobiotics. At the DNA transcription level, crosstalk with estrogen-responsive gene promoter regions (Safe 2001) and with a hypoxia-responsive enhancer is assumed (Nie et al. 2001), which may partly explain observed multiple cellular responses.

The CYP2 family is related to induction by phenobarbital (PB), and by more nonplanar PCBs, although the mode of action of this type of induction is less well understood (Nimms and Lubet 1995). Hence, exposure to different types of contaminants (or mixtures) results in different induction patterns of the several CYP450s. The use of alkoxyresorufin-O-dealkylase (AROD) biomarkers as indicators has been well validated (Bosveld and Van den Berg 1994), and is an established tool. Inhibition is the opposite of induction. In this case, enzymatic activity is blocked, possibly due to a strong binding or complex formation between the enzyme and the inhibitors.

Phase II biotransformation enzymes

The second phase of metabolism involves a conjugation of the xenobiotic parent compound or its metabolites with an endogenous ligand (Figure 3.1, route I). Conjugations are addition reactions in which large and often polar chemical groups or compounds such as sugars and amino acids are covalently added to xenobiotic chemical compounds and drugs (Lech and Vodicnik 1985). The majority of the phase II type enzymes, such as glutathion-S-transferases (GSTs) and uridine diphosphate glucuronyl-transferases (UDPGTs), catalyze these synthetic conjugation reactions, thus facilitating the excretion of chemicals by the addition of more polar groups (such as glutathione and glucuronic acid) to the molecule (Commandeur et al. 1995; Mulder et al. 1990). In addition to the phase I CYP1A genes, the Ah gene battery also comprises phase II genes like NADPH menadione oxidoreductase, aldehyde dehydrogenase, UDPGT, and GST (Nebert et al. 1990; Celander 1993). The mechanism of induction for most forms of phase II enzymes is, therefore, probably regulated via the Ah receptor as well (Sutter and Greenlee 1992; Hayes and Pulford 1995). Apart from their essential functions in intracellular transport and biosyntheses, phase II enzymes play a critical role in the cellular defense against oxidative damage and peroxidative products of DNA and lipids (George 1994).

Oxidative stress parameters

Many environmental contaminants (or their metabolites) have been shown to exert toxic effects related to oxidative stress (Winston and Di Giulio 1991). Oxygen toxicity is defined as injurious effects due to cytotoxic reactive oxygen species (ROS), also referred to as reactive oxygen intermediates (ROIs), oxygen-free radicals, or oxyradicals (Di Giulio et al. 1989). Of particular interest are the reduction products of molecular oxygen (O_2), such as the superoxide anion radical ($O_2^- \bullet$), hydrogen peroxide (H_2O_2), and the hydroxyl radical ($\bullet OH$). These are extremely potent oxidants that are capable of reacting with critical cellular macromolecules, possibly leading to enzyme inactivation, lipid peroxidation, DNA damage, and ultimately cancer or cell death (Winston and Di Giulio 1991). Defense systems that tend to inhibit oxyradical formation include the antioxidant enzymes such as superoxide dismutase (SOD), catalase (CAT), glutathione-dependent peroxidase (GPOX), and glutathione reductase (GRED). In addition, numerous low–molecular-weight antioxidants, such as glutathione (GSH), β-carotene (vitamin B), ascorbate (vitamin C), α-tocopherol (vitamin E), and ubiquinol$_{10}$ have been described (Stegeman et al. 1992; Lopez-Torres et al. 1993). At present, the information on induction mechanisms of antioxidants and antioxidant enzymes is limited.

Biotransformation products

Elevated levels of biotransformation products, such as metabolite levels in body fluids or the amount of covalent adducts formed between metabolites of biodegradable chemicals and endogenous cellular macromolecules, may also be considered as biomarkers. The first type of biotransformation products, metabolites of nonpersistent compounds, is generally indicative of recent exposure to the parent compounds (Lin et al. 1996). The possible uses of metabolites of both xenobiotic chemicals (PAHs, chlorinated phenols, resin acids, and so on) and endogenous metabolites (such as vitellogenin, glutathione, porphyrins, and reproductive hormones) as biomarkers for exposure or effects have been reviewed in detail by Melancon et al. (1992). The second type of biotransformation products, reactive metabolites bound to macromolecules, is indicative for both (long-term) exposure and toxic effects (Dunn 1991). The oxidative metabolism of PAHs proceeds via highly electrophilic intermediate arene oxides, some of which bind covalently to cellular macromolecules such as DNA, RNA, and protein (Neff 1985). It is generally accepted that metabolic activation by the MFO system is a prerequisite for PAH-induced carcinogenesis (Van Schooten 1991). DNA adducts are considered in the section on genotoxic parameters.

Stress proteins, metallothioneins, and multixenobiotic resistance

The stress proteins (also called heat-shock proteins [HSPs]) comprise a set of abundant and inducible proteins involved in the protection and repair of the cell against stress and harmful conditions (Sanders 1993). Since these parameters are rather nonspecific they may be induced by a wide variety of chemical and physical stressors, including high or low temperature, ultraviolet light, oxidative conditions, anoxia, salinity stress, heavy metals, and xenobiotics such as teratogens and hepatocarcinogens (Stegeman et al. 1992; Di Giulio et al. 1995). A special group of stress proteins is the metallothioneins (MTs), which constitute a family of low–molecular-weight, cysteine-rich proteins functioning in the regulation of the essential metals Cu and Zn, and in the detoxication of these and other nonessential metals such as Cd and Hg (Roesijadi and Robinson 1994). MTs are inducible by both essential and toxic heavy metals (Stegeman et al., 1992; Viarengo et al. 2000). Stegeman et al. (1992) stated that the quantification of MT is rather elaborate, which raises the question of what is gained by using MT analysis instead of metal analysis. The interpretation of MT levels may be difficult due to natural variation in MT induction, nonspecificity of this induction, and the possibility of resistance of organisms to metal exposure, causing variation in sensitivity. Nevertheless, field studies have shown that MT analysis can be applied as a general marker for metal exposure, while a further analysis of the metals bound to the MT may shed a light on the inducing metal. Another distinct group of stress proteins is the P-glycoproteins (PGPs) of the multixenobiotic resistance (MXR) mechanism, which may be induced or inhibited by a wide variety of chemicals (Bard 2000). The MXR mechanism acts as an energy-dependent pump that removes both endogenous and xenobiotic chemicals from the cell, thus preventing their accumulation and cytotoxic effects (Kurelec 1992; Epel 1998).

Hematological parameters

Several hematological parameters are potential-effect biomarkers. The leakage of specific enzymes (such as transaminases) into the blood may be indicative of the disruption of cellular membranes in certain organs (Moss et al. 1986). Delta-aminolevulinic acid dehydratase (ALAD) is a cytosolic enzyme, active in the synthesis of hemoglobin (Mayer et al. 1992). The assessment of ALAD activity is possible in liver, and it appears to be applicable as a biomarker for lead exposure in field situations. Although less specific, other hematological parameters, like hematocrit, hemoglobin, protein, and glucose, may be sensitive to certain types of pollutants as well. In addition, the blood levels of specific steroid hormones or proteins normally induced by these hormones may be indicative for certain reproductive effects due to endocrine disruption. The latter parameters are considered in the section on reproductive and endocrine parameters.

Immunological parameters

A large number of environmental chemicals have the potential to impair components of the immune system. Both antibody- and cell-mediated immunity may be depressed by certain pollutants, as reviewed by Vos et al. (1989). Although most research on the immune system has been performed on mammalian species, it may also be considered a promising field in which to search for new fish biomarkers (Wester et al. 1994).

Reproductive and endocrine parameters

The impact of xenobiotic compounds on reproductive and endocrine effects has attracted growing interest in recent years. Since a decreased reproductive capability in feral organisms may threaten the survival of a large number of susceptible species, these parameters certainly deserve thorough examination because of their ecotoxicological significance. Hormone regulation may be impaired as a consequence of exposure to environmental pollutants (Spies et al. 1990). An overview of the methods and strategies to monitor the impact of endocrine-disrupting chemicals is given by Sadik and Witt (1999). Cytochrome P450 aromatase (P450arom or CYP19) is a CYP enzyme that transforms testosterone into 17β-estradiol, hence the conversion of C19 androgens to C18 estrogens (Ankley et al. 2002). Changes in the activity of aromatase may therefore result in differential sex expression. The activity of aromatase may be altered by several compounds, including dioxin-like contaminants. Sensitive markers for estrogenic effects in male oviparous vertebrates are the synthesis of the egg yolk protein vitellogenin (VTG) and the eggshell zona radiate protein (ZRP), which are involved in egg production of females (Arukwe and Goksøyr 2003).

Neuromuscular parameters

Esterases were initially used to assess the exposure of spray operators to organophosphorous (OP) and carbamate (CA) agricultural pesticides. B-esterases are a large group of serine hydrolases that are inhibited by organophosphates. They include acethylcholinesterase (ACHE), the target site for many of these compounds, and carboxylesterases (CBEs), a group of enzymes that are present in a range of tissues in both vertebrates and invertebrates and that can hydrolyze a wide array of esters. With respect to neuromuscular functions, recent studies indicated that the "old" biomarker acetylcholinesterase (ACHE) may be responding to low levels of contaminants in the environment (Payne et al. 1996). The ACHE inhibition has been used to assess the nature and extent of the exposure of wildlife to agricultural and forest sprays (Greig-Smith 1991; Zinkl et al. 1991).

Genotoxic parameters

The exposure of an organism to genotoxic chemicals may induce a cascade of events (Shugart et al. 1992): formation of structural alterations in DNA, procession of DNA damage and subsequent expression in mutant gene products, and diseases (such as cancer) resulting from the genetic damage. The detection and quantification of various events in this sequence may be employed as biomarkers of exposure and effects in organisms exposed to genotoxic substances in the environment (Dunn 1991). The study of DNA adducts in human and animal models has been an important part of research on mechanisms of genetic disorders and on evaluating substances for their potential to cause genetic damage. DNA adducts are covalently bound addition products formed when electrophilic chemical species attack the nucleophilic sites in DNA, and extensive experimental data support their role in the initiation of chemical carcinogenesis (Miller and Miller 1981). Of the techniques currently available for the detection of DNA modifications, the ^{32}P-postlabeling assay (Gupta et al. 1982) appears to offer potential for the qualitative and quantitative analysis of these adducts. This ^{32}P-postlabeling technique, however, is an expensive and time-consuming assay.

A more general approach involves the detection of DNA strand breaks that are produced, either directly by the toxic chemical (or its metabolite) or by the processing of structural damage (Shugart et al. 1992). DNA strand breakage has been considered as one of several sensitive indicators of genotoxicity (Mitchelmore and Chipman 1998). The measurement of this damage as a genetic endpoint represents a means of detecting the effect of a wide range of genotoxicants. To date, the most commonly used methods for the determination of DNA damage are those of alkaline elution and alkaline unwinding, although the alkaline single-cell gel electrophoresis (SCGE) or "comet" assay is being increasingly used (Large et al. 2002; Hamoutene et al. 2002). The comet assay is suitable for any nucleated eukaryotic cell type, usually as single-cell suspensions. The protocols for this technique vary among laboratories and require optimization for the specific cell type used. DNA base composition, oncogene activation, cytogenetic effects, and tumorigenesis also have the potential to be used as biomarkers.

Physiological and morphological parameters

The actual measurement of adverse effects or of the consequences of those effects may also be used as biomarkers. Determination of adverse effects can be performed histopathologically, by investigating lesions, alterations, or tumor formation (neoplasms) in fish tissues. So-called gross indices, such as the liver somatic index (LSI) and the condition factor (CF), may be simply determined by measuring weight and length of body and organs (Slooff et al. 1983; Bagenal and Tesch 1978). Scope for growth (SFG) is an integrative measure of the energy status of an organism at a particular time. It is based on the concept that energy in excess of that required for normal maintenance will be available for the growth and reproduction of the organism, and it requires the measurement of the energy absorbed from food and the energy lost via respiration and excretion (Bayne et al. 1985). Lysosomal alterations (enlargement and membrane destabilization) are accepted as a marker of general stress for a diverse range of aquatic vertebrates and invertebrates (Moore 1990). In stable lysosomes, hydrolases are prevented from reacting with substrate by an intact membrane. Theoretically, membrane stability decreases in response to stress as membrane permeability increases.

Proteomics and genomics

New and promising developments in the biomarker field are the so-called genomics and proteomics. Genomics is based upon the application of DNA microarrays that allow the expression of hundreds to many thousands of genes to be monitored simultaneously, thus providing a broad and integrated picture of the way an organism responds to a changing environment (Gracey et al. 2001). The entire protein complement of the genome, the proteome, can now be analyzed for changes associated with specific treatments, using peptide mass profiling, a combination of two-dimensional gel electrophoresis and mass spectrometry (Shepard et al. 2000). Proteomics research provides certain protein expression signatures (PESs), which are specific sets of proteins, present or absent, which indicate specific toxicity profiles.

Invertebrate biomarkers

The historical development of the biomarker approach had a strong link with medicine and vertebrate biology (NRC 1987), thus most of the field studies and applications in the aquatic environment have been focused on fish. Biomarker measurements, however, are equally feasible in invertebrates that represent 95% of all animal species (Depledge and Fossi 1994), and are major components of all ecosystems. Most of the general biomarker criteria (see the introduction to this chapter) are fulfilled by certain invertebrate biomarkers, whose use in field studies have some additional advantages. First, invertebrate populations are often numerous, so that samples can readily be taken for analysis without significantly impacting the population dynamics. Second, the application of biomarkers to some species allows the linkage between biomarkers responses and adverse effects on populations and communities. Currently, increasing knowledge of the biochemistry and physiology of invertebrates (Livingstone 1991) permits a reasonable interpretation of biomarker responses. The most commonly used biomarkers in invertebrates are discussed below.

Phase I enzymes

Although many aspects of biochemistry are the same for all organisms, major differences exist among animal groups, both for general biochemical pathways and for specific pathways and enzymes of biotransformation of pollutants and xenobiotics (Livingstone 1998). The first knowledge of P450-type activity in marine invertebrates was reported by Khan et al. (1972). Since then, multiple forms of cytochrome P450 and P450-type enzymatic activities have been reported in crustaceans, polychaetes, cnidarians, molluscs, porifera, and echinoderms (Livingstone 1994; James and Boyle 1998; Snyder 2000), although the relevance of these proteins for the metabolism of xenobiotics remains questionable. EROD activity (characteristic of vertebrate CYP1A) is either undetectable or only present in low activity (Livingstone 1996), although EROD induction has been observed in some crustacean species from contaminated field sites (Ishizuka et al. 1996; Porte and Escartín 1998). In contrast, benzo[a]pyrene (BaP) hydroxylase activity is widely detectable in aquatic invertebrates. Overall responses are lower than in fish, such as a 3-fold increase in BaP hydroxylase activity in pyloric caeca of the starfish exposed to BaP or PCBs compared to a 100-fold increase for hepatic EROD activity in fish (Den Besten et al. 1993). Despite these relatively small changes in enzyme activity, field studies indicate putative induction of the cytochrome P450 system in some invertebrate species collected from contaminated field sites (Gassman and Kennedy 1992; Solé et al. 1996; Fossi et al. 1998). More recently, specific evidence on the existence of multiple P450 isoenzymes in invertebrates has been obtained by the use of molecular tools (Snyder 2000), and this approach will probably reveal potential CYP biomarker genes in the near future.

Oxidative stress

Basic comparative studies on pro-oxidant and antioxidant processes have indicated strong similarities in oxidative stress-related phenomena in aquatic organisms. However, significant differences are observed between vertebrates and invertebrates regarding antioxidant enzymes. GPOX activities, for instance, are one to two orders of magnitude lower in invertebrates than vertebrates (consistent with differences in many other enzyme activities), whereas similar or higher SOD and CAT activities are found in invertebrates (Livingstone et al. 1992).

From a biomarker point of view, both adaptive responses and toxic effects associated with xenobiotic-mediated enhancements of oxyradical production have been considered. A significant amount of invertebrate data on antioxidant enzyme activities and concentrations of nonenzymatic components have been collected in both laboratory and field studies. The responses, however, are transient and variable for different species, enzymes, and exposures (Livingstone 2001). In field studies, higher, equal, or lower activities of various antioxidant enzyme activities have been observed in polluted areas when compared to cleaner areas (Narbonne et al. 1999; Ringwood et al. 1999; Livingstone and Nasci 2000; Porte et al. 2000). A different approach applied to invertebrates is the recently developed measurement of total oxidant scavenging capacity (TOSC) for specific reactive oxygen species (Regoli 2000). This assay offers the potential of a more quantitative approach to the understanding of antioxidant function and regulation in aquatic organisms.

Biochemical manifestations of oxidative damage have been observed after exposure to single and mixed contaminants (Livingstone 2001). The observed oxidative damage comprises lipid peroxidation (formation of malonaldehyde-like species and 4-hydroxyalkenals), protein oxidation (nonpeptide carbonyl groups), and DNA damage (8-hydroxydeoxyguanosine and other oxidized bases).

Stress proteins, metallothioneins, and multixenobiotic resistance

One well-characterized general stress response in invertebrates is the induction of heat shock proteins (HSPs). These proteins, particularly HSP70 and HSP60, are inducible by a variety of chemical and physical stresses including temperature, UV radiation, heavy metals, and oxidizing agents. Numerous studies have been carried out on stress proteins as biomarkers in different invertebrate species (molluscs, crustaceans, nematodes, amphipods). Stress protein induction was observed after exposure to many pollutants at environmentally realistic levels (Sanders 1990, 1993; Lewis et al. 1999). However, the responses were not consistent in all studies.

Metallothioneins (MTs) are also widely studied in invertebrates. They have been identified in approximately 50 different species of aquatic invertebrates, the majority of which are molluscs or crustaceans (Langston et al. 1998). MT responses have been used successfully as biomarkers of heavy-metal exposure in molluscs (Viarengo et al. 1999; Hamza-Chaffai et al. 2000). However, their induction in response to metal stress varies among different organisms and tissues, and may be influenced by nonmetal endogenous and exogenous factors (temperature, glucocorticoids, oxidative stress), which still prevents their wide use as biomarkers (Viarengo et al. 2001).

The existence of membrane-associated drug efflux pumps, which can protect cells from a range of toxic compounds, has been demonstrated in many invertebrate species (Higgins 1992). The first report of a P-glycoprotein (PGP) mediated toxin resistance in aquatic organisms was in a freshwater mussel (Kurelec and Pivcevic, 1989). Since then, a similar protein has been detected in marine mussels, sponges, snails, clams, and worms (Kurelec et al. 1992, 1996; Toomey and Epel 1993; Smital et al. 2000). The presence of PGP has been established directly by immunoprecipitation, western blotting, and measurement of PGP mRNA; and indirectly by measuring cellular accumulation of radio-labeled or fluorescent substrates (Livingstone et al. 2000). P-glycoproteins respond to a wide range of pollutants, and different expression levels have been measured among sites in different field studies (Kurelec et al. 1996). However, the role of age, season, sex, food availability, and other environmental factors that can alter PGP expression should be elucidated before determining their feasibility as biomarkers.

Reproductive and endocrine parameters

Field evidence on endocrine disruption in invertebrates is dominated by the study of the antifouling paint ingredient tributyltin (TBT) and its effects on mollusc populations worldwide. TBT and other organotin compounds such as triphenyltin (TPT) are known to induce different masculinization phenomena (imposex and intersex), which consists of the gradual and concentration-dependent increase of virilization of females, with the final result of sterility (Matthiessen and Gibbs 1998). In imposex-affected species, the entire female genital system is conserved but superimposed by male organs (penis and/or vas deferens); in the periwinkle, the only intersex-affected species described to date, the female pallial organs are modified toward a male morphological structure and then supplanted by a prostate gland. A large number of field surveys have been described in the literature (reviewed by DeFur et al. 1999). Imposex is a worldwide phenomenon that has been found to occur in more than 150 species of gastropods. Imposex has been demonstrated at TBT concentrations as low as 0.5 ng Sn/l (Mensink et al. 2002). Because of this sensitive response, several snail species have been used as bioindicators of TBT contamination worldwide.

Neuromuscular parameters

In aquatic organisms, there is considerable diversity in the biochemical properties and distribution of cholinesterases, as well as in their sensitivity to anticholinesterase agents (Bocquené et al. 1990). To date, several studies have successfully used the inhibition of cholinesterase activity as a tool to assess exposure to organophosphate pesticides in invertebrates (Escartín and Porte 1996, 1997). Due to inherent variability among species and tissues, it is necessary to include good controls in both laboratory and field studies.

Genotoxic parameters

Detecting and quantifying DNA adducts have been performed in invertebrates and successfully applied to field studies (Kurelec et al. 1990; Solé et al. 1996; Harvey et al. 1999). The quantitative analysis of DNA adducts enables the determination of the biologically relevant levels of exposure to genotoxic chemicals, thus integrating the physiological factors such as absorption, metabolism, and detoxification involved in adduct formation.

Physiological and morphological parameters

The large majority of the scope for growth (SFG) work has been conducted with marine bivalves, and most of the research shows that SFG can be a fairly sensitive indicator of effects and general ecosystem condition (Widdows and Donkin 1992). SFG has been successfully used in several monitoring programs, coupled with chemical analysis of contaminants in the tissue of mussels (Widdows et al. 1995, 2002). This approach serves to complement the established chemical monitoring programs by providing a means of assessing whether the recorded contaminant levels are causing deleterious effects, and whether all relevant toxicants are being measured.

Lysosomal enlargement has been successfully used in a number of field studies using molluscs as sentinel species (Porte et al. 2001; Petrovic et al. 2001; Lowe and Fossato 2000). Very often the responses could be related to higher levels of metals, PAHs, PCBs, or other anthropogenic pollutants accumulated by the mussels. The mechanism causing this alteration in membrane stability is not well understood, but may involve either a direct effect of chemicals on the membrane or the increased frequency of secondary lysosomes in toxicant-stressed cells.

Fish biomarkers

For several reasons, fish species have attracted considerable interest in studies assessing biological and biochemical responses to environmental contaminants (Powers 1989). Monitoring species should be selected from an exposed community on the basis of their relationship to the assessment endpoint as well as by following some practical considerations (Suter 1993). For the assessment of the quality of aquatic ecosystems, both criteria are met for numerous species of fish. Fish can be found virtually everywhere in the aquatic environment and they play a major ecological role in the aquatic food webs because of their function as a carrier of energy from lower to higher trophic levels (Beyer 1996). The understanding of toxicant uptake, behavior, and responses in fish may therefore have a high ecological relevance. Most of the general biomarker criteria (see the introduction to this chapter) appear to be directly transferable to certain fish biomarkers (Stegeman et al. 1992). Among different fish species, however, considerable variation in both the basic physiological features and the responsiveness of certain biomarkers towards environmental pollution may become apparent. Despite their limitations, such as a relatively high mobility, fish are generally considered to be the most feasible organisms for pollution monitoring in aquatic systems. The most commonly used fish biomarkers are discussed below.

Phase I enzymes

The phase I biotransformation enzymes, notably CYP1A, definitely belong to the most sensitive fish biomarkers known at present (Goksøyr and Förlin 1992; Van der Oost et al. 2003). The value and feasibility of these biomarkers have been demonstrated in numerous laboratory and field studies with various fish species (Whyte et al. 2000). Pollution-induced increases in EROD activity and CYP1A levels were observed in the majority of more than 100 field studies with marine and freshwater fish (Van der Oost et al. 2003). Consequently, there is growing interest in using CYP1A induction in fish as a biomarker to indicate the exposure of aquatic organisms to CYP1A-inducing compounds and to evaluate the degree and possible risk of environmental contamination. CYP1A protein levels and EROD activity can be incorporated in monitoring programs, provided that the experimental design considers all intrinsic and extrinsic variables that may potentially influence this parameter, as well as xenobiotics that inhibit CYP1A activity (such as organotins; see Fent et al. 1998). No sensitive pollution-related responses have been observed thus far in fish for other P450 isozymes (such as CYP3A). At present, CYP1A levels and activities have been validated for use in various areas of (environmental) toxicological research, such as research on toxic mechanisms of xenobiotics, identification of exposure to specific compounds, identifying subtle early effects (early warning), and triggering of regulatory action (Van der Oost et al. 1996b, 2003).

Phase II enzymes

Due to the important role of phase II enzymes in detoxification, they should be considered in future biomonitoring programs as well, although their sensitivity towards pollutant exposure is limited (Andersson et al. 1985; George 1994). Significant pollution-related increases in GST and UDPGT activities were observed in less than 50% of almost 50 field studies considered (Van der Oost et al. 2003). An integrated biomarker approach, using combinations of biomarkers such as the biotransformation index (BTI) that reflects the ratio between phase I and phase II activities, may be more useful for monitoring since this index reflects the balance between bioactivation and detoxification (Van der Oost et al. 1998). The BTI may also be indicative of the susceptibility of certain species to toxic xenobiotics with carcinogenic properties. Further studies are required, however, to improve the usefulness of phase II enzymes as biomarkers in monitoring procedures, such as baseline activities, non-pollution–related confounding factors, linkage between enzyme activities and higher levels of organization (such as organ or whole-animal), and the basic chemistry of purified forms. Current research on the latter topic may reveal specific isoforms of GST and UDPGT that are more sensitive indicators of exposure or effects than the measurement of total activities (George 1994).

Oxidative stress parameters

Since many environmental contaminants exert toxic effects related to oxidative stress, this phenomenon may be an important feature for biomarker development. However, antioxidant enzymes are generally less responsive to pollutants than phase I and phase II enzymes and the relationships between response and contaminant exposure are still less well established. Their function in detoxification processes, however, motivates continued research on the potential use of SOD, CAT, and GRED in monitoring programs. Other promising parameters that may be useful for indicating oxidative stress are the GSH:GSSG ratio and lipid peroxidation products such as aldehydes. Species differences in the efficiency of antioxidant defenses may partly explain the prevalence of pathological lesions observed in certain species of fish (Vigano et al. 1998). Significant alterations in SOD activities are generally not observed in laboratory studies, but significant increases in SOD activity were demonstrated in many field studies, such as in the brown bullhead, carp, dab, grey mullet, Nile tilapia, red mullet, sardine, and spot at polluted sites (Van der Oost et al. 2003). In general, CAT, GPOX, and GRED activities cannot yet be considered valid biomarkers since induction and inhibition are rarely observed after exposure to environmental pollutants (Van der Oost et al. 2003). Studies addressing the feasibility of nonenzymatic antioxidants as a biomarker are very scarce for fish. A large number of biochemical and physiological effects have been associated with increased fluxes of oxyradicals. Some biochemical perturbations that seem particularly promising for oxidative stress biomarkers in fish are lipid peroxidation (LPO), the total oxyradical scavenging capacity (TOSC) assay, DNA oxidation, and methemoglobinemia and redox status (Stegeman et al. 1992).

Biotransformation products

Biliary levels of (conjugated) metabolites of easily biodegradable xenobiotics, such as PAHs, chlorinated phenols, and resin acids, have been validated as fish biomarkers of recent exposure and may now be used in monitoring programs. In order to determine the exposure to and possible effects of PAHs in fish, the fate of PAH metabolites should be investigated so as to quantify the PAH flux (uptake and excretion) in the animals (Collier and Varanasi 1991). Some PAHs are excreted as polar metabolites via the gallbladder (in bile), but most PAHs are excreted after conjugation by phase II enzymes (Vermeulen et al. 1992). A significant increase in biliary fluorescent aromatic compound levels (FAC) was observed in most of the 24 field studies considered by Van der Oost et al. (2003). Pulp and paper mill effluents generally contain high concentrations of resin and fatty acids, as well as chlorophenolics. Elevated fish bile levels of these substances have been used as biomarkers (Leppänen et al. 1998). Additional research must be performed to establish the impact of confounding factors and to define clear relationships between biomarker responses and the organisms' health. It has, for instance, been demonstrated that both the PAH metabolite levels in bile and the bile volumes were highly influenced by the feeding status of the fish (Collier and Varanasi 1991; Brumley et al. 1998). In order to reduce variations in PAH metabolite bile levels due to feeding status, researchers proposed a procedure in which the metabolite concentrations are related to the biliary pigment contents. In addition to biotransformation products, the body burdens of certain persistent chemicals, such as PCBs and DDTs, may serve as bioaccumulation markers for the exposure to those compounds (Van der Oost et al. 1996a).

Stress proteins, metallothioneins, and MXR

Several field studies with feral fish have supported the experimental data that MTs sequester heavy metals and that MT levels correlate with tissue levels of heavy metals (Roch et al. 1982; Roch and McCarter 1984; Olsson and Haux 1986; Hylland et al. 1992; Schlenk et al. 1995; Olsvik et al. 2000). The large individual variability in hepatic P-glycoprotein (PGP) levels in feral fish suggests that fish may have variable abilities to respond to multi-xenobiotic resistance (MXR) inducers (Bard et al. 1998). Since the literature regarding these parameters in fish is limited, the feasibility of stress proteins (HSPs), MTs, and MXR as fish biomarkers for environmental monitoring needs to be further evaluated with additional research. This research needs to focus on baseline data of the normal physiological function and on the influences of non–pollution-related confounding factors in the field.

Reproductive and endocrine parameters

Many environmental contaminants are known to have endocrine-disrupting properties. Effects of pollutants on the endocrine system may be of major importance in monitoring programs, since impairments in reproductive capability may have a serious impact on fish populations. In an extensive review, Kime (1995) presented an overview of the effects of sublethal pollution, both industrial and agricultural. Kime includes all aspects of fish reproduction, from gonad development to spawning, together with a discussion of how some of these effects may be a result of a disturbance of the reproductive endocrine system. Production and secretion of hormones of the hypothalamus, pituitary, and gonads is usually inhibited and hormone metabolism by the liver can be altered. There is also considerable literature on the survival of eggs, larvae, and fry, which are particularly susceptible to pollutants, and may have a major impact on population dynamics (Kime 1995; Hugla and Thome 1999). Sensitive assays, such as levels of vitellogenin (VTG) and zona radiata proteins (ZRP) are promising, but they must be further validated for their ecological significance. The synthesis of vitellogenin, a precursor of yolk proteins, is affected by estradiol. It was demonstrated that PCB-exposed female fish were less capable of producing vitellogenin (Spies et al., 1990). An impaired reproductive function due to decreased plasma vitellogenin levels was reported for female rainbow trout exposed to cadmium (Haux et al., 1988). Vitellogenin synthesis can also be induced in male fish exposed to endocrine disrupting chemicals such as alkylphenols, thus leading to feminization of male fish (Gimeno et al. 1996). A pronounced increase in plasma vitellogenin levels in male fish was observed in many field studies (Purdom et al. 1994; Mellanen et al. 1999; Lye et al. 1999; Larsson et al. 1999). The latter field study revealed that a substantial part of the observed estrogenic effects was due to 17-ethinyloestradiol, a synthetic estrogen used in contraceptives that was present in effluents of sewage treatment plants receiving (mainly) domestic wastewater. Another potential biomarker for estrogenic effects in male fish might be the induction of ZRPs, also known as vitelline envelope proteins (Hyllner et al. 1991). Studies with juvenile salmon indicated that the ZRP response was more sensitive to various environmental pollutants than the VTG response (Arukwe and Goksøyr 1997; Arukwe et al. 2000), thus providing a sensitive means of detecting exposure to environmental estrogens. Arukwe and Goksøyr (2003) reviewed the oogenetic, population, and evolutionary implications of endocrine disruption in relation to eggshell and egg yolk proteins in fish.

Concentrations of beta-estradiol (E2) found in the environment can have disruptive effects on key steroidogenic enzyme pathways that control sexual development in fish (Halm et al. 2002). Steroid metabolism may be influenced by the effects on cytochrome P450 aromatase (P450arom), mainly expressed in the brain and the gonads of fish, which has the potential to be used as a reproductive biomarker.

Neuromuscular parameters

The inhibition of acetyl cholinesterase (ACHE) in fish tissues may be a promising biomarker for the assessment of exposure and effects of complex mixtures of contaminants. Evidence for variation of ACHE in the muscle tissues of dab and flounder was demonstrated along a pollution gradient in the North Sea (Galgani et al. 1992). A significant depression of ACHE activity in fish from polluted sites (organophosphate pesticides and parathion) was also observed in muscle tissues of brown trout and flounder (Payne et al. 1996) and three-spined stickleback (Sturm et al. 1999, 2000), and in brain tissue of menhaden and mummichog (Fulton and Key 2001). It was unknown whether inhibition observed in field studies was due to pesticides, other factors, or a combination of both. More research on the impact of confounding factors and the identification of the (classes of) compounds responsible for observed effects in field trials is required before this parameter can be used in monitoring programs.

Genotoxic parameters

The study of genetic toxicology in aquatic systems is mainly focused on carcinogenesis in fish and shellfish. The formation of hepatic DNA adducts in fish is considered to be a valid biomarker for exposure to PAHs and for the assessment of potential genotoxic effects. A significant increase in hepatic DNA adduct levels was observed in most of the 30 field studies considered by Van der Oost et al. (2003). ^{32}P-postlabeling is a very sensitive assay for measuring DNA adducts, but it may be considered too expensive and time-consuming to be applied in routine monitoring programs. Other DNA modifications, such as strand breaks, may also be feasible as biomarkers for genotoxic chemicals (comet assay and DNA unwinding assay). The alkaline unwinding assay can estimate the increase in the level of DNA strand breaks above background that result from exposure to genotoxic chemicals (Shugart

1990). This method is well suited for routine, *in situ* monitoring of feral fish because of its ease, speed, and low cost. Oncogene activation is a DNA-based assay to measure specific nucleotide changes in oncogenes and other appropriate genes of chemically exposed animals. The p53 gene has been sequenced for several fish species with a view to the possible use of mutations in the highly conserved domains of p53 to identify genotoxins in the aquatic environment (Cachot et al. 1998; Bhaskaran et al. 1999).

Physiological and morphological parameters

When physiological, histological, and morphological parameters in fish are affected, this generally indicates irreversible damage or disease. These parameters are therefore not suitable as early-warning signals in monitoring programs. Histopathological parameters are, however, relatively easily determined and valuable to evaluate the ecotoxicological significance of other biomarkers. The feasibility of using histopathological parameters in fish as a biomarker for aquatic pollution has been reviewed by Hinton et al. (1992) and Hinton (1994). Immunohistochemical detection of cellular- and tissue-related abnormalities offer earlier warning signals than are provided by normal pathological and histopathological examination of histological samples. Vethaak and ap Rheinallt (1992) have reviewed the usefulness of epidemiological studies on the occurrence of fish disease in monitoring marine pollution. They concluded that, on a worldwide scale, the most convincing examples of a causal relationship between fish disease and pollution was provided by intensive and detailed studies carried out in North America, particularly on liver pathology (Myers et al. 1994). Gross indices, such as the LSI and the CF, can be easily measured and may be used to assess the condition of the liver and the general health of fish, respectively. Significant increases in LSI and CF values were observed in less than 50% of 30 field studies considered by Van der Oost et al. (2003). Although these parameters are not very sensitive and may be affected by non–pollution-related factors (such as season, disease, and nutritional level), they may serve as an initial screening biomarker to indicate effects or to provide information on energy reserves (Mayer et al. 1992).

Amphibian biomarkers

Numbers of amphibians have been declining for many years, a reason for concern (Corn 2000). Reasons for these declines may differ among locations, but one of the hypotheses is that the decreasing numbers of amphibians may be due to the effects of contaminants. The increased incidence of deformed amphibians in particular is thought to be related to the presence of contaminants, including pesticides, although other mechanisms may also play a role (Ouellet 2000). These findings indicate that amphibians may be sensitive to contaminant exposure, and may therefore be good bioindicators for monitoring water quality. There are several reasons why amphibians may be sensitive to contamination of aquatic ecosystems: (1) the development from egg to adult is hormonally regulated, and thus may be affected by endo-crine-disrupting compounds; (2) amphibians spend a significant part of their life span in the water and are thus exposed to aqueous contaminants; and (3) amphibians are able to absorb chemicals through their skin. In this section several types of biomarkers, relevant for amphibians, are discussed in rela-tion to their applicability as indicators for contaminant exposure.

Phase I enzymes

Several studies report the use of phase I enzymes as indicators for exposure of amphibians to contaminants (Schwen and Mannering 1982; Harri 1980; Huang et al. 1998, 1999, 2001; Ertl and Winston 1998; Harris et al. 1998; Stien et al. 1997). In general EROD activity is the most frequently applied biom-arker for the exposure to dioxin-like compounds, although the responses in amphibians are highly variable. Huang et al. (2001) detected an increase of CYP1A proteins and associated EROD in Northern leopard frogs after exper-imental exposure to PCB126. However, in a field study with frogs from contaminated sites in the Green Bay (in the U.S.), a chemically detected exposure to PCBs and dioxins did not result in induction of EROD activity (Huang et al. 1999). It was thought that the Northern leopard frog has a relatively high ED_{50} for EROD induction, so this parameter may not be a very sensitive biomarker. Induction of CYP450s was related to the inducing compounds and the specific substrates used for the detection of CYP activity in a review by Ertl and Winston (1998). The authors concluded that the extent of CYP induction in amphibians is lower than in mammals, that the induction needs a longer exposure period, and that certain types of induction are not observed in amphibians. As seen in fish, Phenobarbital (PB) type of CYP induction is absent in some amphibians. Furthermore, seasonal differences in CYP induction have been observed regularly. Although EROD induction in amphibians has been applied as a biomarker for exposure to contaminants, a careful data interpretation is needed. Before applying this parameter in a specific case and species, it should be validated that CYP1A is inducible, that the induction is sensitive enough, and that the isoform of the CYP protein is able to metabolize the specific substrate used in the assay.

Phase II enzymes

In some amphibian species the presence of certain GST forms has been confirmed (Angelucci et al. 2001; Bucciarelli et al. 2001; Di Ilio 2001). How-ever, little is known about the mode of action leading to induction of these enzymes in amphibians. Furthermore, no information is available on the sensitivity of phase II induction. Hence, without further investigations it remains unclear whether phase II enzyme activities may be useful as biom-arkers of exposure. Nevertheless, because of their important role in detoxification, the use of amphibian phase II enzymes should be considered in future ERA. Before this may be accomplished, more research is needed on the applicability of phase II enzymes, on baseline data, and on the effects of confounding factors.

Oxidative stress

In anuran amphibians, oxidative stress has been related to the tolerance to severe dehydration (Hermes-Lima and Storey 1998). Similar to fish, GSH:GSSG ratios may indicate oxidative stress, although the use of such a marker still has to be validated. Little is known, for instance, on the type of contaminants that may induce oxidative stress, the sensitivity of responses, and interspecies differences. Nevertheless, the relationship between oxidative stress and dehydration tolerance of amphibians indicates that this chain of effects may reveal important biomarkers of exposure to contaminants. Therefore, efforts to validate this type of biomarker for use in ERA may be useful.

Hematological parameters

ALAD was found to be active in frogs and was inibited in a dose-dependent manner after exposure to lead (Ireland 1977; Stansley and Roscoe 1996). ALAD activity has been analyzed in other organisms such as mammals and birds (Stansley and Roscoe 1996; Franson et al. 2000), and in those species it is considered a useful biomarker of exposure to lead. Hence, in amphibians ALAD activity in liver tissue appears to be applicable as a biomarker for lead exposure in field situations, although validation of baseline data and sensitivity may still be needed. The assessment of steroid hormones and VTG in blood is discussed in the section on reproduction and endocrine parameters.

Stress proteins

Heat shock proteins (HSP) have been detected in amphibians (Ali et al. 1996). HSPs appear to be fairly identical in different groups of organisms (Dunlap and Matsumura 1997), but little is known on the use of HSPs as markers for stress in amphibians. The use of HSPs as indicators of contaminant exposure is therefore still to be elucidated. Similar to the situation in fish, research on HSP in amphibians needs to focus on baseline data, and on factors besides contamination that may interfere with HSP.

Reproduction and endocrine parameters

Endocrine-disruptive effects due to contaminant exposure may severely impact the reproductive output of adult amphibians, as well as the development of embryos and tadpoles. The main groups of hormones affected are the sex-steroid hormones and the thyroid hormones (Hayes 2000). Disruption of the thyriod axis may result in abnormalities and deformities, and is therefore discussed in the section on physiological and morphological parameters.

Changes in the steroid axis may result in changes in gonad differentiation, and consequently in changes in sex ratios (Kloas et al. 1999). The occurrence of some sex–reversed individuals appears to be normal in amphibian populations, but the anthropogenically induced changes in sex ratios may ultimately result in decreased survival rates of (local) populations. Various modes of action for reproduction-associated effects caused by endocrine disruption have been observed. For example, effects of contaminants on aromatase activity have been described both in lab and field studies (Di Fiori et al. 1998; Van den Brink et al. 2003a). Since aromatase activity was affected by various chemicals, it appeared to be an effect marker, rather than a compound-related biomarker. Another expression of endocrine disruption is the production of VTG by male amphibians, which indicates exposure to estrogens (Kloas et al. 1999; Palmer et al. 1998). VTG production has been validated to be applicable in assessment of wildlife exposure to environmental estrogens (Palmer et al. 1998). Steroid hormone levels may also be used as indicators of endocrine disruption, but social signals, like chorus, may also affect levels of circulating steroid hormones (Burmeister and Wilczynski 2000).

It can be concluded that several amphibian parameters may be used as biomarkers to assess endocrine disruption related to development, reproduction, and sex differentiation. These parameters include biochemical and histological analysis, which allows for the application of biomarkers at different integration levels.

Neuromuscular parameters

Exposure of tadpoles to certain pesticides (triclopyr and fenitrothion) resulted in mortality or paralysis (Berrill et al. 1994) at levels that may be reached in forest surface waters after spraying. Fenitrothion is a nonreversible ACHE inhibitor and animals exposed to it expressed a decrease in ACHE activity. Exposure to pyrethroid insecticides caused abnormal behavior in amphibian tadpoles (Berril et al. 1993), which was related to changes in sodium and calcium channels in nerve tissues. Exposure to DDT resulted in mortality at relatively low levels that was clearly induced by damage to the central nervous system (Harri et al. 1979). These results show that amphibians may express changes in neuromuscular parameters after exposure to contaminants, which can be analyzed by behavioral studies. These studies should be performed at specific stages of embryonic and tadpole development, because the sensitivity may change during development.

Physiological and morphological parameters

In the section on reproductive and endocrine parameters, it was pointed out that alterations in certain hormone levels may result in deformations or altered development. Thyroid hormones play an essential role in the development of amphibians, although this role is less clear in some salamander species. In the larval development, deformities of legs and snout have been reviewed (Hayes 2000), and related to, for instance, exposure to retinoic acid (limb deformities) and DDT (snout deformities). It is expected that deformities in adult amphibians are the result of induction in the larval phase, so the assessment of possible endocrine disruptive effects on deformities should be carried out in the proper stages of development.

The detection of morphological parameters in an ERA is relatively straightforward, although the interpretation of the observed malformations may be difficult. In a review, Ouellet (2000) reports several causes that may lead to malformations, such as diseases, parasitic cysts, nutritional deficiencies, UV radiation, and contaminant exposure. Hence, when malformations are observed this may be used as a gross initial indicator of overall stress, although it is very difficult to track deformities back to a certain cause, such as exposure to a specific class of compounds. The value of such a general approach is not clear, since malformations may occur in nonpolluted situations and false positives may also be detected.

Histological observations of amphibian tissues are rarely presented in field-related research. The effects of styrene on the gonad development, however, have been assessed histologically (Ohtani et al. 2001). Most of the male tadpoles treated with styrene showed testicular as well as ovarian structures, while in nontreated males this occurred in only 3% of the specimens. In a study with frogs exposed to corticosterone, a loss of connective tissue, epidermis, and dermis in the snout has been observed (Hayes et al. 1997). A reduction of testis volume and changes in testis organization structure were observed after treatment with atrazin (Tavera-Mendoza et al. 2002). These examples show that the use of histological observations can be essential in relating effects observed in field situations at higher integration levels (such as malformations and behavior) to individual (classes of) compounds.

Mammalian and avian biomarkers

Avian and mammalian species are air-breathing, and as such do not accumulate any contaminants through gills. In most cases the uptake of aqueous contaminants through the skin is also negligible, so the main route of exposure to contaminants is generally through food uptake. Hence, these species are mostly exposed to contaminants that tend to accumulate in the food web. These contaminants have some features in common: they are not easily biodegradable, they have affinity to be stored in biotic tissues, and they are transferable between trophic levels. Examples of such contaminants are (1) organochlorines like DDTs and PCBs, which are resistant to metabolism and can be stored in fat depots of organisms; (2) certain heavy metals like cadmium, which are nondegradable and can be stored in organisms' tissues; and (3) other organohalogens like brominated flame retardants. Therefore, the majority of mammalian and avian biomarkers that are applicable in ERAs are focused on the exposure to organochlorines, pesticides, and heavy metals. In this section mammalian and avian biomarkers are addressed together.

Phase I enzymes

Phase I enzymes have been extensively studied in mammals and birds (Bosveld 1995) and applied in ERA. AROD biomarkers (indicative for CYP450 induction), however, can be applied. These are best used with freshly obtained liver tissue material, and are therefore destructive to the test organism. Considering mammals and birds, the use of destructive methods may be limited from an ethical as well as a scientific point of view (Fossi et al. 1993) and the use of nondestructive alternatives, such as taking skin biopsies, is preferable. Skin biopsy material is suitable for a wide range of chemical and biomarker analyses, since organochlorines and polycyclic aromatic hydrocarbons can be analyzed in subcutaneous fat, and MFO activity, CYP450 isoforms, and DNA damage can be detected in the epidermis (Fossi et al. 1997). Another nondestructive alternative is the application of the PCB pattern analysis (PPA) (Van den Brink et al. 2000). The PPA analysis is based upon the fact that certain phase I enzymes (CYPs) in mammals and birds can metabolize certain PCB congeners, depending on their structure. Hence, induction of CYP activity will alter the levels of specific PCBs in organisms, and consequently alter the PCB patterns. PPA can be done in blood samples, for instance, which can be obtained without further affecting the test organism. It has been shown in common terns and little owls that changes in PCB patterns could be related to elevated levels of PCBs or other contaminants (Van den Brink and Bosveld 2001; Van den Brink et al. 2003b). Hence, the application of PPA appears to be feasible as a nondestructive alternative for AROD biomarkers.

Phase II enzymes

Different types of phase II enzymes can be distinguished in mammals and birds. Among these are GSTs, UDPGTs, sulfotransferase (ST), and epoxide hydrolase (EH) (Stegeman et al. 1992). Phase II enzymes in birds and mammals can be induced or inhibited by a variety of contaminants. In mallard ducks, exposure to Hg resulted in decreased GST activity, while exposure to Se did not alter GST activity (Hoffman and Heinz 1998). In general, phase II enzymes may be induced by organic contaminants and some heavy metals, although, for example, the activity of ST appears not to be inducible by xenobiotics (Stegeman et al. 1992). The use of phase II enzymes as indicators for specific exposure seems to be limited, due to the fact that induction is not compound-specific and varies with sex, age, and temporal changes. However, the importance of the phase II enzymes in detoxification and excretion of foreign compounds implies that phase II enzymes may be of more importance as effect biomarkers than as exposure biomarkers.

Hematological parameters

ALAD is active in most tissues of birds and mammals, and is inhibited in a dose-dependent manner after exposure to lead. Several authors reported the use of ALAD inhibition as a marker for lead exposure in small mammals (Stansley and Roscoe 1996), and birds (Franson et al. 2000; Work and Smith 1996). The assessment of ALAD activity is possible in small blood samples (±50 µl), and it appears to be applicable as a biomarker for lead exposure in field situations.

Stress proteins and metallothioneins

In mammals and birds, two major families of stress proteins are of importance: HSP60 and HSP70. Under normal conditions these proteins are involved in protein homeostasis, while under stress they play a role in protection and repair mechanisms (Welch 1990). Elevated levels of MTs have been associated with exposure to heavy metals in mammals (Solis-Heredia et al. 2000; Teigen et al. 1999) as well as in birds (Elliot and Scheuhammer 1997; Trust et al. 2000). The use of MTs as indicators for contaminant exposure has been discussed by Stegeman et al. (1992). Although promising, more research (especially on sensitivity and non–pollution-related responses) is required before MT induction can be applied as a valid biomarker.

Endocrine parameters

Aromatase is a CYP enzyme involved in the balance between estrogenic and androgenic hormones in mammals and birds (Janssen et al. 1998). In bank voles, a positive correlation has been established between the presence of aromatase and spermatogenesis (Carreau et al. 2002). Aromatase activity may be inhibited by, for instance, TCDD, although this inhibition is not very compound-specific. Nevertheless, the effects of aromatase inhibition may be severe and it appears to be a good marker for general endocrine disruption.

Microsomal testosterone metabolism has been studied in several avian and mammalian studies. For example, testosterone hydroxylase in great blue herons has been studied in relation to exposure to organochlorines (Sanderson et al. 1997). Specific sex differences of testosterone hydroxylation were observed among sites under basal conditions, while exposure to TCDD resulted in different hydroxylation patterns between sexes. In mammals, activities of CYP 2A1, 2B2, 2C, and 3A4 have all been related to testosterone metabolism (Janssen et al. 1998). Testosterone metabolite measurement in blood samples enables the screening of testosterone metabolite patterns, which can be used as a nondestructive biomarker related to the activities of specific CYP enzymes.

Thyroid hormones are involved in the development of organisms (Schuur 1998). Exposure to polyhalogenated aromatic hydrocarbons (PHAHs) generally results in decreased thyroxine (T4) levels, although the levels of the active hormone 3,3',5-triiodothyronine (T3) may not be altered or even be increased after exposure (Schuur 1998). Changes in blood hormone levels can be partly related to exposure to PHAHs and subsequent interaction with the Ah receptor, but the occurrence of hydroxylated PCBs may also interfere with the plasma transport of T4 (Lans 1995). Since an imbalance of the thyroid hormones may imply effects on the growth and development of animals, these hormones may be useful biomarkers as an early warning for effects due to PHAH exposure.

Neuromuscular parameters

ACHE inhibition in birds and mammals has been observed after exposure to organophosphorus compounds (OPs) and carbamates (CAs). In humans, ACHE inhibition by OPs is irreversible, while recovery of ACHE activity occurs within minutes after acute exposure to CAs (Moretto 1998). Total cholinesterase levels in the blood of pigeons that were orally exposed to parathion decreased significantly and recovered within 30 to 48 hours (Bartkowiak and Wilson 1995). In European starlings and red-winged blackbirds, a species- and age-dependent inhibition of cholinesterase activity was evident after exposure to diazinon and terbufos, both OPs (Wolfe and Kendall 1998). In small mammals, ACHE inhibition after exposure to dimethoate (an OP) was related to altered behavior of the animals. It was concluded that ACHE inhibition can be used as a predictor of OP-induced behavioral effects in free-living wild animals (Dell'Omo et al. 1997).

Morphological and histological parameters

Morphological and histological observations may reveal effects of exposure to contaminants at higher structural levels. Classic examples of morphological changes are the bill deformations in birds that are observed after exposure to PCBs. In pied flycatchers, asymmetric development of the tarsus of nestlings was related to contaminant exposure (Eeva et al. 2000). In addition, the asymmetric development correlated with effects on EROD activity and decreased breeding success near the source of contamination, a copper smelter. Effects on asymmetric development of primaries of nestlings of great tits were also detected near the copper smelter, although these were not related to an induced EROD activity. Exposure to cadmium resulted in several morphological effects in ducks, such as tubular changes in the kidney, reduced heart mass in female ducks, increase of kidney and liver mass in males, reduced gonad mass in both sexes, and more histological changes (Hughes et al. 2000). In developing chicks, effects of contaminants on the lymphocyte densities in the bursa of fabrius have been detected (De Roode 2001). These examples indicate that morphological and histological malformations may be indicative for exposure to contaminants. However, it should be regarded as a gross type of observation because (1) little is known on compound-specific malformations (besides the well-documented bill deformations), (2) sex- and species-specific malformations may occur in normal situations, and (3) it remains unclear how sensitive the occurrence of morphological malformations is to the exposure to contaminants. Nevertheless, due to the relatively easy observations and the strong ecological relevance of occurring malformations, observations of morphological and histological parameters may serve as gross markers for contaminant exposure.

Summary, discussion, and conclusions

In the environment, a wide variety of organisms are exposed to complex and changing levels of pollutant mixtures. It is a challenging issue to be able to relate pollutant body burdens to (harmful) effects. Traditionally, the basic approach of classical hazard assessment was to measure the amount of chemicals present in the environment (surrounding waters, sediments, or biota), and then to relate that, via animal experimental data, to the adverse effects caused by this amount of chemical. The limitations of this approach are (1) it has been possible to set critical levels for only a very limited number of compounds, (2) it does not take into account species differences, and (3) mixture toxicity and toxic metabolites are ignored.

A current approach in ecotoxicology is to examine exposed individuals for molecular and biochemical responses that are elicited by toxicants in an effort to assess the status of an impacted environment. The main role of biomarkers in environmental assessment is to determine whether or not, in a specific environment, organisms are physiologically normal (Peakall and Walker, 1994). The use of biomonitoring methods in the control strategies for chemical pollution has several advantages over chemical monitoring (De Zwart, 1995). First, these methods measure effects in which the bioavailability of the compounds of interest is integrated with the concentration of the compounds and their intrinsic toxicity. Second, most biological measurements form the only way of integrating the effects on a large number of individual and interactive processes. In view of the present chemically oriented pollution abatement policies and the need to reveal specific chemical problems, it is most probable that biological-effect analysis will never totally replace chemical analyses. The biomarker approach, therefore, should not be considered as a replacement for conventional assessment techniques, but as an important supplementary approach of great ecological relevance (Depledge and Fossi 1994).

A limitation of most of the biological-effect measurements is that it may be very difficult or impossible to relate the observed effects to specific aspects of pollution or to effects on the level of populations, communities, or ecosystems. Although biomarker research has contributed significantly to the gaining of knowledge on the effects of pollutants in the real environment, still new and innovative approaches are needed to integrate effects across different levels of biological complexity, and to provide a clear understanding of the danger posed by environmental pollution. Admittedly, at present it is extremely difficult to extrapolate from a biomarker response caused by chronic exposure to the effect on a whole organism or an ecosystem, as it is equally difficult to extrapolate from an acute response resulting from a lethal concentration of a stressor to low-level effects occurring over long exposure periods. Since the effects of non–pollution-related confounding factors may interfere with biomarker responses, biomarker data interpretation must always be carefully controlled for false-negative and false-positive results (see the chapter introduction). Due to these limitations there is a controversy on the use of biomarkers, and many scientists do not recognize the value of biomarkers for ERA research. The opinion of the authors of this chapter is that biomarkers can be valuable tools in assessing the risks of a wide range of pollutants that cannot be assessed by chemical analyses alone.

Invertebrates

Some of the invertebrate biomarkers reviewed in the present work are being used worldwide in several pollution monitoring programs, whereas others still require additional development before they can be routinely used. In general, no individual biomarker can give a complete diagnosis of pollution effects in the environment. In this context, lysosomal responses (lysosomal enlargement and lysosomal membrane destabilization), one of the most reliable biomarkers of general stress, combined with more specific responses (such as acetylcholinesterase inhibition, induction of metallothioneins, and certain phase I related activities), offer great potential for environmental risk assessment using invertebrates. Other measurements, such as DNA damage, have been successfully used in acute stress situations (such as after an oil spill), but not in chronically exposed organisms. One of the most widely and successfully used biomarkers is the presence or degree of imposex in gastropods as a specific indicator of exposure and effect of organotin compounds. Additional research in the field of reproductive disorders will hopefully lead to an extremely useful set of biomarkers of endocrine disruption in invertebrates. In addition, invertebrates may play a key role in determining the linkage between biochemical and subcellular responses and adverse effects on populations and communities.

Fish

According to an extensive evaluation of fish biomarker responses, using the six criteria mentioned in the introduction, phase I enzymes (such as hepatic EROD and CYP1A), biotransformation products (biliary PAH metabolites), reproductive parameters (plasma VTG), and genotoxic parameters (hepatic DNA adducts) are currently the most valuable fish biomarkers for environmental risk-assessment purposes (Van der Oost et al. 2003). It has to be emphasized, however, that the value for other biomarkers may be elevated when additional research on certain topics has been performed successfully. The value of certain fish biomarkers will depend upon the purpose of the investigations to which the biomarkers are applied. It is, for instance, not advisable to use the ^{32}P postlabelling assay for hepatic DNA adduct determinations in routine monitoring programs with large numbers of samples, because it is an expensive and time-consuming assay. If, on the other hand, only a small number of samples have to be investigated, such as to confirm the presence of potentially carcinogenic substances in the environment, this assay is quite suitable since the criterion of a cheap and easy assay is less important. The importance of a multiple biomarker approach has been demonstrated using multivariate statistical analyses (Van der Oost et al. 1997). In conclusion, it can be stated that fish biomarkers are promising tools for environmental risk assessment (ERA), as supplements to existing chemical measures.

Amphibians

Although the overview on the application of amphibian biomarkers in ERA may not be fully comprehensive, it has been shown that markers are available for detecting exposure of amphibians to organic compounds and metals. The application of AROD biomarkers (phase I enzymes) to assess the exposure to dioxin-like compounds is promising, although responses are variable, and are substrate- and species-dependent. Markers to assess endocrine disruption are available at different integration levels, including VTG production, induction of aromatase activity, the analysis of steroids, and histological observations on malformations in gonads. Gross observations of malformations can also point to effects of contaminants on, for example, the homeostasis of the thyroid hormone system. The assessment of ACHE inhibition may be used as a marker for exposure to certain pesticides (OP), while the application of the ALAD assay may be used as an indicator for lead exposure. The formation of metallothioneins is a more generic marker for exposure to heavy metals. These biomarkers have all been applied in effect studies on amphibians, but less in ERA. From this overview it is clear that more baseline data are needed. Furthermore, most of these biomarkers have been developed for other species, so further optimization of the methods for use on amphibians may be needed. Biomarkers in general appear to be valuable tools in the assessment of risks of contamination for amphibians.

Birds and mammals

Biomarkers have been applied in ERAs addressing risks of contaminants for birds and mammals. For instance, phase I enzymes measured by AROD markers have been used and validated. Phase II enzymes, ALAD, stress proteins, endocrine and neuromuscular parameters, and morphological observations have also been used, but in general these are less validated as a tool in ERA than are AROD markers. Although biomarkers are useful in ERA, it should be pointed out that a certain marker should be validated under specific conditions before it can be implemented in an ERA, such as the species to be investigated, the expected contamination, and so on. Such validation should include the assessment of the (expected) sensitivity of the marker, its relevance for the interpretation within the ERA, and its methodological applicability. With respect to the last criterion, the use of nondestructive methods should be propagated. Nevertheless, this overview clearly illustrates that a wide range of mammalian and avian biomarkers is available for implementation in ERA programs.

Perspectives and recommendations

At present, the research on most of the biomarkers described in this chapter is either in the initial phase (novel biomarkers) or in the learning phase, according to McCarthy (1990). Peakall and Walker (1994) state that in the 1990s, biomarkers were in the same stage of development as were the analyses of environmental chemicals in the 1960s. Den Besten (1998) discusses the concepts for implementation of biomarkers in environmental monitoring. Although it has been demonstrated that biomarkers are useful monitoring tools, it is clear that more information is needed about the relation between biomarker responses and the health and fitness of organisms, and even more so about the relation between biomarker responses and the risks for the ecosystem. With respect to future biomarker research it is important to realize that different concepts are needed for the specific purposes of environmental monitoring programs, such as first carrying out cost-effective measurements in a stepwise approach, obtaining insights into the cause of observed effects in the field, studying trends in time or spatial variation, or using biomarker responses as signals of negative effects on the ecosystem. The use of biomarkers for risk assessment at the community and ecosystem level is still rather ambitious. This is also because biomarkers in general only address direct effects of contaminants on organisms, and they are not suitable for detecting secondary effects, such as changes in prey availability for predators due to toxic effects on prey. Such ecological (secondary) effects can only be assessed by additional observations, so for the assessment of site-specific risks, information from biomarkers should be used in combination with other biological data (such as species abundance) and chemical data (den Besten 1998). The inability of direct toxicity assessment (DTA) procedures to satisfactorily evaluate chronic and sublethal risks increases the interest in using *in-situ* biomarkers for the fingerprinting of stress-response properties as a means of diagnosing risk assessment for integrated urban runoff management (Ellis 2000).

The ultimate objective for applied environmental research should be to make biomarkers more applicable in environmental risk assessment. Therefore, Van der Oost et al. (2003) proposed several guidelines for ecotoxicological research in order to actually incorporate biomarkers in ERA monitoring, such as:

- Efforts have to be made to design a set of biomarkers, covering the exposure and early effects of the entire spectrum of potentially hazardous substances present in the environment.
- All biomarkers used to assess exposure to and effects of environmental pollutants have to be objectively evaluated, according to six biomarker criteria (see the chapter introduction).
- It is essential that more research be carried out to demonstrate relationships between biomarker responses and effects on pathology, survival, growth, or reproduction; knowledge of these relationships should be used to design numerical standards for biomarkers.
- Standard procedures should be developed with regard to sampling, sample treatment, assay conditions, and so on, in international programs such as the European BEEP (biological effects of environmental pollutants on marine organisms), CITY FISH (ecological quality of urban rivers related to effects on fish populations), and BEQUALM (biological-effects quality assurance in monitoring programs) projects.
- Additional research has to be performed on the potential impact of non–pollution-related confounding factors on certain biomarker responses, in order to achieve a more reliable data interpretation.
- In order to reduce the number of animals that have to be killed for biomarker research, emphasis has to be put on the development of nondestructive and noninvasive biomarkers. Artificial devices, such as passive sampling (like SPMD) combined with cell-line bioassays, may be used to monitor the water quality as an alternative to animal biomarkers.
- New and promising developments in the biomarker field are the so-called genomics and proteomics, providing an integrated picture of the way an organism responds to a changing environment.

In conclusion, it can be stated that biomarkers are promising tools for ERA as supplements to existing chemical measures. Much work has to be done, however, in order to test and interpret biomarker responses and to develop acceptable QA procedures. Only when both scientific and legal credibility of this information is established can the biomarker techniques be fully applied in routine monitoring programs. It seems obvious that chemical monitoring alone is insufficient for a reliable classification of the overall water quality. Therefore, the efforts to incorporate biological compounds to ERA research will eventually be worthwhile.

References

Ali, A., L. SalterCid, M.F. Flajnik and J.J. Heikkila, 1996. Isolation and characterization of a cDNA encoding a Xenopus 70-kDa heat shock cognate protein, Hsc70.I. *Comp. Biochem. Physiol.* B 113: 681–687.

Andersson, T., M. Pesonen and C. Johansson, 1985. Differential induction of cytochrome P-450-dependent monooxygenase, epoxide hydrolase, glutathione transferase and UDP glucuronosyl transferase activities in the liver of rainbow trout by ß-naphthoflavone or Clophen A50. *Biochem. Pharmacol.* 34: 3309–3314.

Angelucci, S., A. De Luca, P. Moio, R. Casadei, A. Grilli, M. Felaco, F. Amicarelli and C. Di-Ilio, 2001. Purification and characterization of glutathione transferases from the frog (*Xenopus laevis*) liver. *Chem. Biol. Interactions* 133: 228–230.

Ankley, G.T., M.D. Kahl, K.M. Jensen, M.W. Hornung, J.J. Korte, E.A. Makynen and R.L. Leino, 2002. Evaluation of the aromatase inhibitor fadrozole in a short-term reproduction assay with the fathead minnow (*Pimephales promelas*). *Toxicol. Sci.* 67: 121–130.

Arukwe, A.E. and A. Goksøyr, 1997. Fish zona radiata (egg shell) protein — a sensitive biomarker for environmental estrogens. *Environ. Health Perspect.* 105: 418–422.

Arukwe, A.E., T. Celius, B.T. Walther and A. Goksøyr, 2000. Effects of xenoestrogen treatment on *zona radiata* protein and vitellogenin expression in Atlantic salmon (*Salmo salar*). *Aquatic Toxicol.* 49: 159–170.

Arukwe, A.E. and A. Goksøyr, 2003. Eggshell and egg yolk proteins in fish: hepatic proteins for the next generation: oogenetic, population, and evolutionary implications of endocrine disruption. *Comp. Hepatol.* 2: 4–25.

Bagenal, T.B. and F.W. Tesch, 1978. Methods for assessment of fish production in fresh waters, in T.B. Bagenal, Ed., *Age and growth*. Blackwell Scientific Publications, Oxford, 101–136.

Bard, S.M., B. Woodin and J.J. Stegeman, 1998. Induction of the multixenobiotic resistance transporter and cytochrome P450 1A in intertidal fish exposed to environmental contaminants. *Toxicol. Sci.* 42: 16–26.

Bard, S.M., 2000. Multixenobiotic resistance as a cellular defense mechanism in aquatic organisms. *Aquatic Toxicol.* 48: 357–389.

Bartkowiak, D.J. and B.W. Wilson, 1995. Avian plasma carboxylesterase activity as a potential biomarker of organophosphate pesticide exposure. *Environ. Toxicol. Chem.* 14: 2149–2153.

Bayne, B.L., D.A. Brown, K. Burns, D.R. Dixon, A. Ivanovici, D.A. Livingstone, D.M. Lowe, M.N. Moore, A.R.D. Stebbing and J. Widdings, 1985. *The effects of stress and pollution on marine animals.* Praeger, New York.

Berril, M., S. Bertram, A. Wilson, S. Louis, D. Brogham and C. Stromberg, 1993. Lethal and sublethal inpacts of pyrethroid insecticides on amphibian embryos and tadpoles. *Environ. Toxicol. Chem.* 12: 525–539.

Berril, M., S. Bertram, L. McGillivray, M. Kolohorn and B. Pauli, 1994. Effects of low concentrations of forest-use pesticides on frog embryos and tadpoles. *Environ. Toxicol. Chem.* 13: 657–664.

Beyer, J., 1996. *Fish biomarkers in marine pollution monitoring; evaluation and validation in laboratory and field studies.* Academic thesis, University of Bergen, Norway.

Bhaskaran, A., D. May, M. Rand-Weaver and C. Tyler, 1999. Fish p53 as a possible biomarker for genotoxins in the aquatic environment. *Environ. Mol. Mutagen.* 33: 177–184.

Bocquené, G., F. Galgani and P. Truqet, 1990. Characterization and assay conditions for use of AChE activity from several marine species in pollution monitoring. *Mar. Environ. Res.* 30: 75–89.

Bosveld, A.T.C. and M. Van den Berg, 1994. Biomarkers and bioassays as alternative screening methods for the presence and effects of PCDDs, PCDFs and PCBs. *Fresenius J. Anal. Chem.* 348: 106–110.

Bosveld, A.T.C., 1995. *Effects of polyhalogenated aromatic hydrocarbons on piscivorous avian wildlife.* Ph.D. thesis, University of Utrecht, the Netherlands.

Brumley, C.M., V.S. Haritos, J.T. Ahokas and D.A. Holdway, 1998. The effects of exposure duration and feeding status on fish bile metabolites: implications for biomonitoring. *Ecotoxicol. Environ. Saf.* 39: 147–153.

Bucciarelli, T., R. Petruzzelli, S. Melino, L. Cornelio, F. Amicarelli, C. Di Ilio and P. Sacchetta, 2001. Organ distribution of glutathione transferase isoenzymes in common toad (*Bufo bufo*). *Chem. Biol. Interactions* 133: 322–324.

Bucheli, T.D. and K. Fent, 1995. Induction of cytochrome P450 as a biomarker for environmental contamination in aquatic ecosystems. *Crit. Rev. Environ. Sci. Technol.* 25: 201–268.

Burmeister, S. and W. Wilczynski, 2000. Social signals influence hormones independently of calling behavior in the treefrog (*Hyla cinerea*). *Horm. Behav.* 38(4): 201–209.

Cachot, J., F. Galgani and F. Vincent, 1998. Production of polyclonal antibody raised against recombinant flounder p53 protein. *Comp. Biochem. Physiol.* 120C: 351–356.

Carreau, S., D.S. Bourguiba, S. Lambard, D.I. Galeraud, C. Genissel and J. Levallet, 2002. Reproductive system: aromatase and estrogens. *Mol. Cell. Endocrinol.* 193: 137–143.

Celander, M., 1993. *Induction of cytochrome P450 in teleost fish; with emphasis on the CYP1 gene family.* Academic thesis, University of Göteborg, Sweden.

Chapman, P.M., 1990. The sediment quality Triad approach to determining pollution-induced degradation. *Sci. Tot. Environ.* 97/98: 815–825.

Collier, T.K. and U. Varanasi, 1991. Hepatic activities of xenobiotic metabolizing enzymes and biliary levels of xenobiotics in English sole (*Parophrys vetulus*) exposed to environmental contaminants. *Arch. Environ. Contamination Toxicol.* 20: 462–473.

Commandeur, J.N.M., G.J. Stijntjes, and N.P.E. Vermeulen, 1995. Enzymes and transport systems involved in the formation and disposition of glutathione S-conjugates. Role in bioactivation and detoxication mechanisms of xenobiotics. *Pharmacol. Rev.* 47: 271–330.

Corn, P.S., 2000. Amphibian declines: review of some current hypotheses, in D.W. Sparling, G. Linder, C.A. Bishop, (Eds.), *Ecotoxicology of amphibians and reptiles.* SETAC technical publications series, Society of Environmental Toxicology and Chemistry (SETAC), Pensacola, FL, 663–696.

De Kruijf, 1991. Extrapolation through hierarchical levels. *Comp. Biochem. Physiol.* 100C: 291–299.

De Roode, D.F., 2001. *The chicken embryo bioassay as a tool to assess the possible toxic effects of persistent organic pollutants (POPs).* Ph.D. thesis, Wageningen University, the Netherlands.

De Zwart, D., 1995. *Monitoring water quality in the future, volume 3: biomonitoring.* National Institute of Public Health and Environmental Protection (RIVM), Bilthoven, the Netherlands.

DeFur, P.L., M. Crane, C.G. Ingersoll and L.J. Tattersfield, (Eds.), 1999. *Endocrine disruption in invertebrates: endocrinology, testing, and assessment.* Workshop on Endocrine Disruption in Invertebrates: Endocrinology, Testing, and Assessment, 12–15 Dec. 1998, Noordwijkerhout, the Netherlands. Published by the Society of Environmental Toxicology and Chemistry (SETAC).

Dell'Omo, G., R. Bryenton and R.F. Shore, 1997. Effects of exposure to an organophosphate pesticide on behavior and acetylcholinesterase activity in the common shrew, *Sorex araneus. Environ. Toxicol. Chem.* 16: 272–276.

Den Besten, P.J., P. Lemaire, D.R. Livingstone, B. Woodin, J.J. Stegeman, H.J. Herwig and W. Seinen, 1993. Time-course and dose-response of the apparent induction of the cytochrome P450 monooxygenase system of pyloric caeca microsomes of the female sea star *Asterias rubens* L. by benzo[a]pyrene and plychlorinated biphenyls. *Aquatic Toxicol.* 26: 23–40.

Den Besten, P.J., 1998. Concepts for the implementation of biomarkers in environmental monitoring. *Mar. Environ. Res.* 46: 253–256.

Depledge, M.H. and M.C. Fossi, 1994. The role of biomarkers in environmental assessment (2). *Ecotoxicology* 3: 161–172.

Di Fiore, M.M., L. Assisi and V. Botte, 1998. Aromatase and testosterone receptor in the liver of the female green frog, *Rana esculenta. Life Sci.* 62: 1949–1958.

Di Giulio, R.T., P.C. Washburn, R.J. Wenning, G.W. Winston and C.S. Jewell, 1989. Biochemical responses in aquatic animals: a review of determinants of oxidative stress. *Environ. Toxicol. Chem.* 8: 1103–1123.

Di Giulio, R.T., W.H. Benson, B.M. Sanders and P.A. Van Veld, 1995. Biochemical mechanisms: metabolism, adaptation, and toxicity, in G.M. Rand, (Ed.), *Fundamentals of aquatic toxicology: effects, environmental fate, and risk assessment,* second edition. Taylor and Francis Ltd., London, 523–562.

Di Ilio, C., 2001. Amphibian and bacterial glutathione transferases. *Chem. Biol. Interactions* 133: 207–210.

Dunlap, D.Y. and F. Matsumura, 1997. Development of broad spectrum antibodies to heat shock protein 70s as biomarkers for detection of multiple stress by pollutants and environmental factors. *Ecotoxicol. Environ. Safety* 37: 238–244.

Dunn, B.P., 1991. Carcinogen adducts as an indicator for the public health risks of consuming carcinogen-exposed fish and shellfish. *Environ. Health Perspect.* 90: 111–116.

ECETOC, 1993. *Environmental hazard assessment of substances.* European Centre for Ecotoxicology and Toxicology of Chemicals (ECETOC), Technical Report No. 51, Brussels.

Eeva, T.S., Tanhuanpaa, C. Rabergh, S. Airaksinen, M. Nikinmaa and E. Lehikoinen, 2000. Biomarkers and fluctuating asymmetry as indicators of pollution-induced stress in two hole-nesting passerines. *Function. Biol.* 14: 235–243.

Elliott, J.E. and A.M. Scheuhammer, 1997. Heavy metal and metallothionein concentrations in seabirds from the Pacific Coast of Canada. *Mar. Poll. Bull.* 34: 794–801.

Ellis, J.B., 2000. Risk assessment approaches for ecosystem responses to transient pollution events in urban receiving waters. *Chemosphere* 41: 85–91.

Epel, D., 1998. Use of multidrug transporters as first lines of defense against toxins in aquatic organisms. *Comp. Biochem. Physiol.* A 120: 23–28.

Ertl, R.P. and G.W. Winston, 1998. The microsomal mixed function oxidase system of amphibians and reptiles: components, activities and induction. *Comp. Biochem. Physiol.* C. 121: 85–105.

Escartín, E. and C. Porte, 1996. Acetylcholinesterase inhibition in the crayfish *Procambarus clarkii* exposed to fenitrothion. *Ecotoxicol. Environ. Saf.* 34: 160–164.

Escartín, E. and C. Porte, 1997. The use of cholinesterase and carboxylesterase activities from *Mytilus galloprovincialis* in pollution monitoring. *Environ. Toxicol. Chem.* 16: 2090–2095.

Fent, K., B.R. Woodin and J.J. Stegeman, 1998. Effects of triphenyltin and other organotins on hepatic monooxygenase system in fish. *Comp. Biochem. Physiol.* 121C: 277–288.

Fossi, M.C., C. Leonzio and D. Peakall, 1993. The use of nondestructive biomarkers in the hazard assessment of vertebrate populations, in M.C. Fossi and C. Leonzio, (Eds.), *Non-destructive biomarkers in vertebrates*. CRC/Lewis Publishers, Boca Raton, FL, 3–36.

Fossi, M.C. and L. Marsili, 1997. The use of nondestructive biomarkers in the study of marine mammals. *Biomarkers* 2: 205–216.

Fossi, M.C., C. Savelli, L. Marsili, S. Casini, B. Jimenez, M. Junin, H. Castello and J.A. Lorenzani, 1997. Skin biopsy as a nondestructive tool for the toxicological assessment of endangered populations of pinnipeds: preliminary results on mixed function oxidase in *Otaria flavescens*. *Chemosphere* 35: 1623–1635.

Fossi, M.C., C. Savelli and S. Casini, 1998. Mixed function oxidase induction in *Carcinus aestuarii*: field and experimental studies for the evaluation of toxicological risk due to Mediterranean contaminants. *Comp. Biochem. Physiol.* 121 C: 321–331.

Franson, J.C., T. Hollmen, R.H. Poppenga, M. Hario, M. Kilpi and M.R. Smith, 2000. Selected trace elements and organochlorines: some findings in blood and eggs of nesting common eiders (*Somateria mollissima*) from Finland. *Environ. Toxicol. Chem.* 19: 1340–1347.

Fulton, M.H. and P.B. Key, 2001. Acetylcholinesterase inhibition in estuarine fish and invertebrates as an indicator of organophosphorus insecticide exposure and effects. *Environ. Toxicol. Chem.* 20: 37–45.

Galgani, F., G. Bocquene and Y. Cadiou, 1992. Evidence of variation in cholinesterase activity in fish along a pollution gradient in the North Sea. *Mar. Ecol. Prog. Ser.* 13: 77–82.

Gassman, N.J. and C.-J. Kennedy, 1992. Cytochrome P-450 content and xenobiotic metabolizing enzyme activities in the scleractinian coral, *Favia fragum* (Esper.). *Bull. Mar. Sci.* 50: 320–330.

George, S.G., 1994. Enzymology and molecular biology of phase II xenobiotic-conjugating enzymes in fish, in D.C. Malins and G.K. Ostrander, (Eds.), *Aquatic toxicology; molecular, biochemical and cellular perspectives*. CRC/Lewis Publishers, Boca Raton, FL, 37–85.

Gimeno, S., A. Gerritsen, T. Bowmer and H. Komen, 1996. Feminization of male carp. *Nature* 384: 221–222.

Goeptar, A.R., H. Scheerens and N.P.E. Vermeulen, 1995. Oxygen reductase and substrate reductase activity of cytochrome P450. *Crit. Rev. Toxicol.* 25: 25–65.

Goksøyr, A. and Förlin, L., 1992. The cytochrome P450 system in fish, aquatic toxicology and environmental monitoring. *Aquat. Toxicol.* 22: 287–312.

Gracey, A.Y., J.V. Troll and G.N. Somero, 2001. Hypoxia-induced gene expression profiling in the euryoxid fish *Gillichthys mirabilis*. *Proc. Nat. Acad. Sci.* 98: 1993–1998.

Greig-Smith, P.W., 1991. Use of cholinesterase measurements in surveillance of wild-life poisoning in farmland, in P. Mineau, (Ed.), *Cholinesterase-inhibiting insecticides, chemicals in agriculture,* vol. 2. Elsevier Press, Amsterdam, 127–150.

Gupta, R.C., M.V. Reddy and K. Randerath, 1982. 32P-postlabelling analysis of non-radioactive aromatic carcinogen DNA adducts. *Carcinogenesis* 3: 1081–1092.

Halm, S., N. Pounds, S. Maddix, M. Rand-Weaver, J.P. Sumpter, T.H. Hutchinson and C.R. Tyler, 2002. Exposure to exogenous 17beta-oestradiol disrupts p450aromB mRNA expression in the brain and gonad of adult fathead minnows *(Pimephales promelas). Aquatic Toxicol.* 60: 285–299.

Hamoutene, D., J.F. Payne, A. Rahimtula and K. Lee, 2002. Use of the comet assay to assess DNA damage in hemocytes and digestive gland cells of mussels and clams exposed to water contaminated with petroleum hydrocarbons. *Mar. Environ. Res.* 54: 471–474.

Hamza-Chaffai, A., J.C. Amiard, J. Pellerin, L. Joux and B. Berthet, 2000. The potential use of metallothionein in the clam *Ruditapes decussatus* as a biomarker of *in situ* metal exposure. *Comp. Biochem. Physiol.* 127C: 185–197.

Harri, M.N.E., J. Laitinen and E.L. Valkama, 1979. Toxicity and retention of DDT in adult frogs, *Rana temporaria. L. Environ. Pollut.* 20: 45–55.

Harri, M.N.E., 1980. Hepatic mixed function oxidase (MFO) activities during the seasonal cycle of the frog, *Rana temporaria. Comp. Biochem. Physiol.* C. 67: 75–78.

Harris, M.L., C.A. Bishop, J. Struger, M.R. Van den Heuvel, G.J. Van der Kraak, G. Dixon, B. Ripley and J.P. Bogart, 1998. The functional integrity of northern leopard frogs (*Rana pipiens*) and green frog (*Rana clamitans*) populations in orchard wetlands: 1 genetics, physiology, and biochemistry of breeding adults and young-of-the-year. *Environ. Toxicol. Chem.* 17: 1338–1350.

Harvey, J.S., B.P. Lyons, T.S. Page, C. Stewart and J.M. Parry, 1999. An assessment of the genotoxic impact of the Sea Empress oil spill by the measurement of DNA adduct levels in selected invertebrate and vertebrate species. *Mutat. Res.* 441: 103–114.

Haux, C., B.T. Björnsson, L. Förlin, Å. Larsson and L.J. Deftos, 1988. Influence of cadmium exposure on plasma calcium, vitellogenin and calcitonin in viteleogenic rainbow trout. *Mar. Environ. Res.* 24: 199–210.

Hayes, J.D. and D.J. Pulford, 1995. The glutathione S-transferase supergene family: regulation of GST and the contribution of the isoenzymes to cancer chemo-protection and drug resistance. *Crit. Rev. Biochem. Molec. Biol.* 30: 445–600.

Hayes, T.B., T.H. Wu and T.N. Gill, 1997. DDT-like effects as a result of corticosterone treatment in an anuran amphibian: is DDT a corticoid mimic or a stressor? *Environ. Toxicol. Chem.* 16:1948–1953.

Hayes, T.B., 2000. Endocrine disruption in amphibians, in D.W. Sparling, G. Linder, C.A. Bishop, (Eds.), *Ecotoxicology of amphibians and reptiles.* SETAC technical publications series, Society of Environmental Toxicology and Chemistry (SETAC), Pensacola, FL, 573–594.

Hermes-Lima, M. and K.B. Storey, 1998. Role of antioxidant defenses in the tolerance of severe dehydration by anurans: the case of the leopard frog *Rana pipiens. Mol. Cell. Biochem.* 189 (1–2): 79–89.

Higgins, C.F., 1992. ABC transporters: from microorganisms to man. *Annu. Rev. Cell. Biol.* 8: 67–113.

Hinton, D.E., P.C. Baumann, G.C. Gardner, W.E. Hawkins, J.D. Hendricks, R.A. Murchelano and M.S. Okihiro, 1992. Histopathologic biomarkers, in R.J. Huggett, R.A. Kimerly, P.M. Mehrle Jr. and H.L. Bergman, (Eds), *Biomarkers: biochemical, physiological and histological markers of anthropogenic stress.* CRC/Lewis Publishers, Chelsea, MI, 155–210.

Hinton, D.E., 1994. Cells, cellular responses, and their markers in chronic toxicity of fishes, inD.C. Malins and G.K. Ostrander, (Eds.), *Aquatic toxicology: molecular, biochemical and cellular perspective.* CRC/Lewis Publishers, Boca Raton, FL, 207–240.

Hoffman, D.J. and G.H. Heinz, 1998. Effects of mercury and selenium on glutathione metabolism and oxidative stress in Mallard ducks. *Environ. Toxicol. Chem.* 17: 161–166.

Huang, Y.-W., M.J. Melancon, R.E. Jung and W.H. Karasov, 1998. Induction of cytochrome P450-associated monooxygenases in northern leopard frogs, *Rana pipiens*, by 3,3',4,4',5-pentachlorobiphenyl. *Environ. Toxicol. Chem.* 17: 1564–1569.

Huang, Y.-W., W.H. Karasov, K.A. Patnode and C.R. Jefcoate, 1999. Exposure of northern leopard frogs in the Green Bay ecosystem to polychlorinated biphenyls, polychlorinated dibenzo-*p*-dioxins, and polychlorinated dibenzofurans is measured by direct chemistry but not hepatic ethoxyresorufin-*O*-deethylase activity. *Environ. Toxicol. Chem.* 18: 2123–2130.

Huang, Y.-W., J.J. Stegeman, B.R. Woodin and W.H. Karasov, 2001. Immunohistochemical localization of cytochrome P4501A induced by 3,3',4,4'5-pentachlorobiphenyl in multiple organs of northern leopard frogs, *Rana pipiens*. *Environ. Toxicol. Chem.* 20: 191–197.

Hughes, M.R., J.E. Smits, J.E. Elliott and D.C. Bennett, 2000. Morphological and pathological effects of cadmium ingestion on Pekin ducks exposed to saline. *J. Toxicol. Environ. Health*, Part A, 61(7): 591–608.

Hugla, J.L. and J.P. Thome, 1999. Effects of polychlorinated biphenyls on liver ultrastructure, hepatic monooxygenases and reproductive success in the barbel. *Ecotox. Environ. Saf.* 42: 265–273.

Hylland, K., C. Haux and C. Hogstrand, 1992. Hepatic metallothionein and heavy metals in dab *Limanda limanda* from the German Bight. *Mar. Ecol. Prog. Ser.* 91: 89–96.

Hyllner, S.J., D.O. Oppen-Berntsen, J.V. Helvik, B.T. Walther and C. Haux, 1991. Oestradiol-17b induces major vitelline envelope proteins in both sexes in teleosts. *J. Endocrin.* 131: 229–236.

Ireland, M.P., 1977. Lead retention in toad *Xenopus laevis* fed increasing levels of lead-contaminated earthworms. *Environ. Pollut.* 12: 85–91.

Ishizuka, M., H. Hoshi, N. Minamoto, M. Masuda, A. Kazusaka, and S. Fujita, 1996. Alterations of cytochrome P450-dependent monooxygenase activities in *Eriocheir japonicus* in response to water pollution. *Environ. Health Perspect.* 104: 774–778.

James, M.O. and S.M. Boyle, 1998. Cytochromes P450 in crustacea. *Comp. Biochem. Physiol.* 121C: 157–172.

Janssen, P.A.H., J.H. Faber and A.T.C. Bosveld, 1998. (Fe)male *IBN scientific contributions 13.* DLO Institute for Forestry and Nature Research, Wageningen, the Netherlands, ISSN:1385–1586.

Khan, M.A.Q., A. Kamal, R.J. Wolin and J. Runnels, 1972. *In vivo* and *in vitro* epoxidation of aldrin by aquatic food chain organisms. Bull. Environ. Contam. Toxicol. 8: 219–228.

Kime, D.E., 1995. The effects of pollution on reproduction in fish. *Rev. Fish Biol. Fish.* 5: 52–96.

Kloas, W.T., I. Lutz and R. Einspanier, 1999. Amphibians as a model to study endocrine disruptors: II. estrogenic activity of environmental chemicals *in vitro* and *in vivo*. *Sci. Total Environ.* 225: 59–68.

Kurelec, B., and B. Pivcevic, 1989. Distinct glutathione-dependent enzyme activities and a verapamil-sensitive binding of xenobiotics in a fresh-water mussel *Anodonta cygnea*. *Biochem. Biophys. Res. Commun.* 164: 934–940.

Kurelec, B., A. Garc, S. Krca and R.C. Gupta, 1990. DNA adducts in the marine mussel *Mytilus galloprovincialis* living in polluted and unpolluted environments, in J.F. McCarthy, L.R. Shugart, (Eds.), *Biomarkers of environmental contamination*, CRC Press, Boca Raton, FL, 217–227.

Kurelec, B., 1992. The multixenobiotic resistance mechanism in aquatic organisms. *Crit. Rev. Toxicol.* 22: 23–43.

Kurelec, B., S. Krca, B. Pivcevic, D. Ugarkovic, M. Bachmann, G. Imsiecke, and W.E.G. Muller, 1992. Expression of p-glycoprotein in marine sponges: identification and characterization of the 125-kDa drug-binding glycoprotein. *Carcinogenesis* 13: 69–76.

Kurelec, B., S. Krca and D. Lucic, 1996. Expression of multi-xenobiotic resistance mechanism in a marine mussel *Mytilus galloprovincialis* as a biomarker of exposure to polluted environments. *Comp. Biochem. Physiol.* 113C: 283–289.

Langston, W.J., M.J. Bebianno and G. Burt, 1998. Metabolic pathways in marine invertebrates, in W.J. Langston and M.J. Bebianno, (Eds.), *Metal metabolism in aquatic environments*. Chapmann and Hall, London, 209–283.

Lans, M.C., 1995. *Thyroid hormone binding proteins as novel targets for hydroxylated polyhalogenated aromatic hydrocarbons (PHAHs): possible implications for toxicity.* Ph.D. thesis, Wageningen University, the Netherlands.

Large, A.T., J.P. Shaw, L.D. Peters, A.D. McIntosh, L. Webster, A. Mally and J.K. Chipman, 2002. Different levels of mussel (*Mytilus edulis*) DNA strand breaks following chronic field and acute laboratory exposure to polycyclic aromatic hydrocarbons. *Mar. Environ. Res.* 54: 493–497.

Larsson, D.G.J., M. Adolfsson-Erici, J. Parkkonen, M. Pettersson, A.H. Berg, P.-E. Olson and L. Förlin, 1999. Ethinyloestradiol — an undesired fish contraceptive? *Aquatic Toxicol.* 45: 91–97.

Lech, J.J. and M.J. Vodicnik, 1985. Biotransformation, in G.M. Rand and S.R. Petrocelli, (Eds.), *Fundamentals of aquatic toxicology; methods and applications*. Hemisphere Publishing Corporation, New York, USA, 526–557.

Legler, J., C.E. van den Brink, A. Brouwer, A.J. Murk, P.T. van der Saag, A.D. Vethaak and B. van der Burg, 1999. Development of a stably transfected estrogen receptor-mediated luciferase reporter gene assay in the human T47D breast cancer cell line. *Toxicol. Sci,* 48: 55–66.

Leppänen, H., S. Marttinen and A. Oikari, 1998. The use of fish bile metabolite analyses as exposure biomarkers to pulp and paper mill effluents. *Chemosphere* 36: 2621–2634.

Lewis, S., R.D. Handy, B. Cordi, Z. Billinghurst and M.H. Depledge, 1999. Stress proteins (HSP's): Methods of detection and their use as an environmental biomarker. *Ecotoxicology* 8: 351–368.

Lin, E.L., S.M. Cormier and J.A. Torsella, 1996. Fish biliary polycyclic aromatic hydrocarbon metabolites estimated by fixed-wavelength fluorescence: comparison with HPLC-fluorescent detection. *Ecotoxicol. Environ. Saf.* 35: 16–23.

Livingstone, D.R., 1991. Organic xenobiotic metabolism in marine invertebrates, in Gilles, R. (Ed.), *Advances in comparative and environmental physiology*, vol. 7. Springer-Verlag, Berlin, 46–185.

Livingstone, D.R., 1994. Recent developments in marine invertebrate organic xenobiotic metabolism. *Toxicol. Ecotoxicol. News* 1: 88–94.

Livingstone, D.R., 1996. Cytochrome P-450 in pollution monitoring. Use of cytochrome P-450 1A (CYP1A) as a biomarker of organic pollution in aquatic and other organisms, in M. Richardson (Ed.), *Environmental xenobiotics*. Taylor and Francis, Rickmansworth, UK, 143–160.

Livingstone, D.R., 1998. The fate of organic xenobiotics in aquatic ecosystems: quantitative and qualitative differences in biotransformation by invertebrates and fish. *Comp. Biochem. Physiol.* 120 A: 43–49.

Livingstone, D.R., F. Lips, P. García-Martínez and R.K. Pipe, 1992. Antioxidant enzymes in digestive gland of the common mussel, *Mytilus edulis.* Lett. Mar. Biol. 112: 265–276.

Livingstone, D.R. and C. Nasci, 2000. Biotransformation and antioxidant enzymes as potential biomarkers of contaminant exposure in goby (*Zosterissor ophiocephalus*) and mussel (*Mytilus galloprovincialis*), in P. Lasserre and A. Marzollo, (Eds.), *The Venice Lagoon ecosystem; inputs and interactions between land and sea.* Man and the Biosphere Series, Vol. 25, The Parthenon Publishing Group, Carnforth, UK, 357–373.

Livingstone, D.R., J.K. Chipman, D.M. Lowe, C. Minier, C.L. Mitchelmore, M.N. Moore, L.D. Peters and R.K. Pipe, 2000. Development of biomarkers to detect the effect of organic pollution on aquatic invertebrates: recent molecular, genotoxic, cellular and immunological studies on the common mussel (*Mytilus edulis* L.) and other mytilids. *Int. J. Envir. Pollut.* 13: 56–91.

Livingstone, D.R., 2001. Contaminant-stimulated reactive oxygen species production and oxidative damage in aquatic organisms. *Mar. Pollut. Bull.* 42: 656–666.

Lopez-Torres, M., R. Perez-Campo, S. Cadenas, C. Rojas and G. Barja, 1993. A comparative research of free radicals in vertebrates — II. Nonenzymatic antioxidants and oxidative stress. *Comp. Biochem. Physiol.* 105: 757–763.

Lowe, D.M. and V.U. Fossato, 2000. The influence of environmental contaminants on lysosomal activity in the digestive cells of mussels (*Mytilus galloprovincialis*) from the Venice Lagoon. *Aquatic Toxicol.* 48: 75–85.

Lye, C.M., C.L.J. Frid, M.E. Gill, D.W. Cooper and D.M. Jones, 1999. Estrogenic alkylphenols in fish tissues, sediments, and water from the U.K. Tyne and Tees estuaries. *Environ. Sci. Technol.* 33: 1009–1014.

Mantel, N., 1967. The detection of disease clustering and a generated regression approach. *Cancer Res.* 27: 200–209.

Matthiessen, P. and P. Gibbs, 1998. Critical appraisal of evidence for tributyltin-mediated endocrine disruption in mollusks. *Environ. Toxicol. Chem.* 17: 37–43.

Mayer, F.L., D.J. Versteeg, M.J. McKee, L.C. Folmar, R.L. Graney, D.C. McCume and B.A. Rattner, 1992. Metabolic products as biomarkers, in R.J. Huggett, R.A. Kimerly, P.M. Mehrle, Jr., and H.L. Bergman, (Eds.), *Biomarkers: biochemical, physiological and histological markers of anthropogenic stress*. CRC/Lewis Publishers, Chelsea, MI, 5–86.

McCarthy, J.F., 1990. Implementation of a biomarker-based environmental monitoring program, in J.F. McCarthy and L.R. Shugart, (Eds.), *Biomarkers of environmental contamination*. CRC/Lewis Publishers, Boca Raton, FL, 429–440.

McCarthy, J.F. and L.R. Shugart, 1990. Biological markers of environmental contamination, in J.F. McCarthy and L.R. Shugart, (Eds.), *Biomarkers of environmental contamination*. CRC/Lewis Publishers, Boca Raton, FL, 3–16.

McCarthy, J.F., R.S. Halbrook and L.R. Shugart, 1991. *Conceptual strategy for design, implementation, and validation of a biomarker-based biomonitoring capability.* Publication No. 3072, ORNL/TM-11783, Environmental Sciences Division, Oak Ridge National Laboratory, Tennessee.

Mekenyan, O.G., G.D. Veith, D.J. Call and G.T. Ankley, 1996. A QSAR evaluation of Ah receptor binding of halogenated aromatic xenobiotics. *Environ. Health Perspect.* 104: 1302–1310.

Melancon, M.J., R. Alscher, W. Benson, G. Kruzynski, R.F. Lee, H.C. Sikka and R.B. Spies, 1992. Metabolic products as biomarkers, in R.J. Huggett, R.A. Kimerly, P.M. Mehrle Jr. and H.L. Bergman, (Eds.), *Biomarkers: biochemical, physiological and histological markers of anthropogenic stress*. CRC/Lewis Publishers, Chelsea, MI, 87–124.

Mellanen, P., M. Soimasuo, B. Holmbom, A. Oikari and R. Santti, 1999. Expression of the vitellogenin gene in the liver of juvenile whitefish *(Coregonus lavaretus L. s.l.)* exposed to effluents from pulp and paper mills. *Ecotoxicol. Environ. Saf.* 43: 133–137.

Mensink, B.P., J. Kralt, A.D. Vethaak, C.C. ten Hallers-Tjabbes, J.H. Koeman, B. van Hattum and J.P. Boon, 2002. Imposex induction in laboratory reared juvenile *Buccinum undatum* by tributyltin (TBT*). Environ. Toxicol. Phar.* 11: 49–65.

Miller, E.C. and J.A. Miller, 1981. Mechanisms of chemical carcinogenesis. *Cancer Res.* 47: 1055–1064.

Mitchelmore, C.L. and J.K. Chipman, 1998. DNA strand breakage in aquatic organisms and the potential value of the comet assay in environmental monitoring. *Mutat. Res.* 399: 135–147.

Moore, M.N., 1990. Lysosomal cytochemistry in marine environmental monitoring. *Histochem. J.* 22: 187–191.

Moretto, A., 1998. Experimental and clinical toxicology of anticholinesterase agents. *Toxicol. Lett.* 103: 509–513.

Moss, D.W., A.R. Henderson and J.F. Kochmar, 1986. Enzymes: principles of diagnostic enzymology and the aminotransferases, in N.W. Tietz, (Ed.), *Textbook of clinical chemistry*. W.B. Saunders Company, Philadelphia, 663–678.

Mulder, G.J., M.W.H. Coughty and B. Burchell, 1990. Glucuronidation, in G.J. Mulder, (Ed.), *Conjugation reactions in drug metabolism: an integrated approach*. Taylor and Francis, London.

Murk, A.J., M.S. Denison, J.P. Giesy, C. van der Guchte and A. Brouwer, 1996. Chemical-activated luciferase gene expression (CALUX): a novel *in vitro* bioassay for Ah receptor active compounds in sediments and pore water. *Fund. Appl. Toxicol.* 33: 149–160.

Myers, M.S., C.M. Stehr, O.P. Olsen, L.L. Johnson, B.B. McCain, S.L. Chan and U. Varanasi, 1994. Relationships between toxicopathic hepatic lesions and exposure to chemical contaminants in English sole *(Pleuronectus vetulus)*, starry flounder *(Platichthys stellatus)*, and white croaker *(Genyonemus lineatus)* from selected marine sites on the Pacific coast, USA. *Environ. Health Perspect.* 102: 200–215.

Narbonne, J.F., M. Daubèze, C. Clérandeau and P. Garrigues, 1999. Scale of classification based on biochemical markers in mussels: application to pollution monitoring in European coasts. *Biomarkers* 4: 415–424.

Nebert, D.W., D.D. Peterson and A.J. Fornace, 1990. Cellular responses to oxidative stress: the [Ah] gene battery as a paradigm. *Environ. Health Perspect.* 88: 13–25.

Neff, J.M., 1985. Polycyclic aromatic hydrocarbons, in G.M. Rand and S.R. Petrocelli, (Eds.), *Fundamentals of aquatic toxicology: methods and applications*. Hemisphere Publishing Corporation, New York, 416–454.

Nie, M., A.L. Blankenship and J.P. Giesy, 2001. Inteactions between aryl hydrocarbon receptor (AhR) and hypoxia signaling ppathways. *Environ Toxicol Pharmacol.* 10: 17–27.

Nimms, R.W. and R.A. Lubet, 1995. Induction of cytochrome P-450 in the Norway rat *Rattus norvegicus*, following exposure to potential environmental contaminants. *J. Toxicol. Environ. Health* 46: 271–292.

NRC (National Research Council), 1987. Committee on biological markers. *Environ. Health Perspect.* 74: 3–9.

Ohtani, H., Y. Ichikawa, E. Iwamoto and I. Miura, 2001. Effects of styrene monomer and trimer on gonadal sex differentiation of genetic males of the frog *Rana rugosa*. *Environ. Res.* 87: 175–180.

Olsson, P.E. and C. Haux, 1986. Increased hepatic methallothionein content correlates to cadmium accumulation in environmentally exposed perch *(Perca fluviatilus)*. *Aquatic Toxicol.* 9: 231–242.

Olsvik, P.A., P. Gundersen, R.A. Andersen and K.E. Zachariassen, 2000. Metal accumulation and metallothionein in two populations of brown trout, *Salmo trutta*, exposed to different natural water environments during a run-off episode. *Aquatic Toxicol.* 50: 301–316.

Ouellet, M., 2000. Amphibian deformities: current state of knowledge, in D.W. Sparling, G. Linder and C.A. Bishop, (Eds.), *Ecotoxicology of amphibians and reptiles*. SETAC technical publications series, Society of Environmental Toxicology and Chemistry (SETAC), Pensacola, FL, 617–661.

Palmer, B.D., L.K. Huth, D.L. Pieto and K.W. Selcer, 1998. Vitellogenin as a biomarker for xenobiotic estrogens in an amphibian model system. *Environ. Toxicol. Chem.* 17: 30–36.

Payne, J.F., A. Mathieu, W. Melvin and L.L. Fancey, 1996. Acetylcholinesterase, an old biomarker with a new future? Field trials in association with two urban rivers and a paper mill in Newfoundland. *Mar. Pollut. Bull.* 32: 225–231.

Peakall, D.W. and C.H. Walker, 1994. The role of biomarkers in environmental assessment (3). *Ecotoxicology* 3: 173–179.

Peakall, D.W., 1994. Biomarkers: the way forward in environmental assessment. *Toxicol. Ecotoxicol. News* 1: 55–60.

Petrovic, S., B. Ozretic, M. Krajnovic and D. Bobinac, 2001. Lysosomal membrane stability and metallothioneins in digestive gland of mussels (*Mytilus galloprovincialis Lam.*) as biomarkers in a field study. *Mar. Pollut. Bull.* 42: 1373–1378.

Porte, C. and E. Escartín, 1998. Cytochrome P450 system in the hepatopancreas of the red swamp crayfish *Procambarus clarkii*: a field study. *Comp. Biochem. Physiol.* 121 C: 333–338.

Porte, C., X. Biosca, M. Solé and J. Albaigés, 2000. The *Aegean Sea* oil spill on the Galician coast (NW Spain). III: The assessment of long-term sublethal effects on mussels. *Biomarkers* 5: 436–446.

Porte, C., M. Solé, V. Borghi, M. Martínez, J. Chamorro, A. Torreblanca, M. Ortíz, A. Orbea, M. Soto and M.P. Cajaraville, 2001. Chemical, biochemical and cellular responses in the digestive gland of the mussel *Mytilus galloprovincialis* from the Spanish Mediterranean coast. *Biomarkers* 6: 335–350.

Power, M. and L.S. McCarty, 1997. Fallacies in ecological risk assessment practices. *Environ. Sci. Technol.* 31: 370 A–375 A.

Powers, D.A., 1989. Fish as model systems. *Science* 246: 352–358.

Purdom, C.E., P.A. Hardiman, V.J. Bye, N.C. Eno, C.R. Tyler and J.P. Sumpter, 1994. Estrogenic effects of effluents from sewage treatment works. *Chem. Ecol.* 8: 275–285.

Regoli, F., 2000. Total oxyradical scavenging capacity (TOSC) in polluted and translocated mussels: a predictive biomarker of oxidative stress. *Aquatic Toxicol.* 50: 351–361.

Ringwood, A.H., D.E. Conners, C.J. Keppler and A.A. Dinovo, 1999. Biomarkers studies with juvenile oysters (*Crassostrea virginica*) deployed *in situ*. *Biomarkers* 4: 400–414.

Roch, M., J.A. McCarter, A.T. Matheson, M.J.R. Clark and R.W. Olafson, 1982. Hepatic metallothionein in rainbow trout *(Salmo gairdneri)* as an indicator of metal pollution in the Campbell River system. *Can. J. Fish. Aquatic Sci.* 39: 1596–1601.

Roch, M. and J.A. McCarter, 1984. Hepatic metallothionein production and resistance to heavy metals by rainbow trout *(Salmo gairdneri)* held in a series of contaminated lakes. *Comp. Biochem. Physiol.* 77C: 77–82.

Roesijadi, G. and W.E. Robinson, 1994. Metal regulation in aquatic animals: mechanisms of uptake, accumulation and release, in D.C. Malins and G.K. Ostrander, (Eds.), *Aquatic toxicology: molecular, biochemical and cellular perspectives.* CRC/Lewis Publishers, Boca Raton, FL, 387–420.

Sadik, O.A. and D.M. Witt, 1999. Monitoring endocrine-disrupting chemicals. *Environ. Sci. Technol.* 33: 368A–374A.

Safe, S., 2001. Molecular biology of the Ah receptor and its role in carcinogenesis. *Toxicol. Lett.* 120: 1–7.

Sanders, B.M., 1990. Stress proteins: potential as multitiered biomarkers, in L. Shugart and J. McCarthy, (Eds.), *Biomarkers of environmental contamination.* CRC/Lewis Publishers, Boca Raton, FL, 165–191.

Sanders, B.M., 1993. Stress proteins in aquatic organisms: an environmental perspective. *Crit. Rev. Toxicol.* 23: 49–75.

Sanderson, J. T., D.M. Janz, J. Bellward and J.P. Giesy, 1997. Effects of embryonic and adult exposure to 2,3,7,8-tetrachlorodibenzo-p-dioxin on hepatic microsomal testosterone hydroxylase activities in great blue herons (*Ardea herodias*). *Environ. Toxicol. Chem.* 16: 1304–1310.

Sawyer, T. and S. Safe, 1982. PCB isomers and congeners: induction of aryl hydrocarbon hydroxylase and ethoxyresorufin-O-deethylase enzyme activities in rat hepatoma cells. *Toxicol. Lett.* 13: 87–94.

Schlenk, D., Y.S. Zhang and J. Nix, 1995. Expression of hepatic metallothionein messenger RNA in feral fish and caged fish species correlates with muscle mercury levels. *Ecotoxicol. Environ. Saf.* 31: 282–286.

Schuur, A.G., 1998. *Interactions of polyhalogenated aromatic hydrocarbons with thyroid hormone metabolism.* Ph.D. thesis, Wageningen University, the Netherlands.

Schwen, R.J. and G.J. Mannering, 1982. Hepatic cytochrome P450-dependent monooxygenase systems of the trout, frog and snake — III: induction. *Comp.Biochem. Physiol.* 71B (3): 445–453.

Shepard, J.L., B. Olsson, M. Tedengren and B.P. Bradley, 2000. Protein expression signatures identified in *Mytilus edulis* exposed to PCBs, copper and salinity stress. *Mar. Environ. Res.* 50: 337–340.

Shugart. L.R., 1990. Biological monitoring: testing for genotoxicity, in J.F. McCarthy and L.R. Shugart, (Eds.), *Biomarkers of environmental contamination.* CRC/Lewis Publishers, Boca Raton, FL, 205–216.

Shugart, L.R, J. Bickham, G. Jackim, G. McMahon, W. Ridley, J. Stein and S. Steinert, 1992. DNA alterations, in R.J. Huggett, R.A. Kimerly, P.M. Mehrle, Jr., and H.L. Bergman, (Eds.), *Biomarkers: biochemical, physiological and histological markers of anthropogenic stress.* CRC/Lewis Publishers, Chelsea, MI, 155–210.

Sijm, D.T.H.M. and A. Opperhuizen, 1989. Biotransformation of organic chemicals by fish: enzyme activities and reactions, in O. Hutzinger, (Ed.), *Handbook of environmental chemistry, vol. 2E, reactions and processes.* Springer Verlag, Berlin, 163–235.

Slooff, W., C.F. Van Kreijl and A.J. Baars, 1983. Relative liver weights and xenobiotic-metabolizing enzymes of fish from polluted surface waters in the Netherlands. *Aquatic Toxicol.* 4: 1–14.

Smital, T., R. Sauerborn, B. Pivevi, S. Kra and B. Kurelec, 2000. Interspecies differences in P-glycoprotein mediated activity of multixenobiotic resistance mechanism in several marine and freshwater invertebrates. *Comp. Biochem. Physiol.* 126C: 175–186.

Snyder, M.J., 2000. Cytochrome P450 enzymes in aquatic invertebrates: recent advances and future directions. *Aquatic Toxicol.* 48: 529–547.

Solé, M., C. Porte, X. Biosca, C.L. Mitchelmore, J.K. Chipman, D.R. Livingstone and J. Albaigés, 1996. Effects of the *Aegean Sea* oil spill on biotransformation enzymes, oxidative stress, and DNA-adducts in digestive gland of the mussel (*Mytilus edulis* L.). *Comp. Biochem. Physiol.* 113C: 257–265.

Solis-Heredia, M.J., B. Quintanilla Vega, A. Sierra Santoyo, J.M. Hernandez, E. Brambila, M.E. Cebrian and A. Albores, 2000. Chromium increases pancreatic metallothionein in the rat. *Toxicology* 142: 111–117.

Spies, R.B., J.J. Stegeman, D.W. Rice, Jr., B. Woodlin, P. Thomas, J.E. Hose, J.N. Cross and M. Prieto, 1990. Sublethal responses of *Platichtus stellatus* to organic contamination in San Francisco Bay with emphasis on reproduction, in J.F. McCarthy and L.R. Shugart, (Eds.), *Biomarkers of environmental contamination.* CRC/Lewis Publishers, Boca Raton, FL, 87–122.

Stansley, W. and D.E. Roscoe, 1996. The uptake and effects of lead in small mammals and frogs at a trap and skeet range. *Arch. Environ. Contam. Toxicol.* 30: 220–226.

Stegeman, J.J., M. Brouwer, T.D.G. Richard, L. Förlin, B.A. Fowler, B.M. Sanders and P.A. Van Veld, 1992. Molecular responses to environmental contamination: enzyme and protein systems as indicators of chemical exposure and effect, in R.J. Huggett, R.A. Kimerly, P.M. Mehrle, Jr., and H.L. Bergman, (Eds.), *Biomarkers: biochemical, physiological and histological markers of anthropogenic stress.* CRC/Lewis Publishers, Chelsea, MI, 235–335.

Stegeman, J.J. and M.E. Hahn, 1994. Biochemistry and molecular biology of monooxygenase: current perspective on forms, functions, and regulation of cytochrome P450 in aquatic species, in D.C. Malins and G.K. Ostrander, (Eds.), *Aquatic toxicology; molecular, biochemical and cellular perspectives.* CRC/Lewis Publishers, Boca Raton, FL, 87–206.

Stien, X., C. Risso, M. Gnassia-Barelli, M. Roméo and M. Lafaurie, 1997. Effect of copper chloride *in vitro* and *in vivo* on the hepatic EROD activity in the fish *Dicentrarchus labrax*. *Environ. Toxicol. Chem.* 16: 214–219.

Sturm, A., J. Wogram, P.-D. Hansen and M. Liess, 1999. Potential use of cholinesterase in monitoring low levels of organophosphates in small streams: natural variability in three-spined stickleback *(Gasterosteus aculeatus)* and relation to pollution. *Environ.Toxicol. Chem.* 18: 194–200.

Sturm, A., J. Wogram, H. Segner and M. Liess, 2000. Different sensitivity to organophosphates of acetylcholinesterase and butyrylcholinesterase from three-spined stickleback *(Gasterosteus aculeatus)*: application in biomonitoring. *Environ. Toxicol. Chem.* 19: 1607–1615.

Suter II, G.W., 1993. *Ecological risk assessment*. CRC/Lewis Publishers, Boca Raton, FL.

Sutter, T.R. and W.F. Greenlee, 1992. Classification of members of the Ah gene battery. *Chemosphere* 25: 223–226.

Tavera-Mendoza, L., S. Ruby, P. Brousseau, M. Fournier, D. Cyr and D. Marcogliese, 2002. Response of the amphibian tadpole *Xenopus laevis* to atrazine during sexual differentiation of the ovary. *Environ. Toxicol. Chem* 21: 1264–1267.

Teigen, S.W., R.A. Andersen, H.L. Daae and J.U. Skaare, 1999. Heavy metal content in liver and kidneys of grey seals *(Halichoerus grypus)* in various life stages correlated with metallothionein levels: some metal-binding characteristics of this protein. *Environ. Toxicol. Chem.* 18: 2364–2369.

Toomey, B.H. and D. Epel, 1993. Multixenobiotic resistance in *Urechis caupo* embryos: protection from environmental toxins. *Biol. Bull.* 185: 355–364.

Trust, K.A., K.T. Rummel, A.M. Scheuhammer, I.L. Brisbin and M.J. Hooper, 2000. Contaminant exposure and biomarker responses in spectacled elders *(Somateria fischeri)* from St. Lawrence Island, Alaska. *Arch. Environ. Cont. Toxicol.* 38: 107–113.

Van den Brink, N.W. and A.T.C. Bosveld, 2001. Spatial and seasonal trends of PCB concentrations and effects in common terns *(Sterna hirundo)* in the Netherlands. *Mar. Pol. Bull.* 42: 280–285.

Van den Brink, N.W., A.T.C. Bosveld, E.M. De Ruiter-Dijkman, S. Broekhuizen and P.J.H. Reijnders, 2000. PCB pattern analysis (PPA): a potential nondestructive biomarker in vertebrates for exposure to cytochrome P_{450} inducing organochlorines. *Environ. Toxicol. Chem.* 19: 575–581.

Van den Brink, N.W., M.B.E. Lee-de Groot, P.A.F. De Bie and A.T.C. Bosveld, 2003a. Enzyme markers in frogs *(Rana spec)* for monitoring of risks of aquatic pollution. *Aquatic Ecosystem Health Manage.* 6(4): 441–448.

Van den Brink, N.W., N.M. Groen, J. De Jonge and A.T.C. Bosveld, 2003b. Ecotoxicological suitability of floodplain habitats in the Netherlands for the little owl *(Athene noctua vidalli)*. *Environ. Pollut.* 122(1): 127–134.

Van der Oost, R., A. Opperhuizen, K. Satumalay, H. Heida and N.P.E. Vermeulen, 1996a. Biomonitoring aquatic pollution with feral eel *(Anguilla anguilla)*: I. Bioaccumulation: biota-sediment ratios of PCBs, OCPs, PCDDs and PCDFs. *Aquatic Toxicol.* 35: 21–46.

Van der Oost, R., A. Goksøyr, M. Celander, H. Heida and N.P.E. Vermeulen, 1996b. Biomonitoring aquatic pollution with feral eel *(Anguilla anguilla)*: II. biomarkers: pollution-induced biochemical responses. *Aquatic Toxicol.* 36: 189–222.

Van der Oost, R., E. Vindimian, P. van den Brink, K. Satumalay, H. Heida and N.P.E. Vermeulen, 1997. Biomonitoring aquatic pollution with feral eel (*Anguilla anguilla*): III. Statistical analyses of relationships between contaminant exposure and biomarkers. *Aquatic Toxicol.* 39: 45–75.

Van der Oost, R., S.C.C. Lopes, H. Komen, K. Satumalay, R. van den Bos, H. Heida and N.P.E. Vermeulen, 1998. Assessment of environmental quality and inland water pollution using biomarker responses in caged carp (*Cyprinus carpio*): use of a bioactivation:detoxication ratio as biotransformation index (BTI). *Mar. Environ. Pollut.* 46: 315–319.

Van der Oost, R., J. Beyer and N.P.E. Vermeulen, 2003. Fish bioaccumulation and biomarkers in environmental risk assessment: a review. *Environ. Toxicol. Pharmacol.* 13: 57–149.

Van Gestel, C.A.M. and T.C. van Brummelen, 1996. Incorporation of the biomarker concept in ecotoxicology calls for a redefinition of terms. *Ecotoxicology* 5: 217–225.

Van Schooten, F.J., 1991. Polycyclic aromatic hydrocarbon-DNA adducts in mice and humans. Academic thesis, State University of Leiden, the Netherlands.

Vermeulen, N.P.E., G. Donné-Op den Kelder and J.N.M. Commandeur, 1992. Formation of and protection against toxic reactive intermediates, in B. Testa, E. Kyburz, W. Fuhrer, and R. Giger, (Eds.), *Perspectives in medicinal chemistry.* Verlag Helvetica Chimica Acta, Basel, Switzerland, 573–593.

Vethaak, A.D. and T. ap Rheinallt, 1992. Fish disease as a monitor for marine pollution: the case of the North Sea. *Rev. Fish Biol. Fish.* 2: 1–32.

Viarengo, A., B. Burlando, F. Dondero, A. Marro and R. Fabbri, 1999. Metallothioneins as a tool in biomonitoring programmes. *Biomarkers* 4: 455–466.

Viarengo, A., B. Burlando, N. Ceratto and I. Panfoli, 2000. Antioxidant role of metallothioneins: a comparative overview. *Cell. Mol. Biol.* 46: 407–417.

Viarengo, A., B. Burlando, V. Evangelisti, S. Mozzone and F. Dondero, 2001. Sensitivity and specificity of metallothionein as a biomarker for aquatic environment biomonitoring, in P. Garrigues, H. Barth, C.H. Walker, and J.F. Narbonne, (Eds.), *Biomarkers of environmental contamination: a practical aproach.* Elsevier Press, Amsterdam, 29–43.

Vigano, L., A. Arillo, F. Melodia, P. Arlanti and C. Monti, 1998. Biomarker responses in cyprinids of the middle stretch of the river Po, Italy. *Environ. Toxicol. Chem.* 17: 404–411.

Vos, J., H. van Loveren, P. Wester and D. Vethaak, 1989. Toxic effects of environmental chemicals on the immune system. *Trends Pharmacol. Sci.* 10: 289–292.

Welch, W.J., 1990. The mammalian stress response: cell physiology and biochemistry of stress proteins, in R. Moromoto, A. Tissieres, and C. Georgopoulos, (Eds.), *The role of the stress response in biology and disease.* Cold Spring Harbor Laboratory Press, Cold Spring Harbor, NY, 1063–1081.

Wester, P.W., D. Vethaak and W.B. van Muiswinkel, 1994. Fish as biomarkers in immunotoxicology. *Toxicology* 86: 213–232.

WHO (World Health Organization) International Programme on Chemical Safety (IPCS), 1993. Biomarkers and risk assessment: concepts and principles. Environmental Health Criteria, 155, World Health Organization, Geneva.

Whyte, J.J., R.E. Jung, C.J. Schmitt and D.E. Tillitt, 2000. Ethoxyresorufin-O-deethylase (EROD) activity in fish as a biomarker of chemical exposure. *Crit. Rev. Toxicol.* 30: 347–570.

Widdows, J. and P. Donkin, 1992. Mussels and environmental contaminants: bioaccumulation and physiological aspects, in E. Gosling, (Ed.), *The mussel mytilus.* Elsevier Press, Amsterdam, 383–424.

Widdows, J., P. Donkin, M.D. Brinsley, S.V. Evans, P.N. Salkeld, A. Franklin, R.J. Law and M.J. Waldock, 1995. Scope for growth and contaminant levels in North Sea mussels *Mytilus edulis. Mar. Ecol. Prog. Ser.* 127: 131–148.

Widdows, J., P. Donkin, F.J. Sta, P. Matthiessen, R.J. Law, Y.T. Allen, J.E. Thain, C.R. Allchin and B.R. Jones, 2002. Measurement of stress effects (scope for growth) and contaminant levels in mussels (*Mytilus edulis*) collected from the Irish Sea. *Mar. Environ. Res.* 53: 327–356.

Winston, G.W. and R.T. Di Giulio, 1991. Prooxidant and antioxidant mechanisms in aquatic organisms. *Aquatic Toxicol.* 19: 137–161.

Wolfe, M.F. and R.J. Kendall, 1998. Age-dependent toxicity of diazinon and terbufos in European starlings (*Sturnus vulgaris*) and red-winged blackbirds (*Agelaius phoeniceus*). *Environ. Toxicol. Chem.* 17: 1300–1312.

Work, T.M., and M.R. Smith, 1996. Lead exposure in laysan albatross adults and chicks in Hawaii: prevalence, risk factors, and biochemical effects. *Arch. Environ. Contam. Toxicol.* 31: 115–119.

Zinkl, J.G., W.L. Lockhart, S.A. Kenny and F.J. Ward, 1991. The effects of cholinesterase-inhibiting insecticides on fish, in P. Mineau, (Ed.), *Cholinesterase-inhibiting insecticides, chemicals in agriculture*, Vol. 2. Elsevier Press, Amsterdam, 151–172.

Appendix 1

Overview of all biomarker assays

This appendix gives an overview of the methods used to determine all biomarker assays mentioned in this chapter. More detailed standard operating procedures (SOPs) of indicated methods are given in Appendix 2.

Preparation of microsomes and cytosol (SOP 1) — Liver homogenates are centrifuged at 12.000 G for 30 minutes in order to separate heavy cell organelles and cell-wall parts from the other cell contents. Supernatants are centrifuged at 100.000 G for 1 hour in order to separate microsomes from cytosol. Required equipment: Ultra-Turrax homogenizer, ultracentrifuge, Potter homogenizer.

EROD (SOP 2) — Ethoxyresorufine is biotransformed into resorufin by CYP 1A isozymes with NADPH as an essential cofactor. The resorufin production is measured fluorimetrically at specific emission and excitation wavelengths. Required equipment: ultracentrifuge (microsomes preparation), fluorescence spectrophotometer, or fluorescence platereader.

CYP1A — Cytochrome P450 1A (CYP1A) protein is determined by an enzyme-linked immunosorbent assay (ELISA) with specific antibodies, or by Western blotting with specific antibodies. Required equipment: ultracentrifuge (microsomes preparation), either plate reader (ELISA) or electrophoresis and blotting equipment (Western blotting).

Cyt P450 and cyt b5 (SOP 3) — The absorption spectra of the oxidized forms of both cytochrome P450 and cytochrome b5 are different from those of the reduced forms. The absorption spectra of the microsomes have to be determined for both forms and compared. Cytochrome P450 is oxidized by adding carbon monoxide (CO); cytochrome b5 is oxidized by adding NADH. Concentrations of cytochrome P-450 and b5 are calculated from the differences between the spectra. Required equipment: ultracentrifuge (microsomes preparation), advanced spectrophotometer (wavelength scanning).

UDPGT — Uridine diphosphate glucuronyl transferase (UDPGT) catalyzes the transfer of the glucuronic acid from UDP-glucuronic acid (UDPGA) to a substrate. The most commonly used substrate is p-nitrophenol. The reaction is initiated by adding UDPGA to the microsomes, and followed spectrophotometrically for 25 minutes by measuring the p-nitrophenol decrease (405 nm). Required equipment: ultracentrifuge (microsomes preparation), spectrophotometer or platereader.

GST (SOP 4) — The reduced form of glutathione (GSH) binds to glutathion S-transferase (GST), which causes deprotonation to GS⁻. CDNB binds to the enzyme-GS–complex, forming another complex with a specific absorption at 340 nm. The increase in absorption at 340 nm in time is used as a measure for GST activity. Required equipment: ultracentrifuge (cytosol preparation), spectrophotometer or platereader.

GSH/GSSG — Cytosolic glutathione (GSH) and glutathione disulfide (GSSG) can be determined by a platereader method based upon a first-order recycling reaction, in which the rate of TNB production from the substrate DTNB is dependent on the concentration of GSH and GSSG. The method determines the total glutathione content (GSH + GSSG). Therefore a second analysis is carried out in which the GSH is removed. Required equipment: ultracentrifuge (cytosol preparation), platereader.

SOD (SOP 5) — Superoxide-dismutase (SOD) is a cytosolic enzyme, which converts O_2^- in H_2O_2. The activity of superoxide-dismutase is expressed in units of SOD. The SOD activity that inhibits the reduction of, for instance, nitro blue tetrazolium (NBT) for 50% is defined as 1 unit. Required equipment: ultracentrifuge (cytosol preparation), spectrophotometer or plate reader.

CAT — Catalase (CAT) is a cytosolic enzyme, which converts H_2O_2 in H_2O and O_2. The concentration of H_2O_2 will decrease in a mixture of hydrogen peroxide and a cell suspension or cytosolic fraction containing the enzyme (catalase). The H_2O_2 decrease is measured spectrophotometrically at 240 nm. Required equipment: ultracentrifuge (cytosol preparation), spectrophotometer or plate reader.

AST and ALT — Alanine transaminase (ALT) and aspartate transaminase (AST) in blood plasma are determined using J.T. Baker UV-GPT and UV-GOT kits, respectively. Plasma is added to the reaction mixture, and the decrease in NADH levels is followed spectrophotometrically at 340 nm. One unit of activity (U) is defined as 1 μmole of substrate transformed per minute. Required equipment: table centrifuge (plasma preparation), spectrophotometer or plate reader.

ALAD — Activity of -aminolevulinic acid dehydratase (ALAD) can be measured in blood or other tissues. In blood the procedure is relatively simple, and can be measured as the hydrolysis of -aminolevulinic acid. To hydrolyzed blood a standard amount of -ALA can be added, incubated at 37°C, and the amount of -ALA can be measured by absorbency at 555 nm. Required equipment: spectrophotometer, incubator.

MT — Metallothionein (MT) proteins in liver homogenates can be determined by an enzyme-linked immunosorbent assay (ELISA) with specific antibodies, or by Western blotting with specific antibodies. Required equipment: plate reader (ELISA) or electrophoresis and blotting equipment (Western blotting).

VTG and ZRP — Vitellogenin (VTG) and zona radiata proteins (ZRP) in blood plasma can be determined by an enzyme-linked immunosorbent assay (ELISA) with specific antibodies, or by Western blotting with specific antibodies. Required equipment: plate reader (ELISA) or electrophoresis and blotting equipment (Western blotting).

ACHE (SOP 6) — Tissues are homogenized following a membrane lysis. Acetylcholine esterase (ACHE) activity is measured with the substrate acetylthiocholine that is degraded to thiocholine. In the presence of thiocholine, DTNB is converted to TNB that is measured spectophotometrically at 412 nm. Required equipment: tissue homogenizer, spectrophotometer or plate reader.

PAH metabolites — The easiest assay to measure PAH metabolites in bile semiquantitatively is the fixed-wavelength fluorescence (FF) method. Fluorescence of diluted bile is determined at specific emission and excitation wavelengths for different types of PAH metabolites. Required equipment: spectrophotometer or plate reader.

DNA adducts — DNA is isolated from liver tissues and purified. The DNA is digested and base pairs with adducts are selectively labeled using ^{32}P ATP. ^{32}P-labeled adducts are isolated using thin layer chromatography (TLC), visualized by autoradiography, and quantified by scintillation counting of the excised areas of the TLC chromatograms. Required equipment: laboratory suited to work with highly radioactive materials, TLC equipment.

Comet assay — The comet assay is a sensitive way to determine DNA strand breaks. After membrane lysis and electrophoresis of cells, DNA is stained and fragmentation is visualized using fluorescence microscopy and image analysis. The observed tail moment ("comet") is a measure for DNA damage. Required equipment: electrophoresis equipment, fluorescence microscope, imaging system.

Gross indices — The condition factor (CF) of fish is determined as (body weight [g] × 100) ÷ (length [cm])^3. The liver somatic index (LSI) is determined as: (liver weight [g] × 100) ÷ (whole body weight [g]). Required equipment: balance, ruler.

Lysosomal membrane stability (SOP 7) — Lysosomal destabilization is measured as the increased permeability of the substrate naphthol AS-BI N-acetyl-β-glucosaminide, visualized by the reaction with the enzyme N-acetyl-β-hexosaminidase into the lysosomes in the presence of diazonium salt. The preparation of tissues for the examination of cell structures needs the use of specialized methodology to produce high-quality stained sections. Required equipment: microscope, equipment for histological preparations and staining.

Scope for growth — Scope for growth (SFG) is a measure of the amount of energy available to an organism for somatic growth. Physiological responses (clearance rate, respiration rate, and food absorption efficiency) are measured under standardized laboratory conditions.

Appendix 2

Standard operating procedures of biomarker assays

This appendix gives standard operating procedures (SOPs) of a selected set of biomarker assays mentioned in this chapter.

SOP 1: Isolation of microsomes and cytosol from liver tissues

Solutions

A. Homogenization buffer: 50 mM KP-buffer (pH 7.4) + 0.9 % NaCl and 1 mM EDTA
    K_2HPO_4:
    0.87 g K_2HPO_4 (mol wt 174 g/mol)
    0.9 g NaCl (mol wt 58.44 g/mol)
    0.034 g EDTA (mol wt 340 g/mol)
    Dissolve in 100 ml bidest
    KH_2PO_4:
    0.34 g KH_2PO_4 p.a. (mol wt 136 g/mol)
    0.45 g NaCl p.a. (mol wt 58.44 g/mol)
    0.017 g EDTA p.a. (mol wt 340 g/mol)
    Dissolve in 50 ml bidest
    Add KH_2PO_4 to K_2HPO_4 until pH = 7.4

B. Suspension buffer: 100 mM KP-buffer (pH 7.4), + 0.1% EDTA and 25% glycerol (v/v)
    4.65 G K_2HPO_4 + 0.2 g EDTA in 200 ml bidest
    1.81 G KH_2PO_4 + 0.1 g EDTA in 100 ml bidest
    Add KH_2PO_4 to K_2HPO_4 until pH = 7.4
    Add 25 ml glycerol to each 75 ml of buffer

Further requirements

> pipette P10 ml
> Ultra-Turrax homogenizer
> Potter homogenizer
> ultracentrifuge

Needed per sample

> ±15 ml A
> 4.5 ml B

Procedure

- Work on ice in a cold room (4°C).
- Weigh in about 1 to 2 grams (±0.01 g) of liver tissue and cut it into small pieces.
- Homogenize the liver with ±15 ml homogenization buffer (A), using an Ultra-Turrax homogenizer.
- Centrifuge the homogenate for 25 minutes at 12.000 G (T = 4°C)
- Centrifuge the supernatant (in a fresh tube) for 1 hour at 100.000 G (T = 2°C).
- Isolate supernatant (cytosolic fraction), measure the total volume; store the required amount of cytosol in 1.5-ml Eppendorf cups at -80°C.
- Wash the pellet (microsomes) by resuspending it in ±10 ml homogenization buffer (A) and centrifuge for another hour at 100.000 G (T = 2°C).
- Throw away the supernatant and resuspend the microsomal pellet in 4.5 ml suspension buffer (B).
- Homogenize the suspension, using a Potter apparatus (about five times up and down); take care to keep the suspension cold.
- Store the microsomal suspension in 1.5-ml Eppendorf cups at -80°C.

SOP 2: EROD activity in liver microsomes

Principle

Ethoxyresorufine is transformed into resorufin by CYP 1A isozymes with the aid of NADPH. The resorufin production is measured fluorimetrically.

Solutions

A. 100 mM potassium phosphate buffer, pH 7.8
   3.48 g of K_2HPO_4/200 ml (mol wt 174 g/mol)
   1.36 g of KH_2PO_4/100 ml (mol wt 136 g/mol)
   Add KH_2PO_4 to K_2HPO_4 until pH = 7.8
B. 100 µM ethoxyresorufin in DMSO
C. 50 mM NADPH in A (M = 833.4; 41.67 mg/ml); prepare *fresh* and keep on ice)
D. 1 µM resorufin in ethanol (keep on ice)

Further requirements

fluorescence spectrophotometer
waterbath 26°C
fluorescence cuvettes
pipettes: 3 ∞ P100, P5000
parafilm
crushed ice

Needed per determination (duplo)

50 µl microsomes (MIC), on ice
4 ml A
20 µl B
20 µl C

Procedure

- Place buffer in waterbath at 26°C.
- Start EROD program on fluorescence spectrophotometer; set excitation wavelength at 530 nm and emission wavelength at 586 nm.
- Determine calibration curve with 0, 10, 20, 30 and 40 µl of resorufin in 2 ml buffer.
- Fill cuvette with:
   25 µl of microsomes
   2 ml A
   10 µl B
   10 µl C

Measure the increase in fluorescence for 3 minutes (slope should be linear).

Calculations

- Calculate the slope of the calibration curve, enter points in pmol (1μl (res) = 1 pmol (res)).
- EROD activity (pmol/min/ml MIC) = ΔA/min. (slope sample) / slope calibration curve / microsome volume.
- EROD activity in pmol/min/mg protein is obtained by dividing by the microsomal protein concentration (mg/ml).

SOP 3: Cyt P-450 and Cyt b5 Contents of Liver Microsomes

Principle

The absorption spectra of the oxidized forms of both cytochrome P-450 and cytochrome b5 are different from those of the reduced forms. From both forms an absorption spectrum is determined. Cytochrome P-450 is oxidized by adding CO, cytochrome b5 by adding NADH. Concentrations of cytochrome P-450 and b5 are calculated from the differences between the spectra.

Solutions

A. 100 mM potassium phosphate buffer, pH 7.5
   3.48 g K_2HPO_4/200 ml (mol wt 174 g/mol)
   1.36 g KH_2PO_4/100 ml (mol wt 136 g/mol)
   Add KH_2PO_4 to K_2HPO_4 until pH = 7.5.
B. 2% (w/v) solution NADH in buffer (2 mg/100 μl) — prepare the solution *fresh* on the day of measurement and keep on ice

Further requirements

carbon monoxide gas (CO)
sodium dithionite
spectrophotometer with wavelength scanning options
pipettes: 2 × P20, 2 × P1000
1.5-ml cuvettes
vortex
parafilm
crushed ice

Needed per determination (duplo)

0.75 ml microsomes, on ice
3 ml phosphate buffer A
37.5 μl NADH solution B

Procedure for cyt b5

- Take the microsomes out of the -80°C freezer and put on ice.
- Adjust spectrophotometer:
  Measure between 400 and 500 nm
  Scan speed: 2500 nm/min
- Carefully fill 3 cuvettes (*reference, a,* and *b*) with 1 ml buffer A and 0.25 ml microsomes mix.
- Measure *reference* by 450–500 nm wavelength scanning; record the reference scan for difference measurements with *a* and *b*. Note: the reference cuvette contains *no* NADH (solution B).
- Add 12.5 µl of NADH solution B to two cuvettes (*a* and *b*) and mix carefully.
- Measure cyt b5 in cuvettes *a* and *b* by 400–500 nm wavelength difference scanning against the recorded reference scan; print out the scan.

Procedure for cyt P-450

- Add 12.5 µl NADH to the *reference* cuvette and mix carefully.
- Pass CO through all cuvettes for at least 30 seconds (carefully adjust the CO flow to control foam production).
- Measure *reference* by 450–500 nm wavelength scanning. Record the reference scan for difference measurements with *a* and *b*; note: the reference cuvette contains *no* sodium dithionite.
- Add a spatula pinch of sodium dithionite to cuvettes *a* and *b* and mix carefully.
- Measure cyt P450 in cuvettes *a* and *b* by 400–500 nm wavelength difference scanning against the recorded reference scan; print out the scan.

Calculations

- Determine $\Delta A_{490-450}$ (cyt P450) or $\Delta A_{424-409}$ (cyt b5); note: if the initial slope of the curve (between 400–410 nm) is too steep to use for the cyt b5 measurement, a baseline should be estimated for the $\Delta A_{424-409}$ determination.
- Cytochrome concentration = $\Delta A \times 1000 \times$ dilution factor microsomes / ε (ε = extinction coefficient of cytochrome, 91 and 185 $cm^2/\mu mol$ for P450 and b5 respectively) => cyt P-450 conc. = $\Delta A \times 54.95$ and cyt b5 conc. = $\Delta A \times 27.03$.
- The cytochrome concentration in nmol/mg microsomal protein is obtained by dividing by the protein concentration of the sample (mg/ml).

SOP 4: GST Activity of Liver Cytosol

Principle

Glutathione (GSH) binds to GST, in which process it is deprotonated to GS^-. CDNB binds to this enzyme-GS complex, thereby forming a new complex with a specific absorption at 340 nm. The increase in absorption at 340 nm in time is used as a measure for GST activity.

Solutions

A. 125 mM potassiumphosphate buffer, pH 6.5
   2.18 g K_2HPO_4/100 ml (mol wt 174 g/mol)
   1.70 g KH_2PO_4/100 ml (mol wt 136 g/mol)
   Add K_2HPO_4 to KH_2PO_4 until pH = 6.5
B. 12 mM reduced glutathione (GSH) in buffer (M = 307.3; 3.7 mg/ml, *fresh*, on ice)
C. 15 mM CDNB in acetone (mol wt = 202.6 g/mol; 3.04 mg/ml, on ice)

Further requirements

spectrophotometer
water bath 26°C
3-ml cuvettes
pipettes: 2 × P100, P1000, P5000
parafilm

Needed per determination (duplo)

60 µl cytosol (CYT), on ice
2.4 ml A
500 µl B
100 µl C

Procedure

- Place buffer in water bath at 26°C.
- Adjust recorder speed to 1 inch per minute.
- Determine baseline at 340 nm with cuvette containing homogenization buffer (KP, pH 7.4) instead of cytosol.
- Fill cuvette with:
     1.2 ml A
     250 µl B
     50 µl C

- Start reaction by adding 30 μl of cytosol.
- Mix carefully and measure ΔA_{340} for 1 minute (check linearity, since product inhibition will occur).

Calculations

- Calculate ΔA/min for each sample.
- GST activity = ΔA/min $\times$ V cuvette / (e$\times$ V cytosol) = ΔA/min $\times$ 5.31 (ε = extinction coefficient of CDNB = 9.6 cm^2/μmol).
- Dividing by protein concentration of the sample gives GST activity in μmol/min/mg cytosolar protein.

SOP 5: SOD Activity in Liver Cytosol

Principle

Superoxide dismutase (SOD) is a cytosolic enzyme that converts O_2^- in H_2O_2. The activity of SOD is expressed in units of SOD. The activity of SOD, which will inhibit the reduction of nitro blue tetrazolium (NBT) by 50%, is defined as 1 unit.

Solutions

A. 0.043 M Na$_2$CO$_3$/NaHCO$_3$, pH 10.2 (buffer A)
   1.81 g NaHCO$_3$ /500 ml
   2.28 g Na$_2$CO$_3$ /500 ml
   Add NaHCO$_3$ to Na$_2$CO$_3$ until pH = 10.2
B. Substrate solution in buffer A (prepare *fresh*)
   0.1 mM Xanthine (mol wt = 174.1 g/mol; 1.74 mg/100 ml)
   0.1 mM EDTA (mol wt = 372.24 g/mol; 3.72 mg/100 ml)
   0.05 mg BSA/ml (5.00 mg/100 ml)
   0.025 mM NBT (mol wt = 817.67 g/mol; 2.40 mg/100 ml)
C. Enzyme solution
   Xanthine oxidase (stock = 20 U/ml), on ice

Further requirements

parafilm
spectrophotometer
waterbath 25ºC
1.5-ml microcuvettes
pipettes: P100, P20, P5000

Needed per determination (duplo)

50 µl 10× diluted cytosol
5 µl C
1.93 ml B

Procedure

- Thermostate the spectrophotometer at 25°C.
- Place the substrate solution in waterbath at 25°C.
- Set the spectrophotometer:
  wavelength: 573 nm
  integration time: 1 min.
- Zero set the spectrophotometer, using buffer A.
- Fill cuvette with:
  25 µl A
  2.5 µl C
  965 µl B.
- Mix carefully, measure blank, and record the slope
- Fill cuvette with:
  25 µl 10x diluted cytosol
  2.5 µl C
  965 µl B.
- Mix carefully, measure increase in absorbance, and record the slope.

Calculations

- SOD-activity (units SOD) = ΔA blank / (ΔA sample × 2)
- SOD activity (U/ml) = SOD activity (units SOD) × dilution factor = SOD activity (units SOD) × 397
- SOD activity in U/mg protein is obtained by dividing by the protein concentration of the sample (mg/ml).

SOP 6: ACHE Activity in Tissue Homogenates

Principle

The method of measuring cholin esterase activity uses acetylthiocholin iodide (ACTC) as specific substrate. The method is based on the production of TNB, which is produced when thiocholin reacts with the dithio*bis*nitrobenzoate (DTNB) ion. The method is described for use on a microplate reader. The same assay can be performed in cuvettes, after adjustment of the reaction volumes.

acetyl cholinesterase
Acetylthiocholin → thiocholin + acetic acid
Thiocholin + DTNB → TNB-production (yellow color, absorbance at 412 nm)

Solutions

A. 0.1 M potassium phosphate buffer, pH 7.2
B. 75 mM acetylthiocholin iodide (ATC), 21.67 mg/ml bidistilled water
C. 0.01 M dithiobisnitrobenzoic acid (DTNB), 3.96 mg/ml in 0.1 mol/l potassium phosphate buffer, pH 7.2, containing 1.5 mg/ml sodium hydrogencarbonate
D. ACHE assay reagent: 30 ml potassium phosphate buffer + 0.2 ml ATC solution + 1.0 ml DTNB solution (amount required per microplate)

Further requirements

Eppendorf centrifuge
Eppendorf cups
Potter homogenizer with 5-ml tube

Needed per determination (triplicates in microplate method)

0.1 to 1 G tissue
±5 ml potassiumphosphate buffer A
750 µl ACHE reagent D

Procedure

1. Prepreparation of tissue
   - Disrupt 0.1 to 1 g tissue with a scissor into very small parts.
   - Bring the parts into a Potter tube (5 ml) and add 3 ml of ice-cold potassium phosphate buffer A.
   - Homogenize with a Potter until a very homogenous suspension is obtained.
   - Pour the suspension over into a clean 2-ml cup.
   - Remove debris by centrifugation, 10 min at 10.000 G.
   - Transfer the supernatant into a clean 1.5-ml cup.
   - Store the supernatant in the freezer at -20°C.
   - Up to here all steps are carried out on ice or in a cold chamber (±4°C).
2. Determination of the enzymatic reaction on microplates (in triplicate)
   - Thaw the samples on ice.
   - Dilute the samples 10 times (= DFS in formula 3.3) with ice-cold potassium phosphate buffer A.

- Pipette 50 µl of diluted sample per well; include blanks with 50 µl phosphate buffer.
- Add 250 µl ACHE assay reagent.
- Put the microplate in a plate reader with temperature control set at 30°C.
- Wait 5 min for temperature adaptation.
- Read the plate for 5 min at 414 nm; shake the plate before each reading cycle.
- Calculate the average increases of optical density per minute.

Calculations

- Calculate concentrations of the reaction product using the Lambert-Beer law:

$$c = \frac{\Delta OD}{\varepsilon \times d} \qquad (3.1)$$

where:

$d = 0.9$ cm (path length in microplate well)

$\varepsilon = 1.36 \times 10^4$ ml $\times$ mmol^{-1} $\times$ cm^{-1} (extinction coefficient of the reaction product 5-thio-2-nitro-benzoic acid anion at 414 nm)

Note: include a correction for spontaneous product formation (background color development in blanks) into calculations.

The rate is:

$$\frac{c}{t} = \frac{\Delta OD}{12.24 \times t} \; [\mu\text{mole x ml}^{-1} \text{ x min}^{-1}] \qquad (3.2)$$

where:

OD = increase of optical density during time of measurement
c = concentration product formation
t = time of measurement (min)

- The specific enzyme activity (R) per milligram of protein is calculated as:

$$R = \frac{(\Delta OD / \min.) \; x \; DFEA \; x \; DFS}{12.24 \;\; x \; c\,Pr} = \frac{(\Delta OD / \min.) \; x \; 0.49 \; x \; DFS}{c\,Pr} \qquad (3.3)$$

$$[\mu mole \; x \; \min^{-1} \; x \; mg^{-1}]$$

where:
> OD/min = slope kinetic curve CHE assay
> cPr = protein concentration of sample (mg/ml)
> DFEA = dilution factor of sample in enzymatic assay ($300/50 = 6$)
> DFS = dilution factor of sample (predilution of sample)

SOP 7: Lysosomal Membrane Stability in Cells

Principle

The following protocol is a cytochemical procedure that is based on the determination of lysosomal membrane permeability using lysosomal N-acetyl-β-hexosaminidase. Lysosomal destabilization is measured as the increased release of the enzyme into the cytoplasm, which is visualized by the reaction with its substrate, naphtol AS-BI N-acetyl-β-D-glucosaminidine, followed by diazonium coupling.

Solutions and chemicals

A. 0.1 M Na citrate buffer, 2.5% NaCl w/v, pH 4.5 (lysosomal membrane labilizing buffer)
B. 20 mg naphtol AS-BI N-acetyl-β-D-glucosaminidine dissolved in 2.5 ml methoxyethanol, and 3.5 g of polypep P5115 made up to 50 ml with solution A (substrate incubation medium, to be prepared just before use)
C. 0.1 M Na phosphate buffer, pH 7.4, containing 1 mg/ml of diazonium dye Fast Violet B salts (diazonium dye)
D. Calcium formol: 2% Ca acetate + 10% formaldehyde

Further requirements

> aqueous mounting medium
> liquid nitrogen
> cryostat

Procedure

A. Prepreparation of tissue
   - Excise five small pieces (3 to 4 mm³) of the target organ or tissue obtained from five different samples and rapidily place them on an aluminium cryostat chuck.
   - Immerse the chuck in hexane precooled at –70°C in liquid nitrogen, seal them with parafilm, and store at –80°C.

- Cut serial tissue sections (7 to 10 μm thick) in a cryostat at a cabinet temperature of -25°C; transfer the slides at room temperature to flash-dry them.

B. Enzymatic determination of membrane stability
  - Place the sections in a Hellendal jar containing solution A for different times (0, 2, 5, 10, 20, 30, 40 minutes) in order to find out the range of pretreatment time needed to completely labilize the lysosomal membrane.
  - Transfer the set of slides to solution B and incubate the slides for 20 minutes at 37°C in a Hellendal jar, preferably in a shaking water bath.
  - Rinse the slides in 3.0% NaCl for 2 min at 37°C.
  - Transfer the slides to solution C for 10 min at room temperature.
  - Rinse the slides in running tap water for 5 min.
  - Fix sections for 10 min in solution D.
  - Rinse in distilled water and mount in aqueous mounting medium.

Interpretation of results

The lysosomal membrane stability is estimated as the mean lowest time necessary to reach the maximal activity of β-N-acetyl hexosaminidase in sections. To determine the magnitude of the histochemical reaction, scored with a light microscope, do the following:

- Place the slides under a microscope and examine every section by quarters; lysosomes will stain reddish-purple due to the reactivity of the substrate with β-N-acetyl hexosaminidase.
- The average labilization period corresponds to the average incubation times in buffer A that produces maximal staining reactivity.

chapter four

Molecular methods for gene expression analysis: ecotoxicological applications

A. Lange, M. Maras, and W.M. De Coen

Contents

Abstract 153
Introduction 154
Gene expression methods 156
    Discovering differences in gene expression analysis: suppression
        subtractive hybridization (SSH) 156
    Rapid gene-expression analysis of a limited number of genes:
        real-time PCR 160
    Microarray analysis 165
Future developments: pitfalls and recommendations 169
References 172

Abstract

During the last few decades, the potential of biomarker analysis in environmental toxicology has largely been acknowledged. Only a minority of studies, however, have focused on transcriptional endpoints of toxic stress. Recent developments in biological sciences have generated new technologies and methods that may provide insights into the toxicological mode of action of chemicals. This chapter gives an overview of some of these recent molecular developments and highlights their potential application in ecotoxicological research. Three methods are discussed in detail: a discovery tool to pinpoint those genes of interest affected by the stressor, called suppression subtractive hybridization polymerase chain reaction (SSH-PCR); and two ways to quantify the level of expression once it has been selected by SSH-PCR

— real-time PCR analysis and DNA arrays. Examples of successful application of these methods in toxicological studies are discussed. The limitations as well as the advantages of the genomic approach in ecotoxicology are evaluated. Environmental toxicogenomics allows the identification and characterization of genomic finger prints of environmental toxicants at the mRNA level. Before these new tools can be applied in environmental impact studies or risk assessments, however, their ecological relevance (their link with higher levels of biological organization) must be demonstrated.

Introduction

Biomarker research in ecotoxicology has been driven by the general stress concepts of Selye (1976). A bottom-up description of the stress response of an organism is divided into three distinct phases: alarm, compensation, and exhaustion. These three phases are conceptually translated into effects at the primary (neuroendocrine), secondary (physiological), and tertiary (individual responses) levels of organization within the individual. Initial perturbation caused by toxic exposure generates responses within the homeostatically regulated mechanism of the unit of study (cell, individual, and so on). These toxicant-induced, suborganismal alterations are biologically irrelevant if they do not lead to impaired characteristics at higher levels of biological organization (Huggett et al. 1992). This hierarchical division of (sub)organismal responses remains today the framework of biomarker developments. From a practical point of view, however, these biomarker analyses have focused mainly on biochemical and physiological endpoints, whereas the primary response of an organism at the transcriptional level has always been relatively underevaluated.

Moreover, ecotoxicologists are faced with the potential of adverse effects of exposure to environmental pollutants which, in general, are only poorly described at the toxicological level. This is true not only for individual chemicals, but also for chemical mixtures (regardless of their use) such as those in production or in products, or those distributed in waste or wasted material. As a result of the intensive use of chemicals in daily life, humans and biota are exposed during a full life cycle to a mixture of pollutants, usually at low concentrations (Kannan and Falandysz 1997). In order to assess the adverse effects of chemicals of environmental concern, *a priori* a selection of endpoints is made, which should allow a targeted effect evaluation of a chemical. Unfortunately, this can only be seen as "fishing in the sea, with a small net." Effects that are not measured are not documented, and can remain undiscovered. From a practical point of view, at present, it is impossible to investigate all individual pathways, enzymes, and so on.

Recent developments in biological sciences might offer a breakthrough in environmental toxicology. The introduction of genomic technologies, such as DNA arrays (also called DNA chips and biochips) allows simultaneous assessment of the expression of thousands of genes (Schena et al. 1996). This technology has opened up tremendous new possibilities in the study of gene expression and DNA variation. Such methods of gene-expression analysis from exposure to drugs and toxicants offer a convenient way to unravel the modes of actions, toxicity, efficacy, and side effects of chemicals. A major disadvantage for ecotoxicologists is that the "standard" species, which are commonly used, are not readily available in a commercial DNA array context.

Nevertheless, at present there are several molecular strategies available to obtain a targeted set of cDNA of species; genomes have been poorly studied so far. Certainly since the publication of the differential display technique (Liang and Pardee 1992), a plethora of alternatives has been published for differential gene expression analysis. Such "open" methods (Green et al. 2001) allow the full evaluation of any toxicologically relevant phenomenon for any species. From an environmental toxicology point of view, it is desirable that toxicant-specific expressed sequences be identified and that knowledge of their function and relevance is available. Once these toxicant-specific libraries have been established for different cell types and organisms, the effects of environmental contaminants on global health can be assessed in a more mechanistic and integrative way. Using these expressed sequence profiles, one could link the observed sequences to specific modes of action, suggesting the presence of the responsible (types of) contaminants. If the expressed sequences can be linked to genes with a known medical importance, they can serve as early-warning sensors for future health problems. Certainly, there are drawbacks in using species for which limited genome information is available. Homologies among sequences expressed as E-values provide a good mathematical evaluation system, but do not guarantee a similar function, especially when working with invertebrate species. However, since the public genomic databases are growing exponentially, it is only a matter of time before such uncertainties are resolved.

It should be possible with a moderate genome-sequencing effort to pave the way for exploring toxicological events in an unprecedented way. Within the next decade, ecotoxicologists will start to investigate the effects of stressors on a genomic scale. Environmental toxicogenomics will allow the identification and characterization of genomic fingerprints of environmental toxicants at the mRNA level, and in a further step at the protein expression level. The following sections of this chapter describe several methods of gene-expression analysis that have a high potential in ecotoxicology. Three methods are discussed in detail: a discovery tool to pinpoint those genes of interest affected by the studied stressor, called suppression subtractive hybridization polymerase chain reaction (SSH-PCR); and two tools to easily quantify the level of expression once it has been selected by SSH-PCR — real-time PCR analysis and DNA arrays.

The SSH-PCR method can clearly be seen as a new approach in biomarker analysis. Rather than focusing on a defined set of biomarkers, it evaluates and analyzes toxicant-induced changes in gene expression. Recently, several molecular techniques such as differential display, subtractive hybridization, representational difference analysis (RDA), and so on have been established in order to quantify differences in gene expression and to identify specific gene products. Probably one of the most reliable methodologies for differential gene-expression analysis is the SSH-PCR developed by Diatchenko et al. (1996). It is based on RDA technology, but contains a modification to normalize for mRNA abundance. The SSH-PCR approach reveals, at a fundamental molecular level, which genes are induced in different organisms due to exposure in a pollution gradient.

Once a gene of interest has been pinpointed by a differential screening technique such as SSH, a more elegant method is needed to confirm and quantify its mRNA expression level. This can be done most easily by a highly sensitive approach, quantitative real-time reverse-transcription polymerase chain reaction (RT-PCR). Using this method, the expression level of targeted genes can be more easily and rapidly analyzed compared to the traditional RT-PCR methodology. Finally, to perform the simultaneous analysis of several genes, DNA arrays are the method of choice.

These arrays consist of nitrocellulose membranes (macroarrays) or microscopic glass slides (microarrays) that contain small quantities of oligonucleotides or cDNAs. Each represents parts of a single gene product, and is used to screen for the differential expression of gene transcripts. This technology has opened up tremendous new possibilities in the study of gene expression and DNA variation.

Gene expression methods

Discovering differences in gene expression analysis: suppression subtractive hybridization (SSH)

Subtractive cDNA hybridization is an efficient technique for isolating differentially expressed genes by comparing two populations of mRNA. It involves hybridization of cDNA from a test population to large excess of mRNA/cDNA of another, so-called driver population. Under such conditions, only molecular species that are specific to the tester will not hybridize with molecules present in the driver. Afterwards, the unhybridized tester fraction (which contains differentially expressed genes) is physically separated from the hybridized common sequences (by, for instance, oligo (dT) $_{30}$-latex beads or hydroxylapatite chromatography). Thus, only the population of tester-specific cDNA remains. The obtained differentially expressed sequences are inserted into a cloning vector in order to obtain the subtracted cDNA library.

This method, however, has some significant limitations, several of which are crucial for application in ecotoxicological approaches. First, it requires relatively large quantities of RNA, when there are often only small amounts of starting material (a few cells) available. Second, it is inefficient for obtaining low-abundance transcripts. This is a problem because the minor fraction is particularly of interest, since transcripts for many regulatory proteins fall into this category (Gurskaya et al. 1996). These limitations could be overcome by introducing the SSH-PCR (Figure 4.1) developed by Diatchenko et al. (1996). This method overcomes the problem of differences in mRNA abundance by incorporating a hybridization step that normalizes (equalizes) sequence abundance during the subtraction. The subtraction allows the isolation of transcripts, which are specifically up- or down-regulated under the two compared conditions. This virtually excludes the isolation of false-positive and false-negative clones, and makes it possible to detect low copy-number mRNAs. The high subtraction efficiency as well as the equalized representation of differentially expressed sequences (normalization) is based on a suppression PCR step. This is a specific form of PCR that permits exponential amplification of sequences that differ in abundance while suppressing the amplification of sequences that are identically abundant in both compared populations.

One of the most important advantages of isolating differentially expressed genes by means of SSH is that no prior comprehensive knowledge of which genes are up- or down-regulated after xenobiotic exposure is required, but an almost complete set of ecotoxicologically relevant genes is obtained (Rockett et al. 1999). This is a clear benefit in environmental toxicology, since the major biochemical and molecular questions focus on vital biomarker species (selected *a priori*) or important ecotoxicological model organisms for which little data may be available (Hogstrand et al. 2002). Therefore, the development of open functional genomic methodologies such as the SSH method now enables environmental toxicologists to monitor changes in gene expression that result from xenobiotic exposure, even in animal species whose genomes are still poorly characterized.

So far, SSH has been mainly applied to pathophysiological approaches. This has led to the identification of rarely transcribed TNF-inducible genes in human umbilical arterial endothelial cells involved in the regulation of inflammatory responses (Stier et al., 2000), genes expressed by endothelial cells implicated in distinct aspects of angiogenesis (Glienke et al. 2000), or mesangial cell genes involved in diabetic nephropathy (Murphy et al. 1999). Other biomedical applications can be found in cancer research, where SSH has proven its full potential (Hufton et al. 1999; Zhang et al. 2000; Bangur et al. 2002).

In the meantime, studies using SSH in toxicological approaches have successfully been performed. One possible application is the investigation of the impact of metals on biological systems. Metals are ubiquitous environmental contaminants that cause a variety of biochemical, cellular, and physiological alterations. Using SSH, cellular responses from metal exposure can be characterized and the results may lead to better insight into the mechanisms related to metal toxicity. Such a study has already been performed investigating the response of COS-7 cells to Cd exposure (Lee et al. 2002). Besides the induction of the expression of genes that are already known to be induced under Cd stress (such as MT and hsp 60 and 70), the authors were able to identify further genes (such as hsp 10, 40, or 89α) that were so far unknown to be related to Cd toxicity. Another example of the use of SSH

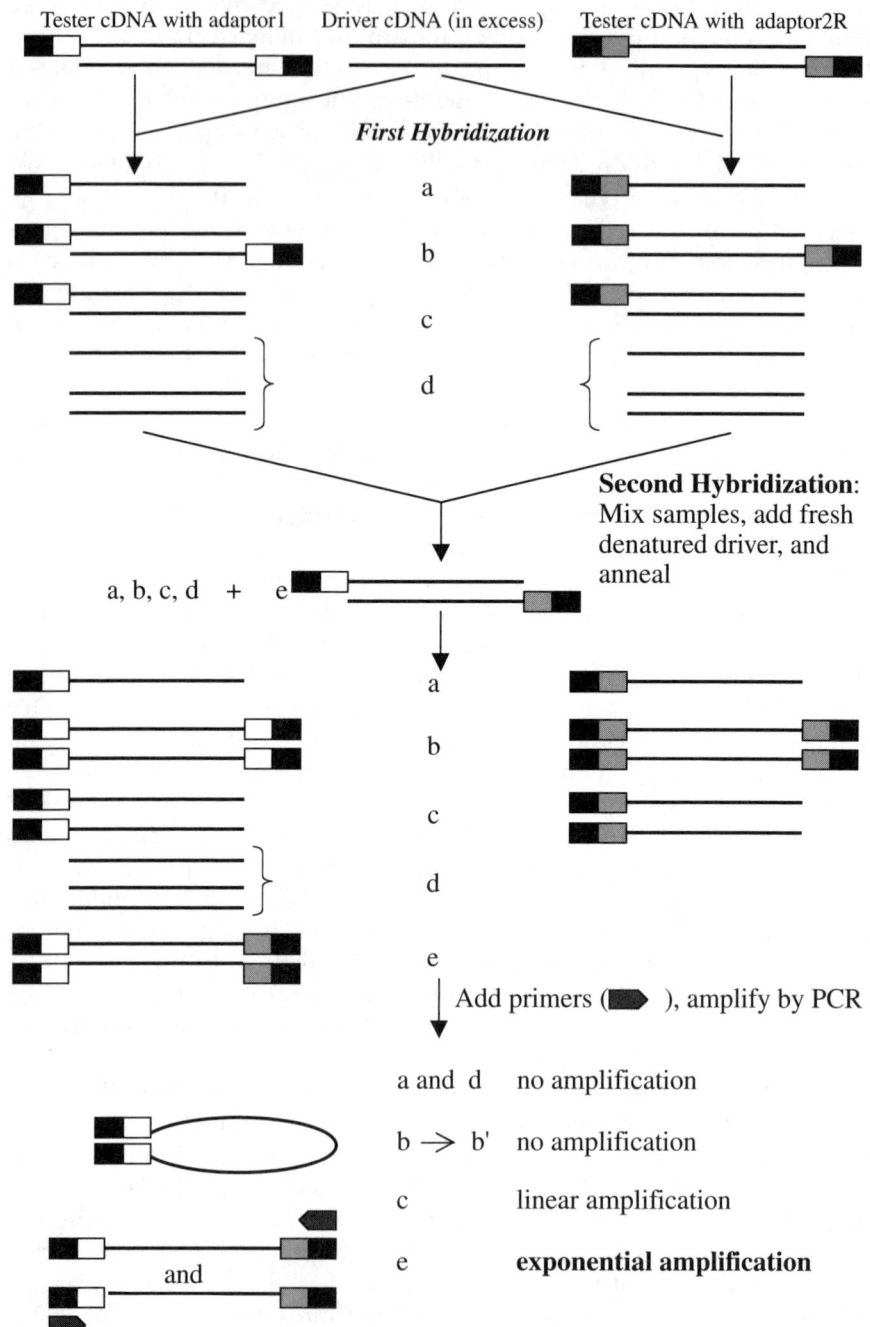

Figure 4.1 Schematic diagram of the SSH method. After cDNA has been synthesized from the two RNA populations being compared, two different cDNA adapters are ligated to the cDNA ends. Both adaptors have stretches of identical sequences to allow annealing of one PCR primer. Two hybridizations are then performed. In the first, an excess of driver is added to each sample of tester. The hybridization kinetics leads to equalization and enrichment of differentially expressed sequences. For the second hybridization, the two primary hybridization samples are mixed together. Only the remaining equalized and subtracted ss tester cDNAs can reassociate and form type e hybrids.

Solid lines represent *Rsa*I digested tester or driver sequences. Solid boxes represent the outer parts of the two adaptors and the corresponding PCR primer sequence. Clear boxes represent the inner part of adaptor1 and the corresponding nested PCRprimer1 sequence. Shaded boxes represent the inner part of adaptor2R and the corresponding nested PCR primer2R sequence.

in toxicological approaches is the identification of aryl hydrocarbon (AhR) receptor-regulated genes that mediate the toxicity of dioxin and dioxin-like compounds (Kolluri et al. 2001). These xenobiotics are of tremendous environmental concern since, among other things, they possess hepatotoxic, embryotoxic, and immunotoxic potential at very low concentrations (Giesy et al. 2002). Uptake and accumulation of these xenobiotics induces a specific cellular response that detoxifies or metabolizes the compound. This response can be characterized by isolating genes that show altered expression in their target tissues or cells following exposure to a chemical stimulus. Taking this into account, it becomes obvious that the construction of subtractive cDNA libraries can be seen as a promising means to generate tools devoted to the monitoring of contamination (Bultelle et al. 2002).

Cell cultures are a common tool in toxicological research that offer several advantages, since the cell is the site of the primary interaction between toxic chemicals and biological systems. This provides the opportunity to study the toxic mechanisms and interactions that underlie complex whole-animal responses (Segner 1998). SSH has already been used in a variety of applications in order to isolate subtractive cDNA libraries from different cell cultures (Ye and Connor 2000; Zhang et al. 2000; Kolluri et al. 2001; Kiss et al. 2003). It is important, however, to keep in mind that when generating subtractive libraries from cells, the results for differentially expressed genes depend on the differentiation stages of the cells and can represent either cell type-specific responses or common responses in many cell types. For this reason, it is wrong to conclude that the results obtained from a cell type-specific library can easily be equated with the gene expression of the whole tissue from which the cell culture is derived. It should always be remembered that a tissue is composed of several cell types that may be responding to an exposure with different gene-expression patterns. Exposure time and concentration are other important factors to consider when trying to transfer in vitro gene-expression results to an in vivo system.

SSH remains applicable only to pair-wise treatment comparisons and must be replicated by switching tester and driver samples in order to identify gene-expression changes in both directions (as well as for the comparison of more than two treatments). Additionally, it does not provide a quantitative measure of expression differences (Moody 2001). Nevertheless, since SSH can be linked to microarray analysis, a combination of both technologies seems to provide a powerful tool for toxicity-related gene-expression profiling. Arrayed SSH repertoires have the advantage that redundant spotting is eliminated and much smaller and efficient DNA chips can be produced (von Stein et al. 1997; Yang et al. 1999). However, one always has to keep in mind that it is not mRNA but protein that makes a cell functional, and due to translational differences or posttranslational modifications, the protein pattern of a cell does not necessarily have to accord with the predicted one (Abbott 1999). Thus, if cellular responses to toxicants are to be characterized, then biochemical, cellular, and physiological studies should be performed in addition to genomic approaches.

Rapid gene-expression analysis of a limited number of genes: real-time PCR

Real-time reverse-transcription PCR determines the amount of a specific template in a very specific, sensitive, and reproducible way. The real-time PCR system is based on the detection and quantitation of a fluorescent reporter molecule. Fluorescence emitted during the reaction is monitored during each PCR cycle (in real time) as a quantitative indication of the amplification process. The increase of fluorescence is directly proportional to the amount of PCR product. Through continuous recording of the signal at each cycle, one can monitor the PCR reaction during the exponential amplification phase and quantify the amplification step at which the first significant increase in the PCR product correlates to the initial amount of target template. Furthermore, after the final amplification process, the fluorescence of the melting curve, which can be used as a control over the specificity of the amplified product, can be monitored.

In contrast to conventional RT-PCR methods, the continuous endpoint detection does not need extra processing (such as gel image analysis) afterwards, which gives the method a higher throughput. Furthermore, real-time PCR also allows measurements within a much wider dynamic range (up to 10^7-fold versus 1000-fold in conventional RT-PCR).

In general, two methods are available for the quantitative detection of the amplicon. The first method is based on the release of an additional fluorescence-labeled oligonucleotide, while the second one involves the detection of fluorescent dsDNA-binding dyes.

The discovery of real-time PCR depended on two important things: (1) the discovery that *Taq* polymerase possesses $5' \rightarrow 3'$-exonuclease activity (Holland et al., 1991), which can be used for degrading specific fluorigenic probes; and (2) the construction of dual-labeled oligonucleotide probes, which emit a fluorescent signal only on cleavage of the probe.

The TaqMan assay (Figure 4.2) combines these two findings. TaqMan probes are dual-labeled oligonucleotides (longer and with a Tm value of 10°C higher than the conventional primers) that contain a fluorescent dye (such as FAM, 6-carboxyfluorescein), usually on the 5' base, and a quenching dye (such as TAMRA, 6-carboxytetramethylrhodamine), typically on the 3' base. When excited at a specific wavelength, the excited fluorescent dye transfers energy to the nearby quenching dye molecule rather than fluorescing (fluorescence resonance energy transfer or FRET). TaqMan probes are designed to anneal to an internal region of a PCR product. Conventional primers, which are not labeled, are used to drive the PCR cycle. When the polymerase replicates a template on which a TaqMan probe is bound, its 5' exonuclease activity cleaves the probe. This ends the activity of quencher (no FRET) and the reporter dye starts to emit fluorescence that increases proportionally to the rate of probe cleavage in each cycle.

Molecular beacons (Figure 4.3) are DNA hybridization probes that are designed to have two arms with complementary sequences, and are thus able to form a stem-and-loop structure from a single-stranded DNA molecule. The loop portion of the molecule is complementary to the target sequence, while the stem is formed by the annealing of complementary arm sequences at both ends of the probe sequence (Tyagi and Kramer 1996). The probes contain a fluorescent dye (such as FAM or TAMRA) at one end of the molecule and a quencher (typically DABCYL, [4-(4'-dimethylaminophenylazo) benzoic acid]) at the other end. Through the hairpin structure that they adopt in free solution, the fluorescent dye and the quencher are positioned in close proximity, which allows FRET to occur. When excited with a specific wavelength, the excited fluorescent dye transfers energy to the nearby quenching dye molecule rather than fluorescing. When the beacon hybridizes to the target during the annealing step, the reporter dye is separated from the quencher and the reporter fluoresces (FRET does not occur). Molecular beacons remain intact during PCR and must rebind to target every cycle for fluorescence emission. This correlates to the amount of PCR product available.

Both TaqMan probes and molecular beacons allow detection of multiple DNA species (multiplexing) by use of different reporter dyes on different probes and beacons. By multiplexing, both the target and endogenous control can be amplified in a single tube. Multiplex assays can be performed using multiple dyes with distinct emission wavelengths as long as the equipment allows their detection.

A cheaper alternative is the double-stranded DNA-binding dye chemistry that quantitates the amplicon production (including nonspecific amplification and primer-dimer-complexes) by the use of a non-sequence-specific fluorescent intercalating agent (such as SYBR Green I or ethidium bromide). SYBR Green I is a minor groove binding dye. It does not bind to ssDNA. The major problem with SYBR-Green-based detection is that nonspecific

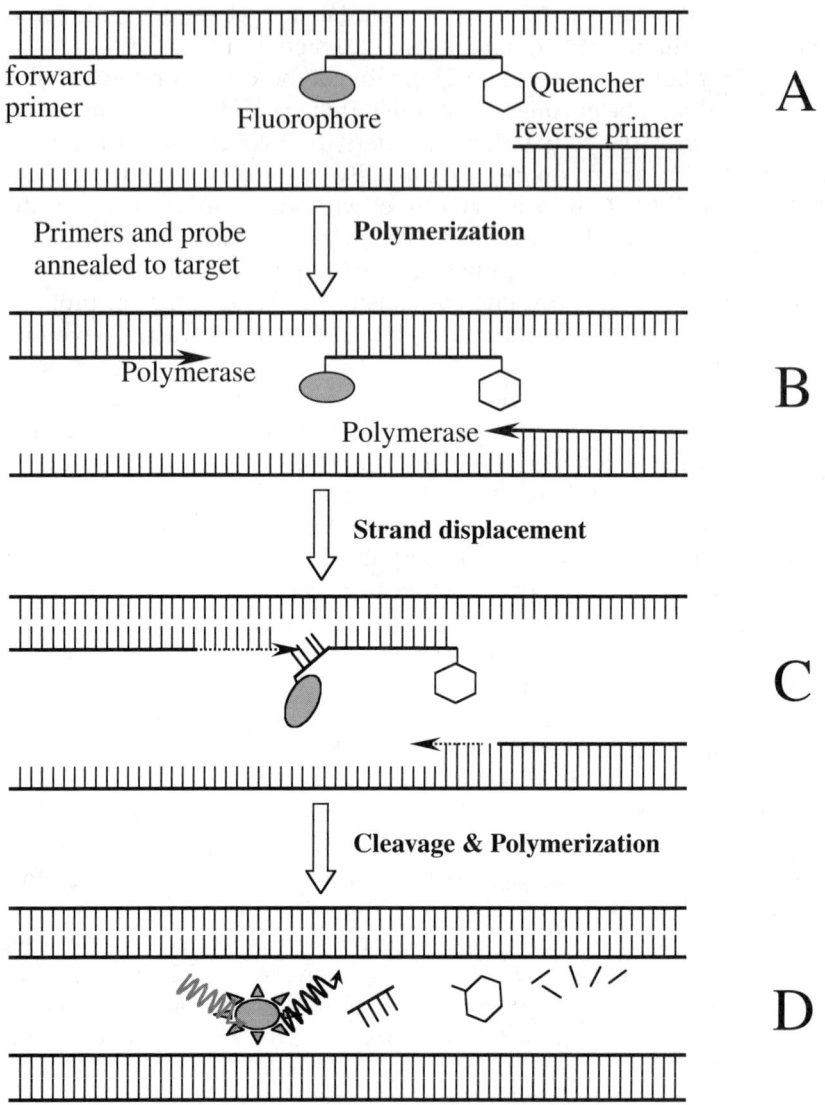

Figure 4.2 The TaqMan assay system. (A): TaqMan probes are dual-labeled oligonu-cleotides that contain a fluorescent dye (usually on the 5′ base) and a quenching dye (typically on the 3′ base). They are designed to anneal to an internal region of a PCR product. Fluorescence is quenched by the proximity of the two dyes. Conventional primers, which are not labeled, are used to drive the PCR cycle. (B and C): When the polymerase replicates a template on which a TaqMan probe is bound, its 5′ exonu-clease activity cleaves to the probe. This terminates the activity of the quencher and the reporter dye starts to emit fluorescence, which in each cycle increases propor-tionally to the rate of probe cleavage.

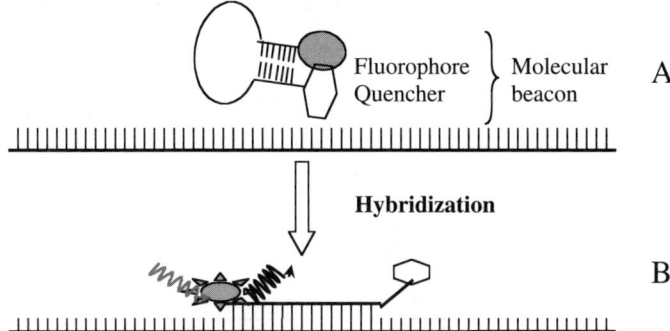

Figure 4.3 The molecular beacon system. (A): Molecular beacons are DNA hybridization probes that are designed to be complementary to a sequence in the middle of the expected amplicon. They have two arms with complementary sequences, so they are able to form a stem-and-loop structure from a single-stranded DNA molecule. The loop portion of the molecule is complementary to the target sequence, while the stem is formed by the annealing of complementary arm sequences at both ends of the probe sequence. The probes contain a fluorescent dye at one end of the molecule and a quencher at the other end. (B): When the probe hybridizes to the target, a linear structure is formed. Fluorophore and quencher are separated and fluorescence is emitted.

amplifications cannot be distinguished from specific amplifications. This can be circumvented by plotting fluorescence as a function of temperature after the full PCR amplification cycles, thus generating a melting curve of the amplicon.

An important parameter in real-time PCR analysis is the threshold cycle or the C_T value (Figure 4.4). This is the cycle at which a significant increase in fluorescence is first detected. The threshold cycle is defined as the point at which the system begins to detect the increase in the fluorescent signal above the background signal associated with an exponential increase in PCR product during the log-linear amplification phase. This phase provides the most useful information about the PCR amplification. The slope of the log-linear phase is a reflection of the amplification efficiency. For the slope to be an indicator of real amplification (rather than signal drift), there must be an inflection point. This is the point on the growth curve when the log-linear phase begins. It also represents the greatest rate of change along the growth curve (signal drift is characterized by gradual increase or decrease in fluorescence without amplification of the product).

The C_T value is important for quantification. The higher the initial amount of template (the copy number of a gene can vary among different samples), the sooner the accumulated product is detected in the PCR process, and the lower the C_T value. The choice of threshold, which will determine the C_T value, is up to the operator and is one of the subjective elements in real-time PCR. It should be placed above any baseline activity and within

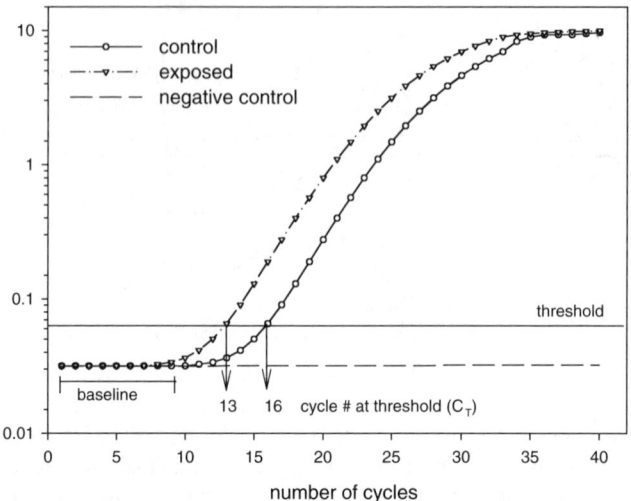

Figure 4.4 Hypothetical real-time PCR amplification plot. The threshold cycle (C_T) is defined as the point at which the system begins to detect the increase in the fluorescent signal (above the background signal) that is associated with an exponential increase in PCR product during the log-linear amplification phase.

the exponential increase phase (which looks linear in the log transformation). Besides being used for quantitation, the C_T value can be used for qualitative analysis as a pass/fail measure.

Several interesting examples demonstrating the usefulness of real-time PCR can be found in the following (eco)toxicological applications:

- Celius et al. (2000) recently showed the potential of using zona radiata (ZR) proteins instead of vitellogenin (VTG) mRNA in rainbow trout (*Oncorhynchus mykiss*) as a marker for estrogen receptor-mediated effects. Using real-time PCR, the authors quantified the induction of ZR and VTG gene in liver after exposure to 17β-estradiol (E2) or α-zearalenol (α-ZEA). Compared to an indirect enzyme-linked immunosorbent assay (ELISA), real-time PCR proved to be more sensitive; ZR was more responsive to low levels of E2 and α-ZEA than VTG.
- Rees et al. (2003) applied real-time PCR for CYP1A screening in Atlantic salmon (*Salmo salar*). CYP1A was detected and was inducible in Atlantic salmon gill, brain, kidney, and liver tissue. Field as well as laboratory experiments were performed. The authors underlined that quantitative PCR analysis of CYP1A expression is useful for studying ecotoxicity in populations of Atlantic salmon in the wild.
- Glue et al. (2002) used real-time PCR to study the effects of phthalates on human cytokine expression in peripheral blood mononuclear cells (PBMCs). Their results suggested that monophthalates had a limited effect on cytokine expression in a monocytic cell line THP-1, and

weak effect on cytokine expression in PBMCs from allergic and non-allergic individuals.

- Apoptosis was studied by Hildebrand et al. (1999) in primary rat hepatocytes through real-time PCR quantification of bax and p53 gene expression. Two toxicants (camptothecin and topotecan) were used, with a clear difference in apoptotic potential.

- Berthiaume and Wallace (2002) used real-time PCR to quantify the proliferation of mitochondria by mitochondrial DNA (mtDNA) copy number. Compounds such as perfluorooctanoic acid (PFOA) that cause peroxisome proliferation in rats and mice could interfere with mitochondrial biogenesis. Perfluorooctanesulfonate (PFOS) and N-ethyl perfluorooctanesulfonamido ethanol (N-EtFOSE) were investigated as peroxisome proliferators and compared to PFOA. Although PFOS and PFOA clearly showed similar potencies as peroxisome proliferators, none of the perfluorooctanoates (PFOS and N-EtFOSE) significantly altered mitochrondrial biogenesis.

- Diaz et al. (2001) indicated tissue-specific changes in the expression of glutamate-cysteine ligase mRNAs in mice exposed to methylmercury. Glutamate-cysteine ligase (GLCL), the rate-limiting enzyme in glutathione (GSH) synthesis (composed of two subunits, a catalytic [GLCLc] and a regulatory subunit [GLCLr]), was down-regulated by MeHg exposure in the liver, and up-regulated in the kidney. Only the catalytic subunit mRNA in the small intestine of female mice was up-regulated.

- Borlak and Thum (2001) showed the usefulness of real-time PCR in evaluating the pleotropic effects of Aroclor 1254 on rat hepatocyte cultures. Induction of nuclear transcription factors, cytochrome P450 monooxygenases, and glutathione S-transferase (alpha) gene expression could easily be quantified. Their results suggested a coordinate genomic response in rat hepatocytes upon exposure to PCBs.

Microarray analysis

DNA microarrays hold great promise for toxicologists who study the effects of chemical compounds on all kinds of living organisms and cells and try to learn more about their modes of action. The latter knowledge is needed for finding correlations between compound exposure and stress, and/or the origin or progression of diseases (Fielden and Zacharewski 2001). DNA arrays allow complete transcriptional signatures, which might shed light on the mechanisms responsible for observed biochemical, physiological, or biological effects upon exposures to potentially hazardous substances. In short, the technique uses "chips" containing several hundreds or thousands of spotted cDNA, PCR fragments, or oligonucleotides, corresponding to a chosen set of genes and/or expressed sequence tags (ESTs) that are templates for hybridizations with dual fluorescent labeled probes (ESTs correspond to fragments of genes for which no functions have been assigned).

For a more thorough explanation of cDNA microarrays, see http://dir.niehs.nih.gov/microarray/figures/background.pdf. The latter probes are synthesized from mRNA that is harvested from compound-treated versus nontreated cells or organisms. Different dyes (mostly Cy3 green versus Cy5 red fluorescing) are used to label the first strand DNAs from respective cell populations. Differential signals from control versus treated samples are then indicative for induction of gene expression by exposure to a toxic compound. A more detailed description of the technique is given by Celis et al. (2000).

Several studies have demonstrated the possibility of using microarray analysis in building new models to further unravel partially known complex biochemical pathways. Hamadeh et al. (2002b) used rats exposed to peroxisome proliferators (clofibrate, Wyeth 14,643, and gemfibrozil) on one hand, and to the enzyme inducer phenobarbital on the other hand. The microarray data corroborated previously described metabolic and toxicological effects of peroxisome proliferators, such as stimulation of triglyceride hydrolysis, fatty acid uptake, conversion of acyl CoA derivatives, and stimulation of the β-oxidation pathway. With phenobarbital, microarray data confirmed previously described upregulation of cytochrome P450 genes (CYP2B2, CYP2C6, CYP3A9), epoxide hydrolase, diaphorase, and several glutathione S-transferases (GSTs). The up- or down-regulation of several novel genes was also detected. Upon Phenobarbital exposure, a higher transcript level was found for carboxylesterase precursor, while carnitine palmitoyl transferase I and Acyl-CoA synthetase were repressed. The latter observation led to the suggestion that an inhibition of fatty acid peroxidation is provoked by Phenobarbital exposure, which was not previously recognized and could be a subject for further research.

Besides discovering potentially new biochemical pathways or revealing previously unknown cross-communications between complex pathways, this technique eventually will be involved in "predictive toxicology" (Pennie et al. 2001). It is expected that each chemical will induce a unique and diagnostic expression profile under a given set of conditions, and this information will allow the predicting of future harmful events or disease. Compounds with similar expression profiles could be subdivided into classes of chemicals with similar pharmacological or toxicological endpoints.

Another important question is whether biomarkers will be easily detected using microarray analysis. In an attempt to recognize chemical-specific signature patterns, Hamadeh et al. (2002b) used hierarchical cluster analysis to conclude that an individual animal could be distinguished by the class of toxicants to which it was exposed. Using principal-component analysis, a close proximity in the gene-expression patterns of animals exposed to clofibrate, Wyeth 14,643, and gemfibrozil was found, while a distinct partition between animals exposed to peroxisome proliferators and Phenobarbital was observed. Biomarkers were suggested to correspond to genes that show time-independent change of expression upon toxicant exposure. As proposed by the authors, the latter genes could be used to develop signatures of the compound classes.

Studies by Waring et al. (2001) again confirmed that gene-expression profiles for compounds with similar toxic mechanisms are related, yet distinguishable: Rat hepatocytes were exposed to 15 different hepatotoxins. Each compound induced unique toxin-specific expression profiles upon screening the 973 genes of the commercial Rat Toxicology U34 Array (from Affymetrix, Inc.). Hierarchical cluster analysis was performed in order to correlate gene responses to fold changes. In accordance with what was previously known about mechanisms of action, close associations were found between Aroclor 1254 and 3-methylcholanthrene, two aromatic hydrocarbons that cause hepatic hypertrophy due to an increase in the hepatocellular content of smooth endoplasmic reticulum. A set of genes was detected that are regulated in common between Aroclor 1254 and 3-methylcholanthrene (for instance, CYP1A1, CYP1B1, CYP2A1). In an analogous way, clusters were detected between indomethacin and carbamazepine (previously shown to produce a decrease in the P450 monooxygenase system), and between carbon tetrachloride, methotrexate, and monocrotaline. The latter two compounds were shown to induce the formation of reactive intermediates. Diethylnitrosamine and etoposide did not form clusters with any of the other hepatotoxins, a result expected from their unique mechanisms of toxicity. While etoposide induces DNA damage by interacting with topoisomerase II, dietylnitrosamine causes DNA damage by alkylation.

Besides *in vivo* models, *in vitro* tissue cultures are used in toxicological studies because of the ease of manipulating the testing material in a reproducible, well-controlled, and relatively cheap way. Microarray analyses with the latter study material were performed by several research groups who were able to identify cytotoxicity-associated gene-expression changes (Amundson et al. 1999; Gore et al. 2000; Puga et al. 2000; Ross et al. 2000; Frueh et al. 2001; Harries et al. 2001; Martinez et al. 2002). As to whether cytotoxicity-related expression changes give additional information on modes of action of chemicals, a lot can be learned from several recent molecular pharmacology studies. In an extensive study by Scherf et al. (2000), 60 different cancer cell lines were exposed to 1400 chemical compounds. First, activity profiles were sought by studying growth inhibition by the different drugs. Clustering of cell lines by using an average-linkage algorithm and a metric based on the growth inhibitory activities (GI_{50}) of the 1400 compounds resulted in 15 distinct branches. Arrays containing 1376 genes were used to generate gene-expression profiles that were further analyzed in relation to the activity profiles of 118 drugs of which the mechanism of action is known. Drugs were then clustered on the basis of Pearson correlation coefficients that related the corresponding activity profiles across 60 cell lines to the expression patterns of the cell lines. For many but not for all drugs, this clustering differed from that based on the known structural or mechanistic features of the drugs. With 5-fluorouracil or L-asparaginase, relationships between variations in transcript levels of particular genes and mechanisms of drug sensitivity and resistance were revealed. Dan et al. (2002) showed a correlation between expression patterns of cancer cells exposed to anticancer drugs and drug-sensitivity profiles. An integrated database approach of gene expression and chemosensitivity profiles was used, and genes were identified that correlated with specific drugs that have similar mechanisms of action. The results of these studies strengthen the belief that microarray analysis could be useful for standard monitoring of the effects of drugs on cell lines and animal models.

Results obtained so far also demonstrate that in the future the microarray technique needs further improvement as a predictive toxicological tool. First of all, as soon as experimental designs are changed by only one factor (such as dose, exposure time, medium composition, age of cells, choice of cell line, and so on), significantly altered expression profiles can be expected. Amundson et al. (1999) demonstrated widely differing responses according to the cell lines used to study genotoxic ionizing radiation. Ross et al. (2000) demonstrated systematic variations in gene-expression patterns according to the human cancer cell lines studied. This means that one has to be very careful when comparing microarray results obtained with different cell lines. Pronounced differences due to different exposure times of carbon tetrachloride to human hepatoma HepG2 cells was nicely demonstrated by Harries et al. (2001). Martinez et al. (2002) demonstrated how changing 2,3,7,8-tetrachlorodibenzo-p-dioxin (TCDD) doses resulted in substantially varying concentration-response patterns in human airway epithelial cells. TCDD exposure was also studied by Puga et al. (2000) and Frueh et al. (2001), but since different gene sets were used for the design of the arrays, these studies are (unfortunately) not comparable. Another problem with *in vitro* cell cultures is their relevance for the *in vivo* model. Indeed, Schuppe-Koistinen et al. (2002) found many significant discrepancies between *in vitro* and *in vivo* gene-expression changes of rat liver treated with aminoguanidine carboxylate 2-{1[hydrazine(imino)methyl]hydrazine} acetic acid. Furthermore, since DNA arrays do not allow distinctions between causative events and adaptive responses, it is necessary to perform microarray analyses in concert with physiological, biochemical, molecular biochemical, and pathological studies, as exemplified by Hamadeh et al. (2002a).

Hamadeh et al. (2002b) suggested that differences in the variations in gene-expression responses (fold inductions) between different similarly treated animals were probably related to the potency of the toxic compounds: the more potent the compound, the less variation in expression could be expected. This has to be taken into account when comparing different chemicals. Genotoxic compounds that cause mutations at random could also provoke difficulties when reproducing expression profiles. Finally, it is generally known that genes can be affected directly by a toxicant, or the expression can be regulated as an adaptive response downstream of xenobiotic insult. Pleiotropic toxicants such as TCDD act by disturbing many different biochemical pathways, probably as a result of many indirect responses. The reproducibility of these responses remains to be studied. The examples mentioned underscore the fact that the microarray technique is still in its infancy and many control experiments are needed to reveal pitfalls and to study reliability and reproducibility. Experiments should be designed to allow easy comparison to other studies from other laboratories. A choice of similar gene arrays, similar cell lines and growth conditions, and similar doses and exposure times of toxicants could be helpful in learning the importance of the intrinsic variability of cell cultures (or animals) from different laboratories in the outcome of experiments. Optimizing the technique will also require further optimization of computational data acquisitions and analyses (Amin et al. 2002; Dobbin and Simon 2002). Obtaining hybridization signals from expressed genes is only one issue. More challenging, however, is developing good arrays, good criteria, and good processing programs to decide which of the differential signals are really toxicologically relevant and how chemical-specific recognition patterns can be found.

Future developments: pitfalls and recommendations

For isolating novel genes that may turn out to become useful biomarkers for a particular state or condition, the application of molecular methods that pinpoint differentially expressed genes (such as SSH) has become very useful for environmental toxicologists. The advantage of "open systems" is that no prior knowledge about specific genes, or about which ones are up- or down-regulated, is required. This allows the isolation of previously unknown genes. In contrast, "closed systems" such as microarrays only allow the assessment of defined genes represented in those systems. However, open and closed systems can act in a complementary way. Once novelty (both absolute novelty and novel orthologs of known genes in less well-characterized systems) has been identified in open systems, these genes can then be used in directed ways in closed systems.

Most work using microarrays has so far focused on genetically well-characterized organisms such as yeast (*Saccharomyces cerevisiae*), the nematode *Caenorhabditis elegans*, mouse (*Mus musculus*), or human (*Homo sapiens*), but not all of them can be considered as being of major interest in the field of ecotoxicology. Here occurs one of the main problems for ecotoxicological applications of gene-expression analysis: there is still a lack of genomic data for ecotoxicological model organisms such as, for example, fish. Nevertheless, gene-expression arrays are being developed for a variety of fish species, frogs, and birds (Neumann and Galvez 2002), which are all of ecotoxicological interest. However, since the complete genomes of most of the ecotoxicologically relevant organisms are so far uncharacterized — except for zebrafish (*Danio rerio*) — it should be a main objective to characterize those genomes first. This can be done using open methodologies such as SSH or differential display RT-PCR. The obtained microarrays can then be used for toxicogenomic-based applications in order to assess the environmental impacts of contaminants on gene expression. The results may be the recognition of genes that represent useful biomarkers. Useful markers are those that are altered in response to chemical exposure and thus are often used as measures for exposure to a contaminant. Results obtained from microarray analysis also allow unraveling the modes of action, toxicity, efficacy, and side effects of chemicals.

An aspect that should always be considered when interpreting genomic data (responses) is the specific nature of the investigated biological system that was used (for example, whether data were obtained for a special cell type or for a tissue that consists of several cell types). As discussed earlier, it is careless to extrapolate conclusions drawn from the responses of cells to the responses of tissues or even whole organisms. Also, the fact that physiological factors such as age, sex, or nutritional status are known to strongly regulate gene expression has to be kept in mind when interpreting gene-expression patterns. Furthermore, microarrays represent a temporally and spatially restricted picture of gene expression, whereas toxicant exposure can elicit complex temporal and spatial changes in gene expression with subsequent altered physiological functions during prolonged exposure (Neumann and Galvez 2002). Since transient and acclimation effects can also occur during long-term sublethal exposure, it is also dangerous to draw conclusions from data obtained for a single time point. To correctly interpret gene-expression patterns obtained from microarrays, dose-response relationships should be studied. These responses are of particular relevance in ecotoxicology because organisms are exposed to a concentration range of toxicants rather than to a stable concentration. It must also be considered that pollutant loading into ecosystems is not continuous but is an intermittent process, subject to meteorological and hydrological conditions and anthropogenic activities (Burton 1999).

Another reason why gene-expression patterns obtained in laboratory experiments for an individual chemical cannot be applied to the environmental situation is that in the environment, organisms are not exposed to individual compounds but to complex chemical mixtures. The concentrations and toxic potencies of compounds present in those mixtures can range over several orders of magnitude. Additionally, interactions between different classes of compounds can modulate the gene-expression pattern and the toxic potential of individual compounds. Accurate predictions of the joint effects of complex mixtures of substances are not yet possible, and a systematic investigation of compound combinations would be necessary for characterizing an environmental response. This is hard to realize in the laboratory, even with a small number of experiments; combinations of stressors, varieties of magnitude, and frequency and duration of each compound would have to be evaluated (Burton 1999).

Once gene-expression responses have been obtained for different cell types or organisms, the question remains whether differentially expressed genes represent a treatment-specific — either differential expression of compound- or damage-specific genes (Farr and Dunn 1999), or "only" a general stress or toxicity response. Gene-expression patterns, however, might support the assessment of the effects of environmental contaminants on global health in a more mechanistic and integrative way. Although gene expression itself is a highly regulated event, several regulatory checkpoints that subsequently regulate the eventual translation of the mRNA into a functional protein product (posttranslational modifications and protein degradation) occur after gene expression (Steiner and Anderson 2000). Since protein and not mRNA (which is investigated with the described genomic approaches) is the functional unit of a cell, subsequent protein functional studies cannot be disregarded. Implying toxicological effects based on gene-transcription profiles, and tentatively attributing a particular gene product to a specific cellular function is risky. This is particularly relevant for nonmodel organisms in which only few functional gene and protein data have been characterized. For this reason, toxicological profiling using microarrays for ecotoxicological purposes requires parallel studies for demonstrating and validating the assumption that altered gene-expression patterns translate into defined toxicological endpoints (Neumann and Galvez 2002). These necessary parallel studies should be conducted with functional and/or physiological endpoints as well as with gene-expression data.

A frequently occurring problem in gene-expression analysis is teh correct selection of an internal control or reference. Obviously, relative gene-expression comparisons are only useful when the gene expression of the chosen endogenous control is more abundant and remains constant in proportion to total RNA. A good endogenous control allows normalization for differences in mRNA amounts, so internal control genes should fulfill two requirements: (1) they should be ubiquitously expressed in the investigated tissue (constitutively expressed housekeeping genes), and (2) their transcription should not be modulated within the investigated experimental context (Stürzenbaum and Kille 2001). The second requirement is especially hard to meet, since even the expression level of control genes is often affected by experimental treatments, stage of development, or cell type. The most commonly used fragments for control genes are the 18S and 28S ribosomal subunits, glyceraldehyde-3-phosphate dehydrogenase (GAPDH), β-actin, cyclophilin, and α– or β-tubulins (Thellin et al. 1999). Suzuki et al. (2000) reviewed the use of internal control genes and stated that GADPH should be used with caution, as increased levels were noted in proliferating cells. According to the same authors, β-actin should be the preferred gene while other authors, such as Drew and Murphy (1997), describe actin expression varying with cell culture conditions or in different cell types within tissues. In the chapter authors' studies, however, 17β-estradiol caused increased β-actin levels in brain tissue of zebrafish and lung tissue of wood mice, while ubiquitin was more stable. The best-suited internal control genes for gene-expression quantification seem to be the ribosomal subunits (Thellin et al. 1999; Stürzenbaum and Kille 2001). Their use, however, is less convenient, since they do not have poly-A tails; random primers rather than oligo-dT primers should be used for cDNA synthesis.

Before using an internal control, every gene should be validated as a normalizer in the tissue of choice. Since it seems that none of these housekeeping genes is expressed in a completely invariant way, at least two types of internal control genes should be used. The safest option, however, is to select a battery of mRNA species. The chosen genes should be proven to be expressed at a constant level at different time points, by both the same individual, as well as by different individuals at the target cell or tissue.

In conclusion, environmental toxicogenomics seems to be a useful tool for the isolation of genes that are differentially expressed under toxicant exposure (SSH-PCR), and for the analysis of gene-expression patterns (microarrays) and quantification of gene-expression levels (quantitative real-time RT-PCR). The main advantage of the first method is that it is designed to amplify genes that demonstrate altered expression. As such, SSH is also useful for the isolation of previously unknown genes that may turn into useful biomarkers for a particular condition. Since methods resulting in the isolation of differentially expressed genes are more demanding, their application on nonmodel or less-common laboratory species will remain difficult until their genomes are fully sequenced. For other species such as humans or mice, toxicogenomic investigations will more and more switch from the application of open to closed systems, such as microarrays, since these are easier and faster to prepare and use. Furthermore, they provide quantitative data and are suitable for high-throughput analysis as well as for the analysis of specific signaling pathways or gene families (Rockett et al. 1999). Once differentially expressed genes have been isolated and identified, however, not only their functional biological but also their ecological significance remains to be evaluated.

References

Abbott, A., 1999. A post-genomic challenge: learning to read patterns of protein synthesis. *Nature* 402, 715–720.

Amin, R.P., Hamadeh, H.K., Bushel, P.R., Bennett, L., Afshari, C.A., Paules, R.S., 2002. Genomic interrogation of mechanism(s) underlying cellular responses to toxicants. *Toxicology* 181–182, 555–563.

Amundson, S.A., Bittner, M., Chen, Y.D., Trent, J., Meltzer, P., Fornace, A.J., 1999. Fluorescent cDNA microarray hybridization reveals complexity and heterogeneity of cellular genotoxic stress responses. *Oncogene* 18, 3666–3672.

Bangur, C.S., Switzer, A., Fan, L., Marton, M.J., Meyer, M.R., Wang, T.T., 2002. Identification of genes over-expressed in small lung carcinoma using suppression subtractive hybridization and cDNA microarray expression analysis. *Oncogene* 21, 3814–3825.

Berthiaume, J. and Wallace, K.B., 2002. Perfluorooctanoate, perflourooctanesulfonate, and N-ethyl perfluorooctanesulfonamido ethanol; peroxisome proliferation and mitochondrial biogenesis. *Toxicol. Lett.* 129, 23–32.

Borlak, J. and Thum, T., 2001. Induction of nuclear transcription factors, cytochrome P450 monooxygenases, and glutathione S-transferase alpha gene expression in Aroclor 1254-treated rat hepatocyte cultures. *Biochem. Pharmacol.* 61, 145–153.

Bultelle, F., Panchout, M., Leboulenger, F., Danger, J.M., 2002. Identification of differentially expressed genes in *Dreissena polymorpha* exposed to contaminants. *Mar. Environ. Res.* 54, 385–389.

Burton G.A., Jr., 1999. Realistic assessments of ecotoxicity using traditional and novel approaches. *Aquatic Ecosyst. Health Manage.* 2, 1–8.

Celis, J.E., Kruhøffer, M., Gromova, I., Frederiksen, C., Østergaard, M., Thykjaer, T., Gromov, P., Yu, J., Pálsdóttir, H., Magnusson, N., Ørntoft, T.F., 2000. Gene expression profiling: monitoring transcription and translation products using DNA microarrays and proteomics. *FEBS Lett.* 480, 2–16.

Celius, T., Matthews, J.B., Giesy, J.P., Zacharewski, T.R., 2000. Quantification of rainbow trout (*Oncorhynchus mykiss*) zona radiata and vitellogenin mRNA levels using real-time PCR after *in vivo* treatment with estradiol-17 beta or alpha-zearalenol. *J. Steroid Biochem.* 75, 109–119.

Dan, S., Tsunoda, T., Kitahara, O., Yanagawa, R., Zembutsu, H., Katagiri, T., Yamazaki, K., Nakamura, Y., Yamori, T., 2002. An integrated database of chemosensitivity to 55 anticancer drugs and gene expression profiles of 39 human cancer cell lines. *Cancer Res.* 62, 1139–1147.

Diatchenko, L., Lau, Y.F.C., Campbell, A.P., Chenchik, A., Moqadam, F., Huang, B., Lukyanov, S., Lukyanov, K., Gurskaya, N., Sverdlov, E.D., Siebert, P.D., 1996. Suppression subtractive hybridization: a method for generating differentially regulated or tissue-specific cDNA probes and libraries. *Proc. Natl. Acad. Sci. USA* 93, 6025–6030.

Diaz, D., Krejsa, C.M., White, C.C., Keener, C.L., Farin, F.M., Kavanagh, T.J., 2001. Tissue specific changes in the expression of glutamate-cysteine ligase mRNAs in mice exposed to methylmercury. *Toxicol. Lett.* 122, 119–129.

Dobbin, K. and Simon, R., 2002. Comparison of microarray designs for class comparison and class discovery. *Bioinformatics* 18, 1438–1445.

Drew, J.S. and Murphy, R.A., 1997. Actin isoform expression, cellular heterogeneity, and contractile function in smooth muscle. *Can. J. Physiol. Pharm.* 75, 869–877.

Farr, S. and Dunn, R.T., 1999. Concise review: gene expression applied to toxicology. *Toxicol. Sci.* 50, 1–9.

Fielden, M.R. and Zacharewski, T.R., 2001. Challenges and limitations of gene expression profiling in mechanistic and predictive toxicology. *Toxicol. Sci.* 60, 6–10.

Frueh, F.W., Hayashibara, K.C., Brown, P.O., Whitlock, J.P., 2001. Use of cDNA microarrays to analyze dioxin-induced changes in human liver gene expression. *Toxicol. Lett.* 122, 189–203.

Giesy, J.P., Hilscherova, K., Jones, P.D., Kannan, K., Machala, M., 2002. Cell bioassays for detection of aryl hydrocarbon (AhR) and estrogen receptor (ER) mediated activity in environmental samples. *Mar. Poll. Bull.* 45, 3–16.

Glienke, J., Schmitt, A.O., Pilarsky, C., Hinzmann, B., Weiss, B., Rosenthal, A., Thierauch, K.H., 2000. Differential gene expression by endothelial cells in distinct angiogenic states. *Eur. J. Biochem.* 267, 2820–2830.

Glue, C., Millner, A., Bodtger, U., Jinquan, T., Poulsen, L.K., 2002. *In vitro* effects of monophthalates on cytokine expression in the monocytic cell line THP-1 and in peripheral blood mononuclear cells from allergic and non-allergic donors. *Toxicol. in Vitro* 16, 657–662.

Gore, M.A., Morshedi, M.M., Reidhaar-Olson, J.F., 2000. Gene expression changes associated with cytotoxicity identified using cDNA arrays. *Functional Integrated Genomics* 1, 114–126.

Green, C.D., Simons, J.F., Taillon, B.E., Lewin, D.A., 2001. Open systems: panoramic views of gene expression. *J. Immunol. Methods* 250, 67–79.

Gurskaya, N.G., Diatchenko, L., Chenchik, A., Siebert, P.D., Khaspekov, G.L., Lukyanov, K.A., Vagner, L.L., Ermolaeva, O.D., Lukyanov, S.A., Sverdlov, E.D., 1996. Equalizing cDNA subtraction based on selective suppression of polymerase chain reaction: cloning of Jurkat cell transcripts induced by phytohemaglutinin and Phorbol 12-Myristate 13-Acetate. *Anal. Biochem.* 240, 90–97.

Hamadeh, H.K., Knight, B.L., Haugen, A.C., Sieber, S., Amin, R.P., Bushel, P.R., Stoll, R., Blanchard, K., Jayadev, S., Tennant, R.W., Cunningham, M.L., Afshari, C.A., Paules, R.S., 2002a. Methapyrilene toxicity: anchorage of pathologic observations to gene expression alterations. *Toxicol. Pathol.* 30, 470–482.

Hamadeh, H.K., P.R., B., Jayadev, S., Martin, K., DiSorbo, O., Sieber, S., Bennett, L., Tennant, R., Stoll, R., Barrett, J.C., Blanchard, K., Paules, R.S., Afshari, C.A., 2002b. Gene expression analysis reveals chemical-specific profiles. *Toxicol. Sci.* 67, 219–231.

Harries, H.M., Fletcher, S.T., Duggan, C.M., Baker, V.A., 2001. The use of genomics technology to investigate gene expression changes in cultured human liver cells. *Toxicol. in Vitro* 15, 399–405.

Hildebrand, H., Kempka, G., Mahnke, A., 1999. Determination of apoptosis in primary rat hepatocytes by real-time quantitative PCR. *Toxicol. in Vitro* 13, 561–565.

Hogstrand, C., Balesaria, S., Glover, C.N., 2002. Application of genomics and proteomics for study of the integrated response to zinc exposure in a non-model fish species, the rainbow trout. *Comp. Biochem. Physiol.* 133B, 523–535.

Holland, P.M., Abramson, R.D., Watson, R., Gelfant, D.H., 1991. Detection of specific polymerase chain-reaction product by utilizing the 5'-3' exonuclease activity of thermus-aquaticus DNA-polymerase. *Proc. Natl. Acad. Sci. USA* 85, 8790–8794.

Hufton, S.E., Moerkerk, P.T., Brandwijk, R., de Bruine, A.P., Arends, J.-W., Hoogenboom, H.R., 1999. A profile of differentially expressed genes in primary corectal cancer using suppression subtractive hybridization. *FEBS Lett.* 463, 77–82.

Huggett, R.J., Kimerie, R.A.S., Mehrie, P.M., Jr., Bergman, H.L. (Eds.), 1992. *Biomarkers: biochemical, physiological, and histological markers of anthropogenic stress.* CRC/Lewis Publishers, Boca Raton, FL.

Kannan, K., Falandysz, J., 1997. Butyltin residues in sediment, fish, fish-eating birds, harbour porpoise, and human tissues from the Polish coast of the Baltic Sea. *Mar. Poll. Bull.* 34, 203–207.

Kiss, C., Nishikawa, J., Dieckmann, A., Takada, K., Klein, G., Szekely, L., 2003. Improved subtractive suppression hybridization combined with high density cDNA array screening identifies differentially expressed viral and cellular genes. *J. Virol. Methods* 107, 195–203.

Kolluri, S.K., Balduf, C., Hofmann, M., Göttlicher, M., 2001. Novel target genes of the Ah (dioxin) receptor: transcriptional induction of N-myristoyltransferase 2. *Cancer Res.* 61, 8534–8539.

Lee, M.J., Nishio, H., Ayaki, H., Yamamoto, M., Sumino, K., 2002. Upregulation of stress response mRNAs in COS-7 cells exposed to cadmium. *Toxicology* 174, 109–117.

Liang, P., Pardee, A.B., 1992. Differential display of eukaryotic messenger-RNA by means of the polymerase chain-reaction. *Science* 257, 967–971.

Martinez, J.M., Afshari, C.A., Bushel, P.R., Masuda, A., Takahashi, T., Walker, N.J., 2002. Differential toxicogenomic responses to 2,3,7,8-tetrachlorodibenzo-p-dioxin in malignant and nonmalignant human airway epithelial cells. *Toxicol. Sci.* 69, 409–423.

Moody, D.E., 2001. Genomics techniques: an overview of methods for the study of gene expression. *J. Anim. Sci.* 79, E128–E135.

Murphy, M., Godson, C., Cannon, S., Kato, S., Mackenzie, H.S., Martin, F., Brady, H.R., 1999. Suppression subtractive hybridization identifies high glucose levels as stimulus for expression of connective tissue growth factor and other genes in human mesangial cells. *J. Biol. Chem.* 274, 5830–5834.

Neumann, N.F., Galvez, F., 2002. DNA microarrays and toxicogenomics: applications for ecotoxicology? *Biotechnol. Adv.* 6206, 391–419.

Pennie, W.D., Woodyatt, N.J., Aldridge, T.C., Orphanides, G., 2001. Application of genomics to the definition of the molecular basis for toxicity. *Toxicol. Lett.* 120, 353–358.

Puga, A., Maier, A., Medvedovic, M., 2000. The transcriptional signature of dioxin in human hepatoma HepG2 cells. *Biochem. Pharmacol.* 60, 1129–1142.

Rees, C.B., McCormick, S.D., Vanden Heuvel, J.P., Li, W., 2003. Quantitative PCR analysis of CYP1A induction in Atlantic salmon (*Salmo salar*). *Aquatic Toxicol.* 62, 67–78.

Rockett, J.C., Esdaile, D.J., Gibson, G.G., 1999. Differential gene expression in drug metabolism and toxicology: practicalities, problems and potential. *Xenobiotica* 29, 655–691.

Ross, D.T., Scherf, U., Eisen, M.B., Perou, C.M., Rees, C., Spellman, P., Iyer, V., Jeffrey, S.S., Van de Rijn, M., Waltham, M., Pergamenschikov, A., Lee, J.C.E., Lashkari, D., Shalon, D., Myers, T.G., Weinstein, J.N., Botstein, D., Brown, P.O., 2000. Systematic variation in gene expression patterns in human cancer cell lines. *Nat. Genet.* 24, 227–235.

Schena, M., Shalon, D., Heller, R., Chai, A., Brown, P.O., Davis, R.W., 1996. Parallel human genome analysis: microarray-based expression monitoring of 1000 genes. *Proc. Natl. Acad. Sci. USA* 93, 10614–10619.

Scherf, U., Ross, D.T., Waltham, M., Smith, L.H., Lee, J.K., Tanabe, L., Kohn, K.W., Reinhold, W.C., Myers, T.G., Andrews, D.T., Scudiero, D.A., Eisen, M.B., Sausville, E.A., Pommier, Y., Botstein, D., Brown, P.O., Weinstein, J.N., 2000. A gene expression database for the molecular pharmacology of cancer. *Nat. Genet.* 24, 236–244.

Schuppe-Koistinen, I., Frisk, A.-L., Janzon, L., 2002. Molecular profiling of hepatotoxicity induced by a aminoguanidine carboxylate in the rat: gene expression profiling. *Toxicology* 179, 197–219.

Segner, H., 1998. Fish cell lines as a tool in aquatic toxicology, in T. Braunbeck, D.E. Hinton, and B. Streit, (Eds.), *Fish Ecotoxicol.*, Vol. 86, 1–38. Birkhäuser Verlag, Basel, Switzerland.

Selye H., (Ed.), 1976. *Stress in health and disease.* Butterworth-Heinemann Publishers, Boston, MA.

Steiner, S., Anderson, N.L., 2000. Expression profiling in toxicology — potentials and limitations. *Toxicol. Lett.* 112–113, 467–471.

Stier, S., Totzke, G., Grünewald, E., Neuhaus, T., Fronhoffs, S., Sachinidis, A., Vetter, H., Schulze-Osthoff, K., Ko, Y., 2000. Identification of syntenin and other TNF-inducible genes in human umbilical arterial endothelial cells by suppression subtractive hybridization. *FEBS Lett.* 467, 299–304.

Stürzenbaum, S.R., Kille, P., 2001. Control genes in quantitative molecular biological techniques: the variability of invariance. *Comp. Biochem. Physiol.* 130B, 281–289.

Suzuki, T., Higgins, P.J., Crawford, D.R., 2000. Control selection for RNA quantitation. *Biotechniques* 29, 332–337.

Thellin, O., Zorzi, W., Lakaye, B., De Borman, B., Coumans, B., Hennen, G., Grisar, T., Igout, A., Heinen, E., 1999. Housekeeping genes as internal standards: use and limits. *J. Biotechnol.* 75, 291–295.

Tyagi, S., Kramer, F.R., 1996. Molecular beacons: Probes that fluoresce upon hybridization. *Nat. Biotechnol.* 14, 303–308.

von Stein, O.D., Thies, W.-G., Hofmann, M., 1997. A high throughout screening for rarely transcribed differentially expressed genes. *Nucleic Acids Res.* 25, 2598–2602.

Waring, J.F., Ciurlionis, R., Jolly, R.A., Heindel, M., Ulrich, R.G., 2001. Microarray analysis of hepatotoxins *in vitro* reveals a correlation between gene expression profiles and mechanisms of toxicity. *Toxicol. Lett.* 120, 359–368.

Yang, G.P., Ross, D.T., Kuang, W.W., Brown, P.O., Weigel, R.J., 1999. Combining SSH and cDNA microarrays for rapid identification of differentially expressed genes. *Nucleic Acids Res.* 27, 1517–1523.

Ye, Z., Connor, J.R., 2000. Identification of iron responsive genes by screening cDNA libraries from suppression subtractive hybridization with antisense probes from three iron conditions. *Nucleic Acids Res.* 28, 1802–1807.

Zhang, L., Cilley, R.E., Chinoy, M.D., Chinoy, M.R., 2000. Suppression subtractive hybridization to identify gene expressions in variant and classic small cell lung cancer cell lines. *J. Surg. Res.* 93, 108–119.

chapter five

Bioassays and biosensors: capturing biology in a nutshell

B. van der Burg and A. Brouwer

Contents

Introduction 177
History 178
Bioassays and biosensors 179
    Definitions 179
    Bioassays 180
        In vivo bioassays 180
        In vitro bioassays 180
        Transgenic animals 182
    Biosensors 184
        Biological recognition elements 184
        Transducers 186
    Biological endpoints 187
Complementary and integrative technologies 187
Validation and application 188
Future perspectives 188
Summary 190
References 190

Introduction

To prevent biological systems in the environment from being damaged by noxious substances, ecotoxicological monitoring depends heavily on chemical-analytical methods. These methods combine high sensitivity, specificity, and the possibility of readily quantifying the compound of interest. These measurements, however, have a major drawback. They are suitable for measuring a limited set of pollutants, selected because they have been found to cause harmful biological effects in experiments directed toward identifying hazardous compounds. This approach was successful at a time when pollution was characterized by high concentrations of a limited number of pollutants with acute biological effects.

The next phase in monitoring is rapidly emerging, succeeding the ongoing and very successful eradication of the release and accumulation of highly noxious materials in the environment. This new phase uses the biological effect itself as an analytical tool. By integrating the effects of a broad spectrum of chemicals at the same biological endpoint, a much more comprehensive testing system may be designed. Three major developments have greatly speeded up the introduction of bioanalytical tools. First, there is an awareness of the environmental spread of an ever-increasing number of chemicals and their metabolites, albeit at relatively low individual levels. This plethora of chemicals hugely increases the possibility of combined effects at the same biological endpoint, thereby causing environmental problems that escape chemical-analytical methods. Second, there has been a rapid advance in the technology that allows using biological endpoints as analytical tools. Third, the new bioanalytical tools have a wide range of applications because they measure endpoints that are not accessible with chemical-analytical methods, and can help replace or reduce animal experimentation in pharmacology, toxicology, drug discovery, and so on.

This chapter gives a broad overview of existing biosensors and bioassays, their principles of action, and their use and applicability, particularly for ecotoxicological purposes. Because of the enormous size of this field of research, the chapter focuses on highlights, novel trends, and recent examples, including those from the authors' own research. Also discussed are different biological systems based on modern technology, such as transgenic animals, as well as the advantages, disadvantages, and possible applications of different approaches.

History

Biological monitoring is not new. It has a long history, going back to crude but effective methods like the use of canaries as early-warning systems for mining gasses such as methane, and using dogs or humans to detect food poisons to protect kings and queens. In ecotoxicology, fish can be used to monitor water quality, and flow-through systems even allow online monitoring. Because of the emergence of new analytical techniques, as well as ethical considerations, most of these methods have disappeared and were gradually replaced by chemical analysis. Even today animal experiments are hard to avoid, and hazard identification of chemicals and pharmaceuticals still greatly depends on *in vivo* determinations in live animals.

However, cell- and molecule-based *in vitro* bioanalytical tools are developing at a dazzling speed and may claim a much more central role in the near future. Rapid technological advances have led to many different types of measuring tools. All of these bioanalytical tools have isolated biological endpoints, such as receptors or key molecules in a particular process, as their analytical hearts. To generate a handy tool, these biological recognition elements are coupled to an easily measurable and quantifiable read-out system. The recognition element in biosensors is directly coupled to a physical or physicochemical transducing system, allowing online measurements.

Direct linkage of a biological recognition element in the form of an enzyme that binds and converts glucose into measurable products led in the early 1960s to the first biosensor, the glucose sensor of Clark and Lyons (1962). The first biosensors were able to measure single compounds that are present in relatively high levels in mixtures such as clinical samples, thereby providing an alternative for chemical measurements (Rogers 2000).

Major technological advances in molecular biology have allowed the identification and isolation of biological receptors, enzymes, and key molecules in biological processes. Within a few decades, molecular identification tools such monoclonal antibodies, subtraction hybridization, differential display PCR, and DNA arrays have been developed. These tools, coupled with such powerful methods as the isolation and cloning of genes, have given us major new insights into molecular processes, biological receptor molecules, and marker and key regulatory genes. These technologies are by no means static, but are continuing to increase in efficiency and accuracy, as discussed below. These advances, together with rapid progress in microtechnology, computer technology, and bioinformatics, has led to the generation of a wealth of new bioanalytical tools, although many have not yet been put to practical use.

Bioassays and biosensors

Definitions

Many biological detection systems consist of a biological recognition element and some kind of transducing system that generates an easily detectable signal. This transducing system can be biological in nature, such as bioassays, or physical, such as biosensors. Because of the possibilities for combining technologies (often from quite distinct scientific fields) in order to create numerous applications, there is a large variation in transducing systems. Consequently, it is difficult to give a uniform definition for the terms bioassay and biosensor (Rogers 2000). The most commonly used definitions in the environmental monitoring field make a functional distinction between the two, mainly based on the read-out system. While a bioassay is a generic term for a wide variety of assays that combine biological recognition elements with a range of biological, biochemical, and molecular biological read-outs, the term biosensor is used exclusively for those systems that include physical and electrochemical transducing systems, and thereby are suitable for online measurements. The distinction between a bioassay and a biosensor is, however, increasingly difficult to characterize. Although bioassays tend to be more complex than biosensors, and the more classical ones generally involve whole animals, in modern biosensors whole organisms like bacteria are sometimes used. The application of nanotechnologies has led to increasingly complex designs of biosensors, thereby creating some overlap with bioassays.

Bioassays

In vivo *bioassays*

Many of the older bioassays, like tests to measure hormone action, use whole animals and relatively straightforward endpoints such as death or the weight of specific organs. For example, the uterotrophic assay, developed more than 70 years ago, determines if a compound mimics the female hormone estradiol in promoting uterine proliferation (Ashby 2001). In this test, female rodents with low estrogen levels (such as prepubertal or ovariectomised animals) are treated with the test compound for several days. Then the increase in uterine weight is compared with control animals, giving a measure of estrogenicity. In this case, both the biological recognition element and the read-out system are to a large extent part of a complex biological system. Although these classical *in vivo* methods have the advantage of taking into account parameters such as toxicokinetics, metabolism, and feedback mechanisms, they are labor-intensive, expensive, and have limited sensitivity, speed, and capacity. Obviously, these types of assays using mammals are not practical for ecotoxicological monitoring. To this end more practical tests have been developed using easy-to-handle organisms that have ecotoxicological relevance, such as daphnia and corophium (Rawash et al. 1975; Hyne and Everett 1998; Keddy et al. 1995). In particular, the daphnia test has been used extensively, and is still being used. Although their relevance is evident, these tests have a rather large degree of variability and labor intensity when compared with *in vitro* assays.

In vitro *bioassays*

New assays for a number of biological endpoints have been developed. These use cultured cells and tissues, thereby reducing animal experimentation (ECVAM Working Group on Chemicals 2002) and cost while increasing the sensitivity, speed, and capacity for screening (Johnston and Johnston 2002). To generate novel *in vitro* bioassays, many cell types from a variety of species are available. This allows generating bioassays with biological endpoints that not only replace *in vivo* assays, but also address endpoints not accessible with *in vivo* assays, such as when the species involved is not suitable as an experimental animal. In particular, the availability of a range of human cell lines, including stem cells able to differentiate *in vitro* (Rizzino 2002), offers many novel bioanalytical possibilities. Read-out systems can be manifold, using endogenously produced marker proteins, enzymes, biochemical reactions, and reporter genes. These reporter genes consist of a gene coding for an easily measurable product, coupled to promoter elements that respond to transcription factors and are modulated when a toxicant is present. The gene codings for firefly luciferase and jellyfish green fluorescent protein are often used in this context. Bioassays using these reporter genes usually have advantages to more conventional assays with respect to sensitivity, reliability, and convenience of use (Naylor 1999).

As an example, methods to measure estrogens were developed that make use of the proliferative response of breast cancer cells towards estrogenic compounds (Soto et al. 1995). This test is known as the E-SCREEN. Through application of reporter-gene technology, more practical, rapid, responsive, and sensitive tests were generated in a variety of cell lines (Balaguer et al. 1999; Legler et al. 1999; Schoonen et al 2000). These assays make use of the knowledge that estrogens enter cells by diffusion, where they bind to intracellular receptors. Upon estrogen binding the receptors become activated, and enter the nucleus to bind to recognition sequences in promoter regions of target genes, known as the estrogen responsive elements (EREs). The DNA-bound receptors then activate transcription of the target genes. This leads to new messenger RNA and protein synthesis, and ultimately to an altered cellular functioning. Reporter genes can be made in which an estrogen-responsive promoter is linked to luciferase. These can be stably introduced in recipient cell lines. When a reporter gene was used with multiple copies of the estrogen responsive elements, and linked to a very minimal promoter and luciferase, an extremely responsive and sensitive cell line was obtained — the ER CALUX® line (Legler et al. 1999; Figure 5.1). This cell line has an EC50 for the main natural ligand 17-estradiol of 6 pM, while the limit of detection is as low as 0.5 pM, allowing precise quantification of estrogenicity of chemicals with low potency but high environmental prevalence (Legler et al. 1999). This assay is more sensitive and gives a better prediction of estrogenicity when compared with another reporter-gene system using yeast cells as a recipient, the so-called YES assay (Legler et al 2002a; Murk et al 2002).

Similarly, reporter-gene systems have been developed for all major classes of steroid receptors (Jausons-Loffreda et al. 1994; Schoonen et al 2000; Sonneveld et al. 2005) including CALUX systems, again using highly responsive and selective reporter genes. These CALUX reporter-gene systems have extremely low detection limits and EC50 values ranging from 3 pM to 500 pM (Sonneveld et al. 2005). Differences between the EC50 values of the assays are in line with known differences in the affinity of the receptors used for their cognate ligands. This set of lines will be integrated into one system to give an overview of the endocrine activity in a given sample. It can be expected that active research in this area, coupled with technological advances, will lead to the development of more *in vitro* bioassays that will address many different biological endpoints.

A very interesting and successful recent application of *in vitro* bioassays is their use as replacements for highly sophisticated chemical-analytical measurements such as gas chromatography/mass spectrometry (GC-MS) to

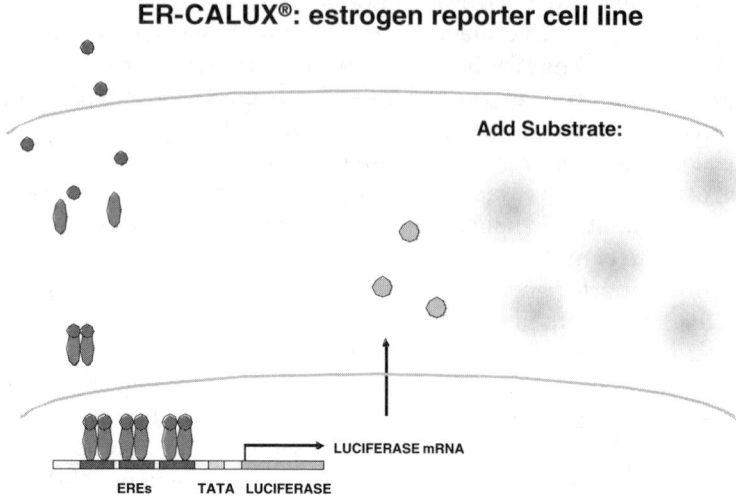

Figure 5.1 Principle of a reporter gene assay — the ER CALUX assay. Upon estrogen binding, the estrogen receptor (ER) becomes activated and binds to recognition sequences in promoter regions of target genes, the so-called estrogen responsive elements (EREs). Three of these EREs have been linked to a minimal promoter element (the TATA box) and the gene of an easily measurable protein (in this case luciferase). The thus-obtained reporter gene was stably introduced in T47D cells. In this way the ligand-activated receptor will activate luciferase transcription, and the transcribed luciferase protein will emit light when a substrate is added. The signal will dose-dependently increase as a result of increasing concentrations of ligand.

detect trace amounts of chemicals. Rather than measuring individual chemicals, these assays measure the net biological effect of receptor-interacting chemicals, thereby giving a better estimate of biological hazard when compared to chemical analysis. An example of a very successful bioassay in this area is the DR CALUX® assay that measures dioxin receptor-interacting compounds. The use of the DR CALUX bioassay for the screening of dioxins and related compounds in food and feed has been accepted in European Union (EU) legislation. Both DR CALUX assays (Behnish et al. 2002; Binderup et al. 2002; Hamers et al. 2000; Koppen et al. 2001; Nyman et al. 2003; Pauwels et al. 2001; Soechitram et al. 2003; Stronkhorst et al. 2002; Van der Heuvel et al. 2002; Vondracek et al. 2001) and ER CALUX assays (Hamers et al. 2003; Legler et al. 2002a, 2002b, 2003; Murk et al. 2002) have been successfully used to measure contamination of a wide variety of environmental matrices.

Transgenic animals

Transgenic animals would classify as *in vivo* bioassays, but because of their special nature are described separately. Two different molecular methods have been developed to modulate the genetic constitution of a number of animal species (called knock-out technologies) to remove or replace genes

from genomes and add genes through transgenesis. These ways to genetically modify animals have led to two basically different possibilities for
generating novel types of bioassays. First, replacement of structural genes
by mutated or inactive versions can lead to novel disease models in which
pharmaceutical and toxic compounds can be tested for their biological
effect. These models also include "humanized" animal models using organisms ranging from mice (Xie et al. 2002) to drosophila (Feany and Bender
2000), in which human genes are introduced that are absent in the animals
or have specific features that make them functionally distinct from their
animal counterparts. Second, marker or reporter genes are introduced,
allowing the sensitive and quantitative measurement of specific biological
processes that are normally difficult to access. In this way methods have
been developed to assess carcinogenicity of compounds more rapidly and
sensitively, avoiding unnecessary animal distress (Thorgeirsson et al. 2000;
Amanuma et al. 2000).

Recently, transgenic models have been developed in which the same
reporter gene was introduced as in the earlier-mentioned ER CALUX *in vitro*
bioassay. This was undertaken because of the concern that estrogenic chemicals may be particularly harmful to developing embryos (Colborn et al.
1993). No methods are available for measuring the activity of estrogen receptors in embryos, and it is uncertain which compounds can reach the embryo
in a biologically active form. Recently, estrogen-responsive reporter gene
expressing mice were generated to allow *in vivo* determination of estrogenicity, in particular with respect to transfer of estrogenic compounds such
as bisphenol A to the embryo. In these animals, noninvasive methods can
be used that allow measurement of luciferase activity (light production) in
intact living embryos, and more quantitative methods using homogenates
of tissues (Ciana et al. 2003; Lemmen et al. 2004).

Using an much more environmentally relevant model, the zebrafish, a
transgenic line has been generated in which rapid determinations of *in vivo*
estrogenicity of compounds present in the aquatic environment can be made
(Legler et al. 2000). With this model, estrogenicity can be determined at all
life stages. Comparison of the response in the zebrafish with the ER CALUX
assay demonstrated that the latter assay is more sensitive and unlikely to
generate false negatives, an essential requirement for an *in vitro* assay that
is to be used as a prescreen for *in vivo* assays. Relatively large quantitative
differences exist, however, between the *in vitro* and *in vivo* assay that seem
largely due to *in vivo* accumulation of lipophilic compounds and metabolism
(Legler et al. 2002b). This makes the transgenic model valuable to complement the *in vitro* tests for estrogenicity. Although this model can also be used
to detect chemical activities in environmental samples, vitellogenin, an
endogenous marker protein for estrogenicity, has been used more extensively
in studies using endemic but also laboratory species (Arukwe and Goksoyr
2003). Transgenic zebrafish strains have also been developed for other applications, including measurements of cadmium and dioxins, and mutational
analysis (Amanuma et al. 2000; Blechinger et al. 2001; Mattingly et al. 2001).

All these vertebrate models will prove to be invaluable for research purposes, providing detailed insight into mechanisms of toxicity. This novel insight can then be used to design simpler and preferably *in vitro* tests. Those replacing chronic tests and those using simple test organisms have great potential as integrative screening models, in which complex biological interactions are taken into account.

Even more simple organisms can be used to generate sentinel models for environmental monitoring. This can be exemplified by the recent generation of *Caenorhabditis elegans* strains using a stress-inducible reporter construct (Candido and Jones 1996), and the earlier-mentioned recombinant bacteria-expressing toxicant-responsive luciferase activity (Keane et al. 2002). Clearly, by varying the organism and reporter construct, specific combinations can be made that have distinct advantages for certain applications.

Biosensors

A biosensor is a combination of a biological recognition element with a physical or physicochemical transducer (reviewed in Brecht and Gauglitz 1995; Nice and Catimel 1999; Rogers 2000; Thevenot et al. 2001). It may be regarded as a specialized type of bioassay, designed for repeated use and online monitoring. Its transducer part converts the binding event of the analyte to the biological recognition element into a measurable signal. For this, binding should lead to a change at the transducer surface, providing a signal to which the transducer responds. In the example of the glucose biosensor, the enzyme glucose oxidase leads to conversion of glucose and oxygen to gluconic acid and hydrogen peroxide. While glucose itself does not generate a signal, a decrease in oxygen or an increase in the reaction products hydrogen peroxide and gluconic acid can do so when brought into the vicinity of a suitable transducer material (an oxygen, pH, or peroxide sensor respectively). Clearly, close proximity and often direct spatial contact between the recognition element and the electrochemical transduction sensor is essential in a biosensor. Through this design the electrochemical biosensor is a self-contained integrated device that can be used repeatedly, and that requires no additional processing steps (such as reagent addition) to be operational (Brecht and Gauglitz 1995; Thevenot et al. 2001). In recent years, a variety of biological recognition elements and transducers have been used in biosensors. Combining these basic elements using various coupling technologies, together with variations in the assay format and read-out, has led to an enormous number of biosensors in a very active field of research. Below is a brief review of some of the basic principles used.

Biological recognition elements

The sensitivity and specificity of a biosensor is determined to a large extent by the biological recognition element and its affinity to the analyte. Without proper biological recognition there is no way to discriminate between ligands. Several types of recognition elements are used, most notably antibodies and enzymes (Table 5.1).

Table 5.1 Major Classes of Components Used in Different Types of Biosensors

Components of Biosensors	
Biorecognition Element	**Physical Transducer**
Enzyme	Electrochemical
Antibody	Optical-electronic (SPR)
DNA	Optical
Receptor	Acoustic
Microorganism	Thermal
Eukaryotic cell[a]	Mass
Tissue[a]	

[a]Laboratory-confined prototypes only.

Enzymes were used in the first biosensors, and direct measurement of their conversion products with the transducing system generated relatively simple devices. These systems, however, tend to be suitable for measuring compounds that are present in relatively high concentrations, and by no means reach the extremely high sensitivity that is needed to measure most biologically active substances. The use of antibodies greatly expanded the range of analytes that can be measured. Again, direct coupling of the biorecognition element to the transducing system is a prerequisite in biosensors for allowing rapid measurements. This distinguishes them from other antibody-based technologies like ELISA and RIA, which use extensive washing procedures and much longer incubation periods.

Antibodies have also been used to couple bacteria to the sensor, while a second, labeled antibody is used to provide the signal to the transducer (Keane et al. 2002). In this case the microbe is not the biorecognition element, but the analyte. Several improvements and amplification steps have improved the sensitivity of the biosensors. In this way the detection limit of 2,4-D has been lowered almost five orders of magnitude using similar antibodies (Rogers 2000). The drawback of these improvements is that they tend to make the sensor technology and the handling more complex, reducing online applicability, and often also increase the time to measure. High sensitivity is needed, however, in systems to measure compounds interfering with major high-affinity biological receptor systems, like those used in the endocrine system. Using the receptors themselves, together with a relatively novel transducing system, surface plasmon resonance (SPR) sensitivity was reached in the range of 100 pM for binding of 17-estradiol to the estrogen receptor (Hock et al. 2002). It should be noted that although this sensitivity is high it still is about two orders of magnitude lower than that reached with reporter-gene systems in eukaryotic cells, such as the ER CALUX system (Legler et al. 1999). This relatively low sensitivity restricts the practical applicability of many biosensors, since detection of ligands interfering with high-affinity receptors (such as the estrogen and dioxin receptors) even now necessitates extraction and concentration methods when using the highly sensitive CALUX systems or GC-MS. Therefore, online measurement with current biosensors is not feasible. Enhancement of sensitivity (for example, by increasing affinity to the analyte) will be a critical factor in biosensor development. Unfortunately, high affinity to the analyte often is difficult to reach and when it is possible tends to reduce reversibility of the binding, decreasing the possibility of reusing the biosensor.

More recently, cells and whole organisms have been used as recognition elements in biosensors. An interesting use of bacteria for environmental monitoring was introduced through the generation of recombinant strains in which the response of bacteria to specific chemicals was used (Keane et al. 2002). Many bacteria have toxicant-responsive genes, the products of which are usually involved in detoxification of the inducing chemical. By fusing the toxicant-responsive regions of such genes to luciferase, bacterial strains can be generated that respond to specific chemicals with light production. Coating suitable sensors with such bacteria generates an interesting class of biosensors that can be used for online measurements such as bioremediation sites.

Whole eukaryotic cells can also be used to couple to transducing surfaces, such as poly-L-lysine (Stenger et al. 2001; Keusgen 2002). The most well-developed versions use neuronal cells and measure ligand-induced electrical signals generated by those cells. In this way, effects on integrated biological pathways downsteam from simple recognition elements can be measured for the first time. Currently, however, no biosensors in the strict sense of the word have been generated and the prototypes still are large, laboratory-bound, and are little more than miniaturized cell biological experiments.

Regardless of the type of biosensor, immobilization of the biorecognition element to the sensor surface is an essential and critical step. This step should be adapted to the kind of recognition element that allows efficient surface coating and preferably leaves the site of ligand recognition unmasked. Particularly when using biological receptors, extreme care should be taken to avoid inactivation and breakdown of these often extremely labile proteins.

Transducers

Many types of transducers, and variations thereof, are used in biosensors (Table 5.1). The most basic types often used in the established enzyme electrodes are the electrochemical (potentiometric, amperometric, or conductometric) type such as pH-sensitive and ion-selective electrodes. Other types of transducers are light-, heat-, or vibration-sensitive. Because of the generic nature of the signals to which the transducers are sensitive, great care should be taken to avoid nonspecific signals. The major means to circumvent such interference are close proximity and a high density of the recognition element at the sensor surface. Because of this, initial biosensors typically have low sensitivities and are subject to nonspecific interference. This latter problem can often be reduced by using a reference transducing system. In addition, modern technologies (such as microfabrication, optoelectronics, and electromechanical nanotechnology) have led to dramatic improvements in design, resulting in increased biosensor sensitivities by orders of magnitude (Hal 2002).

Biological endpoints

The current trend to shift from measuring single compounds using analytical methods toward measuring the effects of complex environmental mixtures using a biological read-out necessitates evaluation and definition of priority effects of ecotoxicological concern. The EU white paper on chemicals defines carcinogenicity, mutagenicity, and reproductive (CMR) toxicity, including developmental toxicity (European Commission 2001) as priority areas for concern. Other areas of concern are immunotoxicity and neurotoxicity. In reproductive toxicity, emphasis has currently been given to chemicals interfering with the nuclear hormone receptor systems activated by androgens, estrogens, and thyroid hormones. From the above it may be clear that current reporter-gene assays, and to a lesser extent biosensors, are suitable for measuring such receptor-mediated events. Some endpoints, like *in vivo* estrogenicity of compounds, show a good correlation with cognate receptor activation (van der Burg et al. [in preparation]). Other *in vitro* bioassays have been developed for acute cytotoxicity and mutagenicity, while models are also being created to predict environmental fate, pharmacokinetics, and metabolism (ECVAM Working Group on Chemicals 2002). However, not all of the relevant endpoints can be readily assessed with a simplified detection system, since there are no simple recognition elements for endpoints such as developmental toxicity, immunotoxicity, neurotoxicity, and more complex endocrine routes, hampering generation of *in vitro* detection systems. In ecotoxicology, another layer of complexity is the presence of multiple species that do not respond similarly to a given chemical. Here, it will be important to generate assays for sentinel species and whenever possible use knowledge of common, conserved routes of toxicity. In this process, more attention is needed to design integrative tests and combinations thereof, leading to a system that can be used for first-line chemical hazard identification and ecotoxicological and epidemiological studies.

Complementary and integrative technologies

To date, bioassays cover a spectrum of relevant toxicological endpoints, and it seems likely that most of the prioritary endpoints will be addressed by new assays in the near future. This will provide good screening tools for initial (tier 1) hazard identification. Adding another level of confidence while aiming to replace most animal experiments is a huge undertaking in which a large panel of assays must be addressed simultaneously. This will necessitate miniaturization, automatization, and a high level of data integration. In all of these areas, technological advance is very rapid, creating great opportunities for future developments. Rapid and efficient screening technologies (so called high-throughput technologies) undergo a major leap forward through huge investments, mainly by pharmaceutical companies, that aim at rapid screening of potential drug candidates from large chemical libraries. For this, miniaturization and robotics are being employed to scale up screening possibilities with bioassays. Another major area of advance is the use of spotted arrays of different gene probes with possible extensions in the biosensor area (McGlennen 2001). The amount of data generated through this approach makes the application of specialized bioinformatics increasingly important. An critical step in developing an integrated system of hazard identification will be the application of pattern- and pathway-recognition software. With such tools, integration of many data with lower specificity can lead to pattern recognition and through this to a much higher specificity. This is the method by which specificity is generated in many biological systems.

Validation and application

Although large amounts of resources have been directed by governments and industries towards development of biosensors, very few have so far come to practical application other than for research purposes (Rogers 2000). The few that have reached commercial application are usually enzyme electrodes used in clinical diagnostics, such as those used for glucose measurement in blood. Applications of *in vitro* bioassays outside the research area are also still limited. Although technical shortcomings (such as low sensitivity or specificity) may play a role for biosensors, another major reason is the huge step that any new analytical system must achieve before entering the market: validation. Validation brings no scientific or commercial merits, and is a major hurdle for academic groups or smaller companies who are often the driving force in the initial research phase that leads to a new system. There is also a large gap between the research phase and the actual market introduction, because the average time requirement for official validation (for example, as an alternative for animal experiments) is about five years (ECVAM Working Group on Chemicals 2002). It should be noted that validation of a method refers to the establishment of the relevance and reliability of the method for a particular purpose. Therefore, when a novel detection system seems suitable for different applications, introduction of a single biodetection system may require several different routes of validation. In this process it is generally advantageous when the system is a variation of an already validated system. If similarity is sufficient, a faster catch-up validation process is also sufficient (ECVAM Working Group on Chemicals 2002). Therefore, the great variation in format of bioassays and biosensors is a handicap at this phase of development.

Future perspectives

Modern molecular and cell biology, nanotechnologies, and bioinformation technologies have led to powerful new bioanalytical tools. These are able to successfully compete with chemical-analytical methods and whole-animal experiments, and can already provide information that stretches beyond that obtained with competing methods. Although the introduction of these *in vitro* methods as alternatives for classical tests is a slow process, biological detection has found its way into a number of applications. Because the level of integration of a bioassay lies between that of a chemical determination and a whole-animal experiment, it can be expected that bioassays and biosensors will claim this central place in a much more prominent manner in the near future (Figure 5.2). Simple biosensors have a promising future for rapid online measurements, but have a relatively low sensitivity compared with bioassays using whole cells. For broader application and higher selectivity, arrays of biosensors seem a promising way to go. Therefore, miniaturization, automatization, and integration are the keys for successful new developments. Integration can be generated with arrays of systems followed by extensive bioinformatics. However, this process also needs a high level of integration that is already present in relevant *in vitro* cell culture systems. Because of this, it is essential to continue to develop biologically relevant and innovative cell culture systems. In this process a merge of cell culture and biosensor technologies can be expected. The aim in any of the fields of application of these model systems is to give a rapid but reliable prediction of pharmacological or ecotoxicological effects. Of course, this task to recapitulate biology in a nutshell is infinitely complex, leading to a never-ending process of constant improvement. This is, however, not different from current

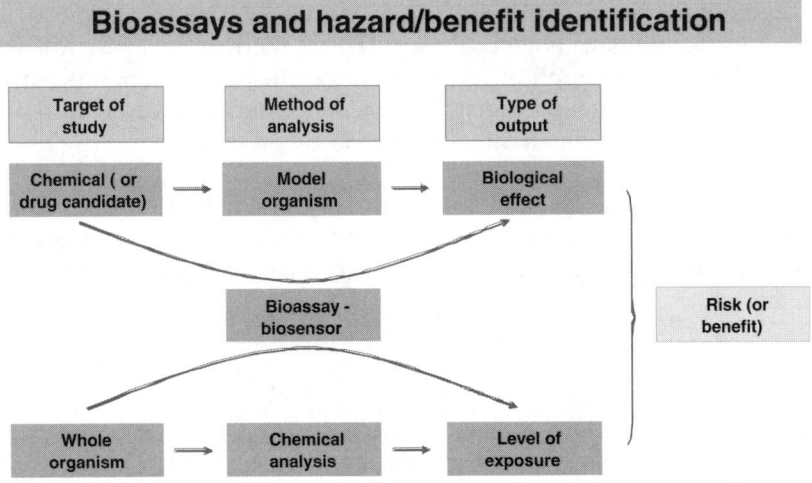

Figure 5.2 Bioassays and biosensors and hazard/benefit identification. Currently, determination of risk (or benefit in case of a pharmaceutical) is determined through analysis of the biological effect (either harmful of beneficial) of the chemical in a model organism. In addition, the level of exposure is determined through chemical analysis in whole organisms or ecosystems. Together, risk (in relation to benefit in case of a drug candidate) is assessed. Through combining the characteristics of an analytical instrument (such as small size, specificity, and sensitivity) and biological relevance, biosensors and bioassays are expected to play an increasingly central role in risk-benefit assessment of chemicals, including pharmaceuticals.

methods; using animal models and chemical analysis we expect that choosing *in vitro* bioassays and biosensors will lead to major advances in analytic power, and will protect humans, animals, plants, and the environment. A reductionalist approach is the only way to incorporate new knowledge and to generate new insights that aim to steer processes in biological systems. This approach will lead to new ways of modulating those systems in a pharmacological manner, and to new insights on how to protect the systems. Clearly, these integrated efforts will require multidisciplinary approaches, technological advances, and above all insight into biological systems.

Summary

Biosensors and bioassays other than the classical invertebrate assays are gradually claiming a prominent place in ecotoxicological monitoring strategies. Modern bioassays also provide alternatives for chemical-analytical monitoring, using the biological effect itself as an analytical tool. Three major developments have greatly speeded up the introduction of bioanalytical tools. First, there is an awareness of the environmental spread of an ever-increasing number of chemicals and their metabolites, albeit at relatively low individual levels. This plethora of chemicals hugely increases the possibility of combined effects at the same biological endpoint, thereby causing environmental problems that escape chemical-analytical methods. Second, there has been a rapid advance in the technology that allows using biological endpoints as analytical tools. Third, the new bioanalytical tools have a wide range of applications because they measure endpoints that are not accessible with chemical-analytical methods, and they can help replace or reduce animal experimentation in pharmacology, toxicology, drug discovery, and so on.

An overview was given in the chapter of the different types of bioanalytical tools and their applications, including recently developed laboratory tools that can be used to measure interference with a number of hormonal systems.

References

Amanuma, K., Takeda, H., Amanuma, H., Aoki, Y., 2000. Transgenic zebrafish for detecting mutations caused by compounds in aquatic environments. *Nat. Biotechnol.* 18, 62–65.

Arukwe, A. and Goksoyr, A. 2003. Eggshell and egg yolk proteins in fish: hepatic proteins for the next generation: oogenetic, population, and evolutionary implications of endocrine disruption. *Comp. Hepatology* 2, 1–21.

Ashby, J., 2001. Increasing the sensitivity of the rodent uterotrophic assay to estrogens, with particular reference to bisphenol A. *Environ. Health Perspect.* 109, 1091–4. Review.

Balaguer, P., Francois, F., Comunale, F., Fenet, H., Boussioux, A.M., Pons, M., Nicolas, J.C., Casellas, C., 1999. Reporter cell lines to study the estrogenic effects of xenoestrogens. *Sci. Total Environ.* 233 (1-3), 47–56.

Behnisch, P.A., Hosoe, K., Brouwer, A., Sakai, S., 2002. Screening of dioxin-like toxicity equivalents for various matrices with wild type and recombinant rat hepatoma H4IIE cells. *Toxicol. Sci.* 69, 125–130.

Binderup, M.L., Pedersen, G.A., Vinggaard, A.M., Rasmussen, E.S., Rosenquist, H., Cederberg, T., 2002. Toxicity testing and chemical analyses of recycled fibre-based paper for food contact. *Food Additives Contaminants* 19, Supplement, 13–28.

Blechinger, S.R., Warren, J.T., Jr., Kuwada, J.Y., Krone, P.H., 2001. Developmental toxicology of cadmium in living embryos of a stable transgenic zebrafish line. *Environ. Health Perspect.* 110, 1041–104.

Brecht, A. and Gauglitz, G., 1995. Optical probes and transducers. *Biosensors Bioelectronics* 10, 923-936.

Candido, E.P. and Jones, D., 1996. Transgenic *Caenorhabditis elegans* strains as biosensors. *Trends Biotechnol.* 14, 125–129.

Ciana, P., Raviscioni, M., Mussi, P., Vegeto, E., Que, I., Parker, M.G., Lowik, C., Maggi, A., 2003. *In vivo* imaging of transcriptionally active estrogen receptors. *Nat. Med.* 9, 82–86.

Clark, L.C. and Lyons, C., 1962. Electrode systems for continuous monitoring in vascular surgery. *Ann. N.Y. Acad. Sci.* 102, 29–45.

Colborn, T., vom Saal, F.S., Soto, A.M., 1993. Developmental effects of endocrine-disrupting chemicals in wildlife and humans. *Environ. Health Perspect.* 101, 378–84.

ECVAM Working Group on Chemicals, 2002. Alternative (non-animal) methods for chemical testing: current status and future prospects. *Am. Theological Libr. Assoc.* 30, Supplement 1, 1–125.

European Commission, 2001. *EU white paper: strategy for a future chemicals policy,* Brussels.

Feany, M.B. and Bender, W.W., 2000. A Drosophila model of Parkinson's disease. *Nature* 404, 394–398.

Hall, R.H., 2002. Biosensor technologies for detecting microbiological foodborne hazards. *Microb. Infect.* 4, 425–432.

Hamers, T., van Schaardenburg, M.D. Felzel, E.C., Murk, A.J., Koeman, J.H., 2000. The application of reporter gene assays for the determination of the toxic potency of diffuse air pollution. *Sci. Total Environ.* 262, 159–174.

Hamers, T., van den Brink, P.J., Mos, L., van der Linden, S.C., Legler, J., Koeman, J.H., Murk, A.J., 2003. Estrogenic and esterase-inhibiting potency in rainwater in relation to pesticide concentrations, sampling season and location. *Environ. Pollut.* 123, 47–65.

Hock, B., Seifert, M., Kramer, K., 2002. Engineering receptors and antibodies for biosensors. *Biosensors Bioelectronics* 17, 239–49.

Hyne, R.V. and Everett, D.A., 1998. Application of a benthic euryhaline amphipod, *Corophium sp.*, as a sediment toxicity testing organism for both freshwater and estuarine systems. *Arch. Environ. Contamination Toxicol.* 34, 26-33.

Jausons-Loffreda, N., Balaguer, P., Roux, S., Fuentes, M., Pons, M., Nicolas, J.C., Gelmini, S., Pazzagli, M., 1994. Chimeric receptors as a tool for luminescent measurement of biological activities of steroid hormones. *J. Bioluminescence Chemiluminescence* 9 (3), 217–21.

Johnston, P.A. and Johnston, P.A., 2002. Cellular platforms for HTS: three case studies. *Drug Discovery Today* 7(6), 353–63.

Keane, A., Phoenix, P., Ghoshal, S., Lau, P.C.K., 2002. Exposing culprit organic pollutants: a review. *J. Microbiol. Methods* 49, 103–119.

Keddy, C.J., Greene, J.C. and Bonnell, M.A., 1995. Review of whole-organism bioassays: soil, freshwater sediment, and freshwater assessment in Canada. *Ecotoxicol. Environ. Safety* 30, 221–251.

Keusgen, M., 2002. Biosensors: new approaches in drug discovery. *Naturwissenschaften* 89, 433–444.

Koppen, G., Covaci, A., Van Cleuvenbergen, R., Schepens, P., Winneke, G., Nelen, V., Schoeters, G., 2001. Comparison of CALUX-TEQ values with PCB and PCDD/F measurements in human serum of the Flanders Environmental and Health Study (FLEHS). *Toxicol. Lett.* 123, 59–67.

Legler, J., Van den Brink, C.E., Brouwer, A., Murk, A.J., Van der Saag, P.T., Vethaak, A.D., van der Burg, B., (1999). Development of a stably transfected estrogen receptor-mediated luciferase reporter gene assay in the human T47-D breast cancer cell line. *Toxicol. Sci.* 48, 55–66.

Legler, J., Broekhof, J.L.M., Brouwer, A., Lanser, P.H., Murk, A.J., Van der Saag, P.T., Vethaak, A.D., Wester, P. Zivkovic, D., van der Burg, B. (2000) A novel *in vivo* bioassay for (xeno)estrogens using transgenic zebrafish. *Environ. Sci. Technol.* 34, 4439–4444.

Legler, J., Dennekamp, M., Vethaak, A.D., Brouwer, A., Koeman, J.H., van der Burg, B., Murk, A.J., 2002a. Detection of estrogenic activity in sediment-associated compounds using *in vitro* reporter gene assays. *Sci. Total Environ.* 293, 69–83.

Legler, J., Zeinstra, L.M., Schuitemaker, F., Lanser, P., Bogerd, J., Brouwer, A., Vethaak, A.D., De Voogt, P., Murk, A.J., van der Burg, B., 2002b. Comparison of *in vivo* and *in vitro* reporter gene assays for short-term screening of estrogenic activity. *Environ. Sci. Technol.* 36, 4410–4415.

Legler, J., Jonas, A., Lahr, J., Vethaak, A.D., Brouwer, A., Murk, A.J., 2002c. Biological measurement of estrogenic activity in urine and bile conjugates with the *in vitro* ER CALUX reporter gene assay. *Environ. Toxicol. Chem.* 21, 473–479.

Legler, J., Leonards, P., Spenkelink, A., Murk, A.J., 2003. *In vitro* biomonitoring in polar extracts of solid phase matrices reveals the presence of unknown compounds with estrogenic activity. *Ecotoxicology* 12(1-4), 239–249.

Lemmen, J.G., Arends, R.J., Van Boxtel, A.L., van der Saag, P.T., van der Burg, B. (2004) Tissue- and time-dependent estrogen receptor activation in estrogen reporter mice. *J. Mol. Endocrinol.* 32, 689–701.

Mattingly, C.J., McLachlan, J.A., Toscano, W.A., Jr., 2001. Green fluorescent protein (GFP) as a marker of aryl hydrocarbon receptor (AhR) function in developing zebrafish (*Danio rerio*). *Environ. Health Perspect.* 109, 845–849.

McGlennen, R.C., 2001. Miniaturization technologies for molecular diagnostics. *Clin. Chem.* 47, 393–402.

Murk, A.J., Legler, J., van Lipzig, M.M., Meerman, J.H., Belfroid, A.C., Spenkelink, A., van der Burg, B., Rijs, G.B., Vethaak, D., 2002. Detection of estrogenic potency in wastewater and surface water with three *in vitro* bioassays. *Environ. Toxicol. Chem.* 21, 16–23.

Naylor, L.H., 1999. Reporter gene technology: the future looks bright. *Biochem. Pharmacol.* 58(5), 749–57.

Nice, E.C. and Catimel, B., 1999. Instrumental biosensors: new perspectives for the analysis of biomolecular interactions. *Bioessays* 21, 339–352.

Nyman, M., Bergknut, M., Fant, M.L., Raunio, H., Jestoi, M., Bengs, C., Murk, A., Koistinen, J., Backman, C., Pelkonen, O., Tysklind, M., Hirvi, T., Helle, E., 2003. Contaminant exposure and effects in Baltic ringed and grey seals as assessed by biomarkers. *Mar. Environ. Res.* 55(1), 73-99.

Pauwels, A., Schepens, P.J., D'Hooghe, T., Delbeke, L., Dhont, M., Brouwer, A., Weyler, J., 2001. The risk of endometriosis and exposure to dioxins and polychlorinated biphenyls: a case-control study of infertile women. *Hum. Reprod.* 16, 2050-2055.

Rawash, I.A., Gaaboub, I.A., El-Gayar, E.M., El-Shazli, A.Y., 1975. Standard curves for nuvacron, malathion, sevin, DDT and kelthane tested against the mosquito *Culex pipiens L.* and the microcrustacean *Daphnia magna (Straus)*. *Toxicology* 4, 133-144.

Rizzino, A., 2002. Embryonic stem cells provide a powerful and versatile model system. *Vitam. Horm.* 64, 1–42.

Rogers, K.R., 2000. Principles of affinity-based biosensors. *Mol. Biotech.* 14, 109–129.

Schoonen, W.G., Deckers, G., de Gooijer, M.E., de Ries, R., Mathijssen-Mommers, G., Hamersma, H., Kloosterboer, H.J., 2000. Contraceptive progestins. various 11-substituents combined with four 17-substituents: 17alpha-ethynyl, five-and six-membered spiromethylene ethers or six-membered spiromethylene lactones. *J. Steroid. Biochem. Mol. Biol.* 74(3), 109–23.

Soechitram, S.D., Chan, S.M., Nelson, E.A., Brouwer, A., Sauer, P.J., 2003. Comparison of dioxin and PCB concentrations in human breast milk samples from Hong Kong and the Netherlands. *Food Additives Contaminants* 20, 65–99.

Sonneveld, E., Jansen, H.J., Riteco, J.A.C., Brouwer, A., van der Burg, B. (2005) Development of androgen- and estrogen-responsive bioassays, members of a panel of human cell line-based highly selective steroid responsive bioassays. *Toxicol. Sci.* 83, 136–148.

Soto, A.M., Sonnenschein, C., Chung, K.L., Fernandez, M.F., Olea, N., Serrano, F.O., 1995. The E-SCREEN assay as a tool to identify estrogens: an update on estrogenic environmental pollutants. *Environ. Health Perspect* 103, Supplement 7, 113–122.

Stenger, D.A., Gross, G.W., Keefer, E.W., Shaffer, K.M., Andreadis, J.D., Ma, W., Pancrazio, J.J., 2001. Detection of physiologically active compounds using cell-based biosensors. *Trends Biotechnol.* 19:304–309.

Stronkhorst, J., Leonards, P., Murk, A.J., 2002. Using the dioxin receptor CALUX *in vitro* bioassay to screen marine harbor sediments for compounds with a dioxin-like mode of action. *Environ. Toxicol. Chem.* 21, 2552–2561.

Terouanne, B., Tahiri, B., Georget, V., Belon, C., Poujol, N., Avances, C., Orio, F., Balaguer, P. and Sultan, C., 2000. A stable prostatic bioluminescent cell line to investigate androgen and antiandrogen effects. *Mol. Cell. Endocrinol.* 160, 39–49.

Thevenot, D.R., Toth, K., Durst, R.A., Wilson, G.S., 2001. Electrochemical biosensors: recommended definitions and classification. *Biosensors Bioelectronics* 1-2, 121–131.

Thorgeirsson, S.S., Factor, V.M., Snyderwine, E.G., 2000. Transgenic mouse models in carcinogenesis research and testing. *Toxicol. Lett.* 112-113, 553–555.

Van Den Heuvel, R.L., Koppen, G., Staessen, J.A., Hond, E.D., Verheyen, G., Nawrot, T.S., Roels, H.A., Vlietinck, R., Schoeters, G.E., 2002. Immunologic biomarkers in relation to exposure markers of PCBs and dioxins in Flemish adolescents (Belgium). *Environ. Health Perspect.* 110, 595–600.

Vondracek, J., Machala, M., Minksova, K., Blaha, L., Murk, A.J., Kozubik, A., Hofmanova, J., Hilscherova, K., Ulrich, R., Ciganek, M., Neca, J., Svrckova, D., Holoubek, I., 2001. Monitoring river sediments contaminated predominantly with polyaromatic hydrocarbons by chemical and *in vitro* bioassay techniques. *Environ. Toxicol. Chem.* 20, 1499–506.

Xie, W., Barwick, J.L., Downes, M., Blumberg, B., Simon, C.M., Nelson, M.C., Neuschwander-Tetri, B.A., Brunt, E.M., Guzelian, P.S., Evans, R.M., 2002. Humanized xenobiotic response in mice expressing nuclear receptor SXR. *Nature* 406, 435–439.

chapter six

Satellite remote sensing in marine ecosystem assessments

T.R. Pritchard and K. Koop

Contents

Introduction 196
Background 196
History and relevance of ocean color 196
    Key satellite-mounted sensors 197
    Ocean color products 199
    Chlorophyll and primary productivity 200
    Optically complex coastal waters (Case 2 waters) 201
Environmental issues and applications 202
    Global scale phenomena: biogeochemical cycles, climate change,
        and El Niño southern oscillation 203
    Regional seas: mesoscale processes and biological variability 208
    Coastal zones: human activity and ecosystem health 211
        Water quality 211
        Algal blooms 213
Fisheries 213
Case study: marine algal blooms in coastal waters off southeast
    Australia 215
    Management issues 215
    Developing a predictive understanding using remote sensed data . 216
    Noctiluca bloom: January 1998 217
    Trichodesmium bloom: March and April 1998 220
Conclusions 222
Acknowledgements 222
References 223

Introduction

Remote sensing technologies range from small-scale, high-frequency devices such as towed video plankton recorders (Davis et al. 1992) to satellite-mounted sensor arrays providing global estimates of primary production (Joint and Groom 2000). This chapter describes a range of applications of satellite-sensed data, especially ocean color and sea surface temperature products, to illustrate how they can be used to develop an understanding of ecosystems and human impacts on them. Global, regional, and local applications are summarized after which a more detailed case study is presented to illustrate how ocean color technology can be employed to develop a predictive understanding of algal bloom development and associated issues in the coastal waters of New South Wales, Australia.

Satellite-borne ocean color products have improved in recent years and many are freely available, so with increased personal computer processing power, applications now fall within the reach of a vast number of potential users.

Background

The world's immense human population exerts profound stresses on aquatic ecosystems at all scales. Direct impacts occur through catchment runoff, discharge of wastes, atmospheric deposition of pollutants, overexploitation, and habitat modification. Further, insidious impacts include the spread of introduced species and manifestations of global warming. Monitoring, predicting, and managing changes within coastal ecosystems are clearly important; remote sensing technologies provide unsurpassed spatial coverage with ever-increasing spatial, temporal, and spectral resolutions to help address these issues.

Although this chapter deals with remote sensing and information technologies that are fast evolving, the type of information needed for assessment and management of aquatic ecosystems remains essentially the same.

History and relevance of ocean color

The color of the ocean can indicate levels of phytoplankton activity. To the casual observer, the color of seawater may vary from the dark green of eutrophic estuarine waters to the deep blue of oligotrophic oceanic waters. Coastal water colorations, however, are often complex with various hues of gray, brown, and yellow due to terrigenous influences such as estuarine plumes, anthropogenic discharges, resuspended sediments, and the presence of dissolved organic substances.

Shipboard and aircraft studies first showed that radiance upwelling from the ocean in the visible region (400 to 700 nm) was related to the concentration of chlorophyll and other plant pigments.

Following this, the first satellite-borne ocean color sensor — the Coastal Zone Color Scanner (CZCS) — was launched in 1978 as a one-year "proof-of-concept" mission. Despite this, CZCS delivered ocean color data for eight years and led to the development of algorithms to estimate primary productivity in our surface oceans (Platt and Sathyendranath 1988). Data from CZCS revolutionized the understanding of phytoplankton distributions and dynamics at a global scale and in many coastal systems (Shannon 1985). Remote sensing provided a synoptic view of large zonal structures that had been overlooked in field studies and ignored in mathematical models because time and length scales were not easily detected by classical field investigations (Nihoul 1984).

After a hiatus of nearly a decade, new ocean color sensors were launched in the middle and late 1990s in response to the need to quantify the carbon cycle, and motivated by increasing concerns about climate change and an appreciation of interactions between climate effects and marine ecosystems.

Key satellite-mounted sensors

Present, future, and past ocean color scanners are summarized in Table 6.1. Information is updated by the International Ocean Color Ocean Coordination Group (IOCCG) at http://www.ioccg.org/sensors/500m.html.

The principal source of published ocean color data presented or referred to in this chapter is the sea-viewing wide field-of-view sensor (SeaWiFS). SeaWiFS was launched in 1997 as the operational successor to the CZCS and was one of the first of a new generation of ocean color satellites (Hooker and McClain 2000; Acker et al. 2002). Much of the processing, quality control, and initial analysis of SeaWiFS data in this chapter were undertaken using the SeaWiFS Data Analysis System (SeaDAS) software (freely available from http://seadas.gsfc.nasa.gov).

Analysis and interpretation of ocean color data is often supported by data from the advanced very-high-resolution radiometers (AVHRRs) aboard the U.S. National Oceanographic and Atmospheric Administration (NOAA) series of satellites. AVHRR scanners deliver four to five channels (depending on the model), including visible and sea surface temperature (SST) images at spatial resolutions comparable to most satellite-borne ocean scanner data (Hastings and Emery 1992). Successive satellites have resulted in a time series of AVHRR data back to 1986.

The launch of the moderate resolution imaging spectroradiometer (MODIS) in December 1999 represented a further leap in ocean color capability compared to SeaWiFS, with more wave bands, higher signal-to-noise ratio, more complex on-board calibration, and the capability of simultaneous observations of ocean color and sea surface temperature (Joint and Groom 2000). MODIS provides global coverage every one to two days. The U.S. National Aeronautics and Space Administration (NASA) provides free and open access to MODIS data, including access to merged data products (Sea-WiFS/MODIS; see http://modis.gsfc.nasa.gov.

Table 6.1 Satellite Mounted Ccean Colour Sensors

Sensor	Agency	Satellite	Launch Date	Swath (km)	Resolution (m)	Number of Bands	Spectral Coverage (nm)
Current Sensors							
COCTS	CNSA (China)	HaiYang-1 (China)	15/05/02	1400	1100	10	402–12500
MERIS	ESA (Europe)	ENVISAT-1(Europe)	01/03/02	1150	300/1200	15	412–1050
MODIS-Aqua	NASA (USA)	Aqua (EOS-PM1)	04/05/02	2330	1000	36	405–14385
MODIS-Terra	NASA (USA)	Terra (USA)	18/12/99	2330	1000	36	405–14385
OCI	NEC (Japan)	ROCSAT-1 (Taiwan)	27/01/99	690	825	6	433–12500
OCM	ISRO (India)	IRS-P4 (India)	26/05/99	1420	350	8	402–885
OSMI	KARI (Korea)	KOMPSAT (Korea)	20/12/99	800	850	6	400–900
SeaWiFS	NASA (USA)	OrbView-2 (USA)	01/08/97	2806	1100	8	402–885
Future Sensors							
S-GLI	NASDA (Japan)	GCOM (Japan)	2007	1600	750	11	412–865
VIIRS	NASA/IPO	NPP	2006	3000	370/740	22	402–11800
VIIRS	NASA/IPO	NPOESS	2009	3000	370/740	22	402–11800
OCM-II	ISRO (India)	IRS-P7 (India)	2005/06	—	—	—	—
KGOCI[a]	Korea	—	2008	3000	500	8	400–865
Past Sensors							
CMODIS	CNSA (China)	Shen Zhou-3 (China)	25/03/02—15/9/02	—	400	34	403–12500
CZCS	NASA (USA)	Nimbus-7 (USA)	24/10/78—22/06/86	1556	825	6	433–12500
CZI	CNSA (China)	HaiYang-1 (China)	15/05/02—1/12/03	500	250	4	420–890
GLI	NASDA (Japan)	ADEOS-II (Japan)	14/12/02—25/10/03	1600	250/1000	36	375–12500
MOS	DLR (Germany)	IRS P3 (India)	21/03/96—early 04	200	500	18	408–1600

Source: International Ocean Color Ocean Coordination Group at http://www.ioccg.org/sensors/500m.html.
[a]KGOCI will be in geostationary orbit. All others are in polar orbits with typical revisit times of 2 to 3 days.

The MODIS sensors, together with the European medium resolution imaging spectrometer (MERIS) launched in March 2002, and the Chinese moderate resolution imaging spectroradiometer (CMODIS) launched in May 2002, provide increased coverage with correspondingly greater opportunities to capture short-duration events.

Ocean color products

Ocean color sensors capture light scattered by the atmosphere and reflected from the sea surface as well as the light radiating from surface waters of the ocean. It is this "water leaving radiance" that carries ecologically important signals. Ocean color algorithms extract this signal and deliver various ocean color products such as those listed in Table 6.2 (derived from Parslow et al. 2000).

Various texts describe the optical properties of ocean and coastal waters and provide the theoretical basis for extracting signals of biological significance (Bukata et al. 1995; Kirk 1994; Mobley 1994).

Satellite-mounted sensors have clear advantages over direct *in situ* observations, but also suffer from some critical limitations mainly due to limited

Table 6.2 Remote Sensed Products

Chlor	Chlorophyll fluorescence as a measure of phytoplankton biomass
ProductionW	Water column primary production using photosynthesis-irradiance relationships, although suspended solids and dissolved organic matter in coastal waters may confound estimates of light attenuation (which is required together with chlorophyll-a and surface irradiance to calculate primary production)
Light	Light attenuation and water color resulting from organic biomass (chlorophyll and other pigments), dissolved substances (yellow), and mineral particles
Pigment/type	Pigment composition and bloom type based on differences in absorption spectra (and perhaps back-scattering spectra) across algal classes
SS	Suspended sediments (particle back-scattering)
Yellow	Yellow substances (colored dissolved organic matter)
Dynamics	Physical dynamics using reflecting optical properties (ocean color) of the upper layer, which are considered better than infrared imagery
Habitat	Bottom depth, benthic reflectance, and habitat for optically shallow coastal waters (using hyperspectral sensor)
ProductionB	Benthic primary production may be derived from bottom light intensity (derived from surface irradiance and attenuation coefficients) and plant biomass distributions

Note: Product identifiers relate to Table 6.3.

light penetration and noise acquired as the signal passes through the water and atmosphere to the satellite.

Cloud cover fundamentally limits the areal extent of coverage, although this can be minimized by extrapolation over time and space through modeling (Aiken et al. 1992) and, in some cases, by compositing successive images if features change slowly with respect to successive or complementary overpasses. Sun glint can also obscure the signal (Lockhart 1994) although optimizing the aspect of the sensor and careful analysis (such as appropriate stray light thresholds) can reduce this.

Another fundamental limitation is limited light penetration through water, which restricts vertical coverage. Ocean color sensors receive radiance from the optical depth (depth of light penetration), which is related to the visible depth. Optical depth ranges from more than 20 m in oligotrophic tropical oceans to 5 to 10 m in typical mesotrophic conditions, and can be as little as 1 to 2 m in high-concentration phytoplankton blooms or sediment-laden waters (Aiken et al. 1992). This can be a critical limitation for subsurface chlorophyll maxima.

Other confounding factors relate to the effects of the water and the atmosphere through which the signal passes. Algorithms must account for the bulk optical properties of the upper water column in order to extract relevant ocean color products (Bukata et al. 1995), and optical effects due to gases and aerosols in the atmosphere must be addressed (Joint and Groom 2000).

The development of inverse modeling techniques for the interpretation of ocean color measurements is an ongoing process. Ground truth data are required to better quantify confidence limits for ocean color products, especially for coastal applications including benthic mapping.

Recognition of these limitations of satellite-borne ocean color data and the need for integrated assessments has led to emphatic recommendations for remote sensing to complement rather than entirely replace *in situ* observations (IOCCG 2000).

Chlorophyll and primary productivity

Ocean color sensors were primarily developed for their potential to monitor chlorophyll and primary production. In general, chlorophyll-a can be measured more accurately *in situ* than from space (Engelsen et al. 2002) but remotely mounted sensors provide synoptic coverage over unparalleled spatial scales and at frequencies unobtainable by any other sampling procedure.

Chlorophyll pigments are among the principal ocean colorants, but estimates of chlorophyll concentrations from satellite data are subject to the nonuniform distribution of chlorophyll concentration with depth. Furthermore, the nonlinear relationship between photosynthetic primary production and photosynthetically available radiance can confound estimations of primary productivity.

Despite these problems, good estimates of open-ocean primary production can be obtained and it is possible to estimate phytoplankton primary production for coastal waters by using algorithms that take local water characteristics into account (Bukata et al. 1995). Standard algorithms for estimating water column primary production are based on photosynthesis-irradiance relationships that rely on remote sensed chlorophyll-a, light attenuation, and estimated surface irradiance. These estimates of primary production are extremely sensitive to light attenuation by substances other than phytoplankton (Platt et al. 1988), which can be problematic in coastal waters where high levels of suspended sediments and dissolved organic matter may be present. Furthermore, remotely sensed surface chlorophyll concentrations must be extrapolated to vertical chlorophyll profiles in order to estimate primary production. Historical *in situ* data, supplementary sea surface temperature data, or physical modeling of mixed layer depths are usually used to extrapolate to chlorophyll profiles (Parslow et al. 2000).

Optically complex coastal waters (Case 2 waters)

Initial applications of ocean color data focused on open ocean systems (case 1 Waters) but with improved sensors, interest has focused on applications in coastal waters that are optically more complex (Case 2 Waters).

Unfortunately, the degree of optical complexity of a natural water body is, in general, directly related to its proximity to land masses (Bukata et al. 1995). In particular, coastal waters contain a variety of absorbing and scattering centers due to distributions of dissolved organic matter, suspended matter, and air bubbles. Algorithms continue to be developed to improve both atmospheric corrections and chlorophyll-a estimates for Case 2 waters. For instance, early atmospheric correction algorithms for open ocean (case 1) waters assumed zero water leaving radiance from red or near-infrared wavelengths; these wavebands were used together with a prescribed aerosol reflectance spectrum to extrapolate and remove aerosol effects. However, the assumption of negligible near-infrared water leaving radiance breaks down for Case 2 waters. Additional wave bands and new algorithms have overcome some of these added complexities (Ruddick et al., 2000), but further room remains for improvements.

The IOCCG reviewed algorithm development for Case 2 waters (IOCCG 2000). The limited number of wavebands on CZCS did not allow the development of elaborate multiwaveband algorithms required for optically complex coastal waters. Significant advances have been made with the advent of the latest generation of satellite-mounted ocean color sensors and associated algorithm development. However, quantitative remote sensing of Case 2 waters will remain challenging because it is fundamentally a multivariable, nonlinear problem. Accuracy of remotely sensed products will improve as the inherent optical properties of coastal waters are better understood. The development of inverse modeling techniques for coastal regions requires precise multispectral radiances, with contemporary optical and concentration measurements of the water constituents (Doerffer et al. 1999). IOCCG (2000) identified a general trend in Case 2 algorithm approaches toward model-based techniques based on the first principles of ocean optics rather than on purely empirical approaches. Regional algorithms, optimized for local conditions, were found to perform well when compared with global algorithms. Considerable scope exists for integration of regional or special-case algorithms within an overarching branching algorithm.

The IOCCG has emphasized a need for further work to ensure that error information is routinely available to avoid inappropriate application of remotely sensed data. The accuracy and precision of remote sensed products varies with conditions and concentrations, due to the nonlinearity of the system and the extreme ranges in the concentrations of individual components that contribute to ocean color. Error estimates can be obtained from sensitivity analysis (models) and comparisons with *in situ* data, recognizing that there may be a mismatch in temporal and spatial scales of *in situ* data.

Environmental issues and applications

Satellite ocean color imagery can provide cause-and-effect indicators at appropriate time and space scales for assessment and management of coastal systems (Parslow et al. 2000). Satellite-mounted ocean color sensors provide complete global coverage, unencumbered by political and military sensitivities that can limit other observing systems, such as aerial photography. Potential and actual applications of ocean color products have been categorized by issue or sector; see Table 6.3. The focus in this chapter is on the top five issues in Table 6.3, because relevant ocean color products are well established and freely available (such as MODIS and research applications using SeaWiFS). Published applications of data from more recent satellite scanners such as COCTS, MERIS, and MODIS-aqua are less numerous than those from SeaWiFS, although recognized applications are equally varied (Doerffer et al. 1999).

Benthic habitat mapping requires spatial and spectral resolutions typically restricted to commercial airborne scanners and experimental satellite-mounted hyperspectral scanners, which are beyond the scope of this chapter. Green et al. (2000) provides general practical guidance on reliability, accuracy, and cost of a wide range of remote sensing products, including habitat mapping with a focus on tropical coastal management.

The examples that follow serve to illustrate the spectrum of existing and potential applications of remote sensed ocean color data. The following applications are considered: at the global scale (hundreds to thousands of kilometers), where emphasis has been on climate change and biogeochemical cycles; at the scale of regional seas (many tens to hundreds of kilometers), where mesoscale systems and processes have been investigated; and within the coastal zone (scales of several to many tens of kilometers), where the effects of human activity on ecosystem health are often most apparent.

Global scale phenomena: biogeochemical cycles, climate change, and El Niño southern oscillation

Early CZCS data revealed significant differences between northern and southern hemispheres. In the northern regions spring blooms dominated distributions of chlorophyll concentration; in the southern ocean, currents and prevailing winds were the dominant factors explaining chlorophyll concentrations (Harris et al. 1993). A comprehensive reanalysis of CZCS data with improved algorithms incorporating *in situ* data now permits quantitative analysis of trends in global ocean chlorophyll spanning two decades (Gregg et al., 2002). CZCS data (1979 to 1986) have been reprocessed for comparison with SeaWiFS data (September 1997 to the present) using the same algorithms (Antoine et al., 2003; data available at http://www.rsmas.miami.edu/groups/rrsl/lpcm-seawifs-CZCS).

The oceans contain approximately 85% of the carbon circulating in the earth's biosphere and provide the main long-term control of atmospheric CO_2 and the strength of the natural greenhouse effect (Aiken et al. 2000). Remotely sensed ocean color has been used with models and other data to estimate carbon removal through the fixation of dissolved carbon by phytoplankton and its subsequent burial in sediment or export to deep ocean waters. Such research has suggested that the global ocean is a major sink for fossil and biogenic carbon released to the atmosphere by human activities (Parslow et al. 2000), while coastal areas appear to act globally as a net source because rivers inject massive quantities of land-derived carbon (Smith and Hollibaugh 1993). There is significant variability, however, among various coastal zones (Smith and Hollibaugh 1993) and through time (Kempe 1995).

Ocean color was used to assess sequestration of carbon to depth following the first *in situ* iron fertilization experiment in the region of intermediate and deep water formation in the southern ocean (Boyd and Law 2001). Iron limitation of phytoplankton growth was confirmed during summer, but SeaWiFs imagery together with modeling suggested no significant downward particulate export of the accumulated phytoplankton. Boyd and Law speculated that mass algal sedimentation may have been prevented by horizontal dispersion of high chlorophyll-a waters to adjacent waters.

SeaWiFS has provided routine global chlorophyll observations since 1997, capturing the response of ocean phytoplankton to major El Niño and La Niña events as well as observing interannual variability unrelated to these phenomena.

SeaWiFS data, such as those presented in Figure 6.1, revealed seasonal chlorophyll distributions across the surface waters of the world's ocean as described by Gregg (2002). High-latitude regions experience a very wide seasonal range of chlorophyll, with a prominent and large local spring and summer bloom and a large die-off in local winter. Mid-latitude regions exhibited much smaller seasonal differences, with local winter maxima. Chlorophyll patterns around India are associated with the northwest monsoon in December and the larger southwest monsoon in July (Gregg,

Table 6.3 Environmental and Management Issues Served by Remote Sensed Products

Issues	Key Products[a]
Global Change and Regional Biogeochemical Cycles The fundamental dynamics of coastal ecosystems and their role in the global carbon cycle will continue to change due to the cumulative effects of: climate-induced changes to sea level, upper ocean temperatures, storm activity and erosion, coastal habitat change, fresh water impoundments, nutrient loading to coastal waters from catchments, sewage, atmospheric sources, and over-fishing. Changes need to be monitored, understood, and, where possible, managed.	*Chlor* *ProductionW* *Dynamics*
Eutrophication Excessive nutrient loadings from catchment and point sources can increase algal biomass and change species composition, often favoring nuisance algae.	*Chlor*
Harmful Algal Blooms Evidence suggests worldwide increase in incidence of harmful algal blooms over the last few decades (Anderson, 1995) possibly due to anthropogenic nutrient loadings, changed flushing regimes, introduced exotic species that can threaten wild and cultivated fisheries, and tourism.	*Chlor* *Pigment/type*
Impacts of Catchment Activities on Estuarine and Coastal Waters Agriculture, forestry, mining, dams, irrigation schemes and urban and industrial development can change patterns of freshwater, sediment, and nutrient and pollutant delivery, and thus impact on coastal waters.	*Light* *Chlor* *SS*
Wild Fisheries Effective management of fisheries requires an ecosystem approach, which in turn requires development of understanding and tools relating to many of the above.	*Light* *Chlor* *Pigment/type* *Dynamics*

	Key Products[a]
Aquaculture	
The rapidly growing aquaculture industry needs appropriate siting and monitoring of environmental impacts of, and on, the industry.	*Light*
	Chlor
	Pigment/type
Macroalgae culture depends on water quality, including light attenuation.	*Habitat*
Shellfish culture depends on phytoplankton biomass and composition (including harmful algae), and particle-bound contaminants.	*ProductionW*
	ProductionB
Crustacean and fish ponds are typically highly eutrophic, so interactions with adjacent waters can be problematic.	*SS*
Fish-cage culture represents a large source of recycled nutrients but requires high water quality and is vulnerable to harmful algal blooms, anoxic sediments, and bottom waters.	*Dynamics*
Maritime Operations	
Navigation, shipping, diving, and hazard detection.	*Light*
	Habitat
	Dynamics
Impacts of Coastal Development on Coastal Habitats and Changes in Flushing Rates	*Light*
Urban and tourist development, port and harbor development, dredging and outfalls can disturb or remove critical habitats, remobilize sediments and pollutants, and change circulation patterns.	*Habitat*
	SS
Conservation	
Effective conservation requires an understanding of the spatial and temporal patterns of environmental forcing and the dynamical response of the marine ecosystem.	*All*
Tourism	
Healthy coastal environments are critical in attracting visitors, especially in high conservation areas, which in turn can be threatened by tourist development.	*Light*
	Chlor
	SS
Integrated Coastal Zone Management	
Issues and uses of remote sensed data (above) interact strongly through coastal ecosystems; core and derived remote sensed products contribute to assessments and a predictive understanding that will facilitate integrated management.	*All*

[a]Key Products relate to Table 6.2.

Source: Parslow, J.S., Hoepffner, N., Doerffer, R., Campbell, J.W., Schlittenhardt, P., and Sathyendranath, S. *Remote sensing ocean color in coastal, and other optically-complex, waters.* Reports of the International Ocean-Color Coordinating Group, No.3, IOCCG, Dartmouth, Novia Scotia, Canada, 2000.

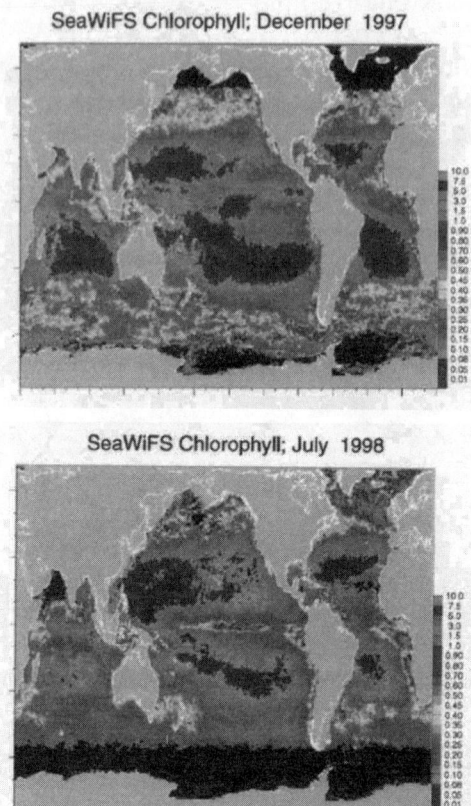

Figure 6.1 Monthly mean SeaWiFS chlorophyll for December 1997 and July 1998. These observations span a major transition from El Niño to La Niña. Areas of the Arabian Sea failed SeaWiFS criteria due to aerosol effects in December 1997. Modified from Gregg 2002.

2002). Elevated chlorophyll levels in the equatorial Atlantic correspond to maximum upwelling (Monger et al. 1997), while high levels during winter (such as in December 1997) are associated with maximum discharge from the Congo River (Gregg 2002).

A major El Niño was underway in September 1997 when SeaWiFS was launched, and it continued until May 1998 when it was succeeded by a La Niña episode in the tropical Pacific. El Niño suppressed upwelling in the equatorial Pacific, resulting in a band of low chlorophyll just above the equator and corresponding to the equatorial counter current (Figure 6.1). During the El Niño, abnormally high wind stresses in the eastern tropical Indian Ocean produced anomalous upwelling that resulted in high chlorophyll levels during December 1997. Reestablishment and intensification of upwelling conditions occurred in the equatorial Pacific when La Niña conditions developed.

A bloom developed rapidly during mid-1998 with a wave pattern centered on the equator, culminating in the highest surface chlorophyll concentrations ever observed in the central equatorial Pacific, more than 1 mg·m^{-3} (McClain et al. 2002). The magnitude and persistence of this bloom is self-evident in the time sequence of estimated primary production shown in Figure 6.2. These data pose as yet-unanswered questions about the mechanism that caused the bloom and how it was maintained for so long. In this region, iron is assumed to be the primary limiting nutrient (Coale et al. 1996), although

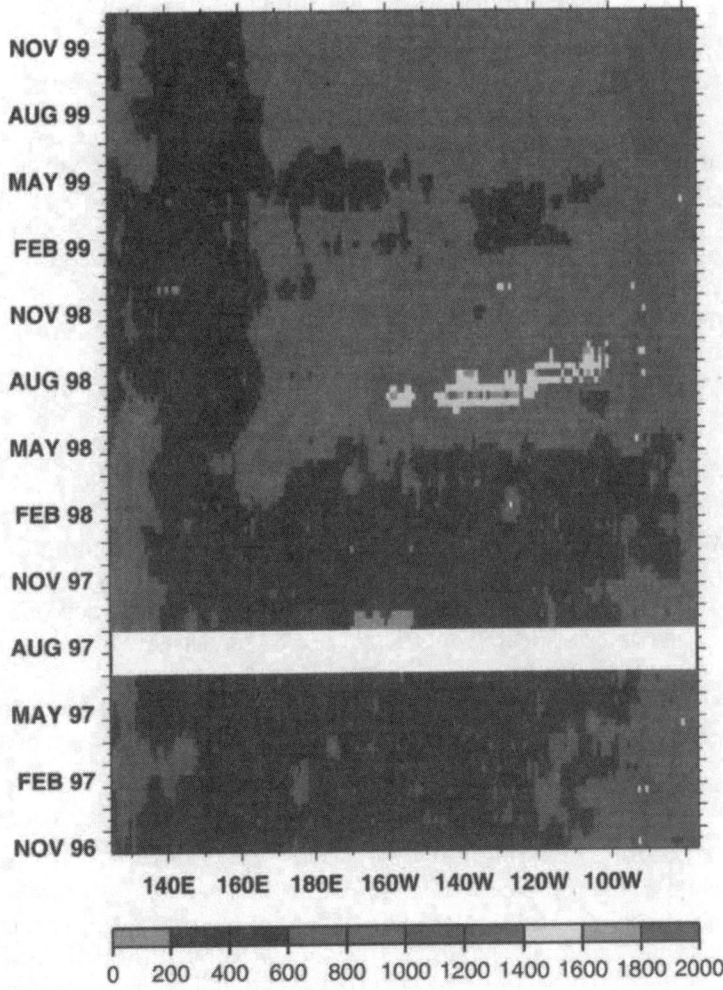

Figure 6.2 Longitude-time plot of primary production (mg C m^{-2} day^{-1}) based on OCTS and SeaWiFS monthly mean chlorophyll from McClain et al. (2002).

wind data appear to discount Ekman upwelling as a source of iron, and atmospheric iron supply remains equivocal (McClain et al. 2002). The persistence of the bloom and the apparent absence of a sustained source of iron suggest efficient retention within the surface layer and ineffective sedimentation over a few weeks or even months.

Recent research has focused on numerical modeling to investigate causal mechanisms and interrelationships of the variability observed in the ocean color data. For example, Gregg (2002) tracked the SeaWiFS record with a coupled physical/biogeochemical/radiative model of the global oceans. Simulations suggested different phytoplankton responses of the Pacific and Indian ocean basins to El Niño. Diatoms were predominant in the tropical Pacific during La Niña, but other groups were predominant during El Niño. The opposite condition occurred, however, in the tropical Indian Ocean.

Other studies have established linkages to meteorological forcing. Follows and Dutkiewicz (2002) used SeaWiFS data to identify meteorological modulation of the spring bloom in the North Atlantic and to examine the implications of decadal changes on biological productivity with a simplified model; Yakov et al. (2001) related seasonal phytoplankton cycles to meteorological factors influencing water stratification of the water column.

SeaWiFS data have also been used to develop and verify ocean general-circulation models (OGCMs), which are critical in global warming assessments. For example, global monthly mean fields of the attenuation of photosynthetic radiation derived from SeaWiFS data have been used to investigate the importance of subsurface heating on surface mixed-layer properties in OCGMs, resulting in a marked increase in the sea surface temperature (SST) predictive skill of the OGCM at low latitudes (Rochford et al. 2002).

SeaWiFS data have also been used together with UV irradiance at the ocean surface (remotely sensed via the total ozone mapping spectrophotometer) to investigate the potential ecological effects of ozone depletion via a model of seawater optical properties in the UV spectral region (Vasilkov et al. 2001).

These studies are examples from a much larger body of work that has employed remote sensed ocean color data to better understand global-scale impacts resulting from human activities.

Regional seas: mesoscale processes and biological variability

Ocean color data have been crucial in relating mesoscale processes to continental shelf ecology through studies of frontal features (Armstrong 1994), eddies (Bardey et al. 1999), upwelling zones (Sathyendranath et al. 1991; Barlow et al., 2001), island wakes (Blain et al. 2001; Caldeira et al. 2002), current patterns (Lee et al. 2001), water mass distributions (Van Der Piepen et al. 1999; Karabashev et al. 2002; Gomes et al. 2000), and various water-quality parameters.

Research has increasingly focused on integration of various remote sensed and *in situ* data. For example, McClain et al. (2002) analyzed chlorophyll concentrations derived from SeaWiFS together with winds (in part from the satellite-mounted scatterometer SeaWinds), sea surface temperature distributions (from AVHRR), and bathymetry data to investigate upwelling phenomena off the west coast of Central America. This region was known for strong upwelling and jets driven by winds that blow from the Atlantic through three narrow mountain passes (McCreary et al., 1989). Synoptic coverage of recent remote sensed data allowed elucidation of interactions between coastal upwelling jets and mesoscale eddies (McClain et al. 2002). Figure 6.3 shows monthly average data for March 1999 when all three upwelling regions were active. High chlorophyll levels (more than 1 mg · m⁻³) extended many hundreds of kilometers offshore from the three mountain passes and were associated with strong offshore wind stress and cool surface waters (1°C to

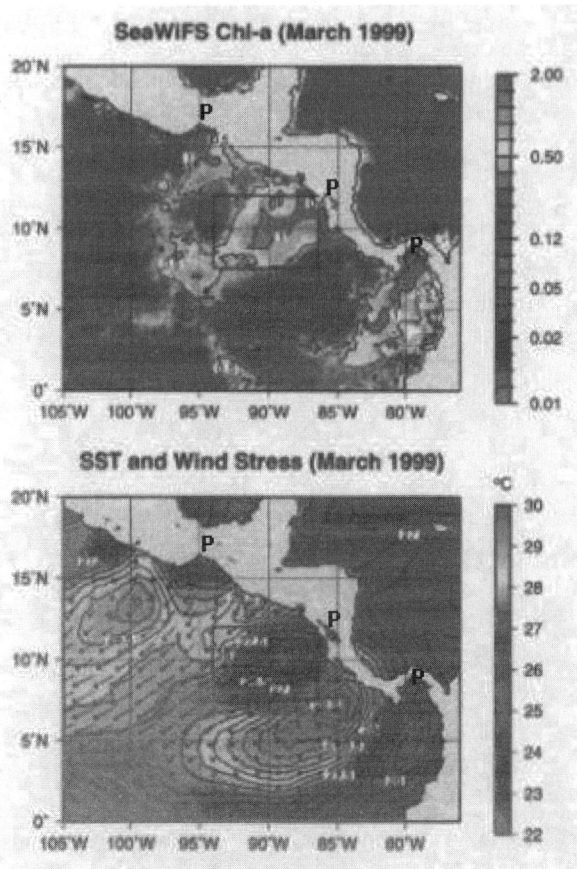

Figure 6.3 Monthly mean SeaWiFS chlorophyll-a (mg·m⁻³) and monthly mean sea surface temperature and wind stress vectors for March 1999. 'P' indicates location of mountain pass. Modified from McClain et al. 2002.

3°C contrast) consistent with jet-driven upwelling. Large mesoscale eddies were spawned by these wind-driven offshore jets (McClain et al. 2002).

A similar multifaceted study used a range of simultaneous remote sensed data to investigate interactions between flow fields and topography/bathymetry around Madeira Island in the northeast Atlantic (Caldeira et al. 2002). AVHRR, CZCS, and SeaWiFS data revealed the following: wind spiral vortices (Von Karman Vortex Street) in the lee of Madeira Island that served to expose the sea surface layer to intense solar radiation compared to cloud covered waters surrounding it; a warm water wake possibly associated with this solar heating (Figure 6.4); geostrophically balanced lee eddies spinning off both flanks of the island including cold core eddies associated with high productivity; localized upwelling and high productivity associated with an underwater ridge; and evidence of the presence of a subtropical front at Madeira's latitude that may influence dispersion.

Semovski et al. (1999) used CZCS chlorophyll estimates together with AVHRR sea surface temperature data, AVHRR channel 1 data as a turbidity indicator, *in situ* data, and modeling to describe the three-dimensional ecosystem structure of mesoscale features in Baltic coastal waters.

A number of studies have used remote sensed ocean color to monitor population dynamics of organisms dependent on phytoplankton. For example, early CZCS studies by Shannon (1985) related ocean color to phytoplankton and pelagic fish distributions. Jaquet et al. (1996) showed that the distribution of sperm whales was strongly correlated with ocean color (chlorophyll) and identified the time (and space) lag between peak chlorophyll concentration and peak sperm whale density with the coefficient of correlation increasing with increasing spatial scales. Polovina et al. (2000)

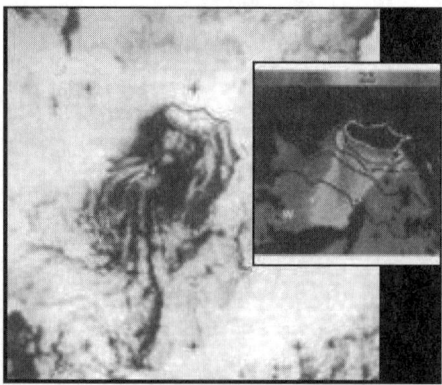

Figure 6.4 AVHRR image showing island mass effects causing interrupted cloud cover and spiral vortices in the lee of Madeira Island, North East Atlantic (19/8/94). An AVHRR sea surface temperature image illustrates typical warm water island wake off Madeira Island (28/7/96) when the wind was north northeast. Modified from Caldeira et al. 2002.

identified an association between loggerhead turtles and frontal zones through analysis of remote sensed sea surface temperature and chlorophyll and geostrophic currents; this conclusion was offered to explain high incidental catches of loggerhead turtles when long-line fishing coincided with frontal zones off Hawaii.

Understanding seasonally high primary productivity can be of great importance in some regions. For example, spring blooms in the Barents Sea provide a strong pulse of energy through the ice-associated and pelagic marine food webs that directly influences the abundance of upper trophic levels, including large marine mammal and sea bird populations (Engelsen et al. 2002). Empirical formulae developed by Engelsen et al. (2002) provided estimates of integrated water column phytoplankton biomass using SeaWiFS data, which held provided that light was the limiting factor.

Together these studies show that a great deal of mesoscale variability can only be observed using satellite remote sensing.

Coastal zones: human activity and ecosystem health

The feasibility of using remote sensing techniques for monitoring water quality in inland and coastal waters was initially limited by their complex optical properties (Kondratyev et al. 1998), but advances in sensors and algorithms deliver a means to discriminate the three main components that account for the optical complexity of case 2 waters: phytoplankton, suspended sediments, and dissolved organic matter. These same components may be used for assessing water quality, algal blooms, and fisheries in the coastal zone.

Water quality
Ocean color (SeaWiFS data) supported by *in situ* observations has been used to investigate outpourings from rivers and coastal catchments. For example, Mertes and Warrick (2001) found that disproportionately large plumes with high concentrations of suspended solids emanated from small coastal Californian catchments compared to large rivers; Siddorn et al. (2001) found an inverse relationship between salinity and yellow substances that could be used to determine the distribution of the Zambezi River plume; Del Castillo (2001) mapped the intrusion of the Mississippi River plume in the West Florida Shelf; and Andrefouet et al. (2002) found that river plumes off Honduras may extend to offshore coral reefs, indicating connectivity of these reefs with the mainland.

Turbid plumes originating from five coastal catchments in southeast Australia after a high rainfall event are shown in Figure 6.5 (from Lee and Pritchard 1999). *In situ* observations during this event confirmed low ocean chlorophyll levels (less than 1 µg/l), thus verifying that the plume images were due to terrigenous matter; the ocean color scale corresponded to log ranges in measured total suspended sediments. A similar logarithmic relationship was found for the Gironde turbid plume in the Bay of Biscay

(Froidefond et al. 2002). Spatial analyses were used in the Australian example to estimate the areal extent of the flood plumes as tabulated in Figure 6.5. The Hunter plume carried an estimated sediment load of about 7000 tons; this figure was based on remotely sensed areal extent and direct observations along offshore transects that indicated a plume-layer thickness of approximately 1 m out to 10 km from the entrance. Significant fallout and dispersion was inferred from the difference between the load carried within the plume and the discharge load estimated at the river mouth.

Woodruff et al. (1999) suggested that photosynthetically available radiation (PAR) attenuation may be estimated from long-term AVHRR satellite data sets as a measure of turbidity; they developed a robust relationship between reflectance observed by AVHRR and light attenuation in the Pamlico Sound estuary in North Carolina in the U.S., although consistent relationships between reflectance and suspended sediment concentrations were elusive due to changing sediment characteristics.

Most studies focus on biological responses (of phytoplankton) to water quality, but Budd et al. (2001) focused on water-quality responses to biological activity (filter feeding). AVHRR reflectance imagery indicated distinct

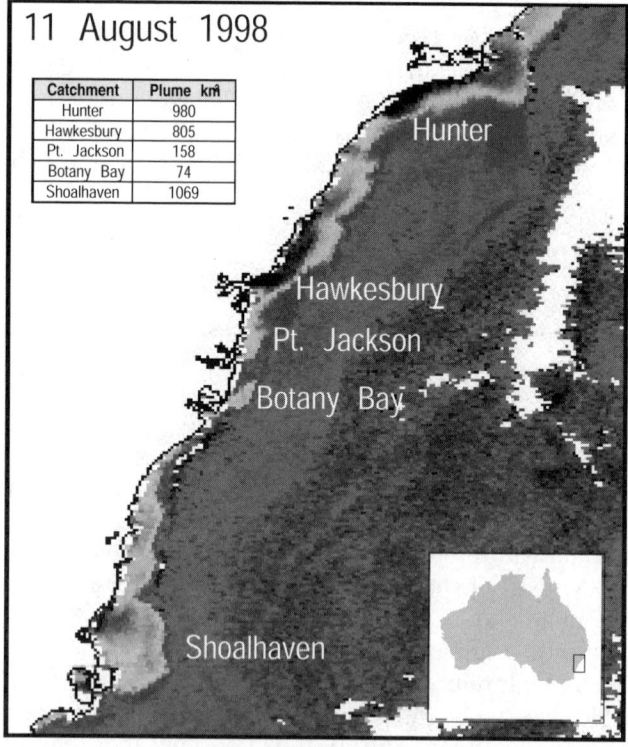

Catchment	Plume km²
Hunter	980
Hawkesbury	805
Pt. Jackson	158
Botany Bay	74
Shoalhaven	1069

Figure 6.5 SeaWiFs image for 11 August 1998 indicated plumes emanating from the Hunter, Hawkesbury, Pt. Jackson, Botany Bay, and Shoalhaven catchments in New South Wales, Australia. Modified from Lee and Pritchard 1999.

and persistent increases in water clarity after zebra mussels (*Dreissena polymorpha*) were discovered in 1991 in Saginaw Bay, Lake Huron, U.S.

Few if any investigations of sewage plumes were found in the international scientific literature because, for satellite-mounted ocean sensors, spatial scales are typically too coarse to resolve sewage plumes. However, untreated sewage discharged from Iraq via a man-made river was implicated as the source of pollution and algal blooms evident in SeaWiFS imagery off the shores of Kuwait in the Persian Gulf (Antonenko et al. 2001).

Algal blooms

The ability to track harmful algal blooms from space can provide coastal communities and seafood harvesting industries with warnings of approaching blooms (Antonenko et al. 2001).

Algorithms are currently unavailable to distinguish between most types of phytoplankton blooms, although SeaWiFS data have been used together with field data to monitor and predict specific harmful algal blooms (such as *Karenia brevis* blooms in the Gulf of Mexico [Stumpf 2001]).

Some bloom types have distinctive ocean color signatures that allow them to be recognized from SeaWiFS data. Examples are the highly reflective coccolithophores that can have a profound effect on the ecosystem, mainly due to extreme reductions in water clarity (Vance et al.,1998; see Figure 6.6), and *Trichodesmium erythraeum*, due to its distinctive spectral response (Subramaniam et al., 2002). Indeed, SeaWiFS-derived trichodesmium chlorophyll concentration has been used for remote estimation of nitrogen fixation by trichodesmium (Hood et al. 2002).

Opportunities exist to use multiple sensors to monitor algal blooms. Lin et al. (1999) attempted to assess the relative performance of nine different types of satellite-mounted ocean color and high-resolution visible sensors to monitor algal blooms, while Rud and Gade (2000) have explored the benefits of using multisensor data (AVHRR, SeaWiFS, Landsat thematic mapper, and ERS synthetic aperture radar) for algal bloom monitoring.

The utility of remote sensed data for diagnostic and prognostic assessment of algal blooms is demonstrated in the case study discussed later in this chapter.

Fisheries

SeaWiFS data were used to demonstrate the relatively clear, pigment-poor surface waters of the Mediterranean with a generally increasing oligotrophy eastwards. Turley et al. (2000) suggested that the combination of low primary production and bacterial dominance of secondary production in the east could account for the low fisheries production, the low vertical flux of material, and low biomass of benthic organisms in this region.

At a finer scale of resolution, Agostini and Bakun (2002) used mean seasonal satellite-sensed ocean color, wind data, and bathymetry to identify

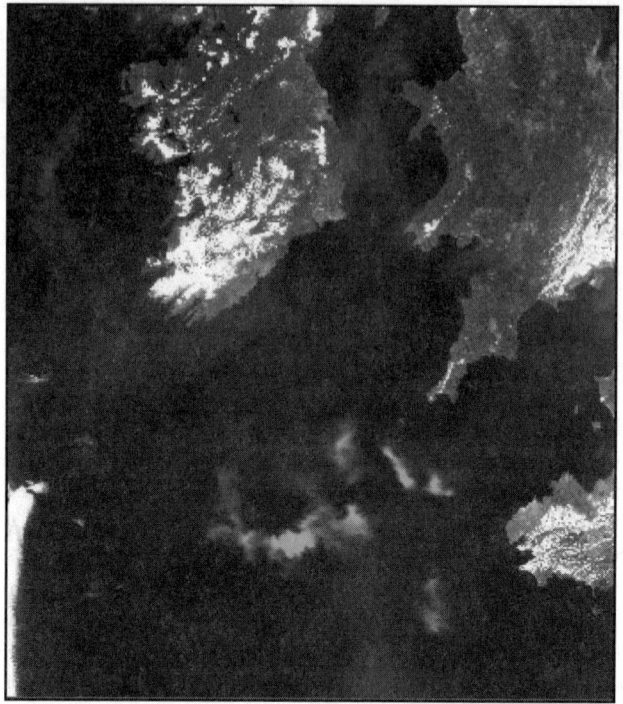

Figure 6.6 Coccolithophore bloom off Cornwall, United Kingdom, on 18/1/1998. True color (Modular Optoelectric Scanner, MOS) from Deutsches Zentrum für Luft- und Raumfahrt, DLR (German Aerospace Centre).

potentially favorable fish reproductive habitats in the Mediterranean based on nutrient enrichment, larval food distributions, and local retention of eggs and larvae.

Platt et al. (2003) used ocean color data from the periods 1979 to 1981 (CZCS), 1997 (POLDER) and 1998 to 2001 (SeaWiFS) to demonstrate that the survival of larval fish (haddock [*Melanogrammus aeglefinus*]) off the eastern continental shelf of Nova Scotia, Canada, depends on the timing of the local spring bloom of phytoplankton. They compared an index of survival (the year-class size at age 1 year, divided by the spawning stock biomass) with anomalies in the timing of spring blooms (the difference in bloom timing from the mean timing for the series). They found that 89% of the variance in larval survival could be accounted for by variation in the timing of the spring bloom. Early spring blooms favored high survival rates, possibly due to greater overlap of spawning and bloom periods. Direct evidence for a putative trophic link such as this is an important factor in analysis of dwindling fish stocks.

Routine synergistic analysis of satellite-borne ocean color and sea surface temperature data sets is currently possible (Solanki et al. 2001) for targeting fishing efforts and monitoring algal bloom development. In the future more frequent coincidence of data from existing and future sensors will deliver synergy among a greater range of remote sensed data including synthetic aperture radar data and data from thermal and optical satellite sensors, as demonstrated by Ufermann et al. (2001).

Parslow et al. (2000) suggest that ocean color data could best contribute to integrated coastal management via diagnostic and prognostic models that also assimilate *in situ* observations and supplementary remote sensed data (such as sea surface temperature via AVHRR, sea surface height via TOPEX/Poseidon, and winds via GEOSAT). At present, integration of ocean color data for the coastal zone with corresponding physical/biogeochemical/radiative models remains a challenge due to the optical complexity of case 2 waters and the requirement for higher spatial resolution compared to open ocean approaches.

Case study: marine algal blooms in coastal waters off southeast Australia

Management issues

Eutrophication has been recognized as a serious threat to the health of coastal ecosystems both globally (Pelley, 1998) and within Australia (Zann 1995). Phytoplankton represent the floating pastures of the ocean, so changes in phytoplankton type and abundance due to eutrophication may profoundly affect the food web. Furthermore, some evidence exists for a worldwide increase in the occurrence of harmful algal blooms (Anderson 1995; Paerl 1997). Some biotoxins selectively kill fish by inhibiting their respiration, while others affect humans generally via seafood.

Visible or harmful algal blooms have the potential to affect tourism in New South Wales (NSW), Australia. Tourism is focused on coastal regions and is worth more than A$6 billion a year. In NSW coastal waters, the magnitude and frequency of "red tides" of the nontoxic dinoflagellate *Noctiluca scintillans* appear to have increased during the last two decades (Ajani et al. 2001a).

Prior to the 1990s, *N. scintillans* appeared as a relatively minor component of the phytoplankton community in NSW coastal waters (Dakin and Colifax 1933), blooming infrequently (Hallegraeff 1995; Ajani et al. 2001b). Since 1990, most red tides in NSW have been due to *N. scintillans* (Figure 6.7). In weekly sampling at Port Hacking off Sydney, Ajani et al. (2001a) found *N. scintillans* in most samples. Major visible blooms of *N. scintillans* have aroused community and media concern in recent years, such as that during January 1998 (see below).

The NSW aquaculture industry, currently worth A$42 to A$45 million a year, is projected to increase to A$250 million a year by 2010. Phytoplankton have been implicated in seafood contamination and fish kills at different times elsewhere in NSW coastal waters (Ajani et al. 2001b). For example, *Dinophysis acuminata*, a producer of diarrhetic shellfish poisoning (DSP), was

Figure 6.7 Spectacular *Noctiluca scintillans* bloom off the popular tourist beach at Manly near Sydney, New South Wales, Australia during 1997. Frontal processes (local convergence) accumulated *Noctiluca* which was then fragmented by the wind into bright red streaks directed shoreward (windrows). Photo courtesy of Beachwatch, NSW EPA.

implicated in the contamination of pipis (edible surf clam, *Donax sp.*) at Ballina, about 700 km north of Sydney (in December 1997), and Newcastle, just south of Port Stephens (in February 1998), with a total of 82 cases of gastroenteritis in consumers.

Regional Algal Coordination Committees have been established by the state government to manage responses to reports of algal blooms while seafood (biotoxin) issues are addressed through the Pipi Biotoxin Management Plan and a SafeFood Marine Algal Biotoxin Contingency/Management Plan. The Pipi Biotoxin Management Plan requires focused, routine monitoring of phytoplankton in water samples while other plans are responsive to alerts (such as visible algal blooms). Prognostic and diagnostic tools would assist in risk management of algal blooms relating to both recreational and seafood issues.

Developing a predictive understanding using remote sensed data

Natural upwelling/uplifting has been identified as the principal driver of marine (offshore) algal blooms in NSW coastal waters, despite significant sewage inputs near major urban centers (Hallegraeff and Reid 1986; Ajani et al. 2001a; Pritchard et al. 2003). This finding together with an understanding of upwelling/uplifting processes provides an opportunity to use remote sensed products together with meteorological data to predict periods of increased risk of marine algal blooms.

The combination of the East Australian Current (EAC) activity on the shelf break (enhancing stratification and bottom stress) and upwelling-favorable winds promotes upwelling (Tranter et al. 1986; Oke and Middleton 1999, 2000; Pritchard et al.,2003). The thermal signatures of the EAC and associated eddies are readily identifiable from remotely sensed sea surface temperature (via NOAA's AVHRR).

Most slope-water intrusions that precede phytoplankton blooms on the NSW continental shelf do not outpour at the surface, although in many instances surface water temperatures are depressed and can be identified on AVHRR images (Cresswell 1994; Pritchard et al. 1999). Phytoplankton responses were found to lag several days behind intrusions of nutrient-rich slope water, so AVHRR images can provide early indications of risk of algal blooms.

Companion synoptic ocean color can indicate oligotrophic EAC waters and monitor phytoplankton responses through time due to nutrient enrichment and cycling, and through space due to advection.

The vast majority of red tide (visible) blooms in NSW marine waters have been due to either *N. scintillans* or *Trichodesmium erythraeum*. Remote sensed data provide a predictive and diagnostic capability, as illustrated by the events described below.

Noctiluca bloom: January 1998

AVHRR SST (Figure 6.8) and SeaWiFS ocean color (Figure 6.9) for January 11 and 12, 1998, identified the warm oligotrophic EAC waters diverging from the coast off Port Stephens, with cool water and high phytoplankton activity on the inside edge of this southward EAC flow. Meteorological observations indicated upwelling-favorable winds during early and mid-January 1998 (Lee et al. 2001). Investigative modeling has shown a tendency for intrusions of cool, nutrient-rich slope water onto the shelf to be associated with the changing shelf configuration to the north of Port Stephens (Oke and Middleton 2000). More localized phytoplankton activity near Jervis Bay (on January 12) is associated with a bathymetric protrusion that has also been shown to favor upwelling (Gibbs et al. 1997). A similar scenario appears to be in operation off Eden on the NSW south coast, where a mesoscale anticyclonic eddy has intensified the divergent flow from the coast.

Regional southward flows on the shelf are indicated by wake effects in the lee of most major changes in the orientation of the coastline (SeaWiFS, January 12, 1998). Time series of ocean color imagery provide greater resolution of flow features than AVHRR SST imagery, although ocean color cannot be regarded as a conservative tracer.

SeaWiFS imagery for January 20 indicates the formation of a cyclonic (clockwise) back eddy inshore of the EAC front in the lee of a major change in shelf orientation near Port Stephens. Baroclinic instabilities such as this eddy also favor upwelling and tend to be associated with along-shelf topographic variability such as that seen near Port Stephens (and Jervis Bay).

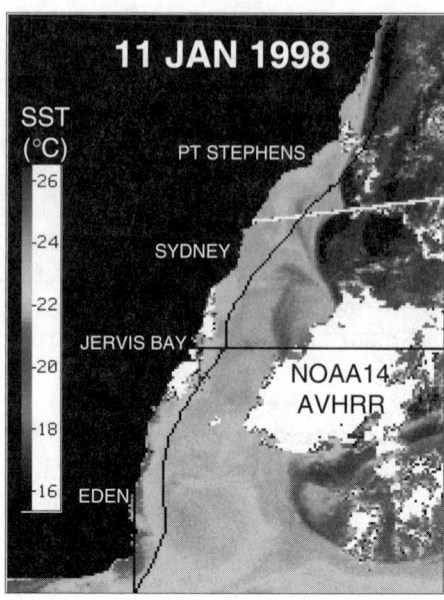

Figure 6.8 Sea surface temperature (SST) image showing separation of the East Australian Current from the shelf off Port Stephens (200 m isobath shelf break indicated). Image courtesy of CSIRO Marine Laboratory.

Cyclonic eddies promote localized upwelling (Ekman pumping) because bottom stress associated with the clockwise rotation promotes convergence of bottom waters (toward the center of the eddy) and consequent upward transport, together with divergence at the surface. Intense phytoplankton activity in this recirculation cell, evident in Figure 6.9 (January 20), is consistent with further localized upwelling. The cell also tends to isolate nutrient-rich waters, incubating phytoplankton that leaks southward with the regional flow on the shelf.

   In situ observations of temperature and chlorophyll-a throughout the water column off Sydney (Figure 6.10) support the notion of a remote source — that is, near-simultaneous arrival of both slope water (nutrients to the euphotic zone) and phytoplankton with no evidence of a lag corresponding to expected phytoplankton response times. The notion of a remote source is consistent with indications of a maturing noctiluca population with increasing southerly extent (Murray and Suthers 1999). Modeling suggests the propensity for the uplifting of slope water north of Port Stephens and subsequent southward transit (Oke and Middleton 2000), and previous observations of EAC-induced upwellings being advected southward as a plume by ambient flows (Cresswell 1994).

   In situ observations (Figure 6.10) were important in verifying SeaWiFS chlorophyll-a distributions with respect to the vertical position of chlorophyll-a maxima. Conductivity, temperature, and depth (CTD) data (not shown) along the transect between PH50 and PH100 on January 15 indicated

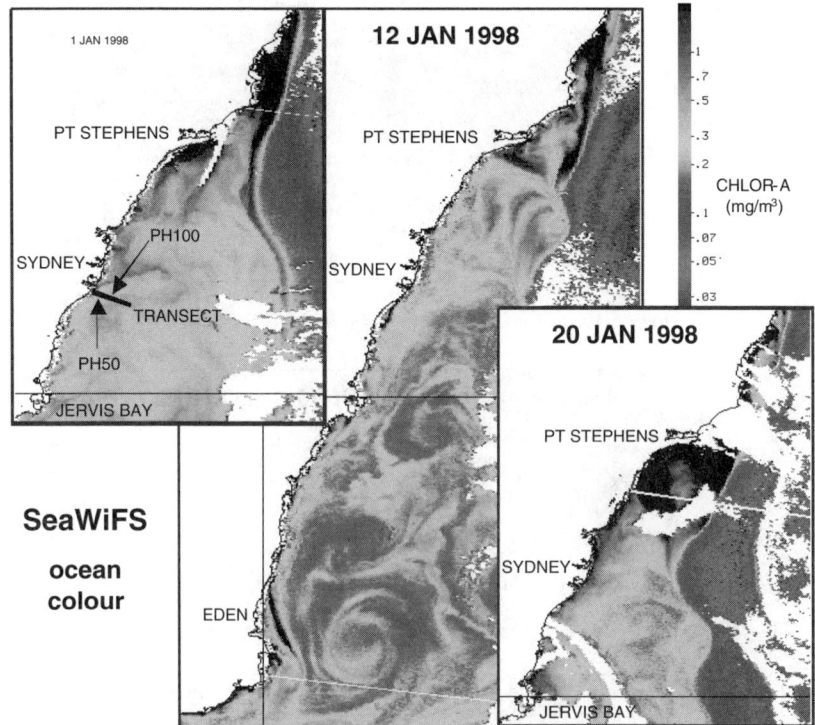

Figure 6.9 SeaWifS chlorophyll-a estimates during January 1998 indicate phytoplankton accumulations along fronts in the lee of major changes in the orientation of the coastline especially along the inner edge of the East Australian Current south of Port Stephens which ultimately formed a plankton-rich cyclonic eddy on January 20, 1998. Images courtesy of CSIRO Marine Laboratory.

prominent shoreward tilting of isotherms, consistent with the vertical distribution of chlorophyll-a at PH100 due to the upwelling forcing. Figure 6.10 shows that phytoplankton blooms were clearly within the upper mixed layer and thus amenable to mapping by satellite-borne ocean color scanners. *In situ* data complements remote sensed data by highlighting the role of thermal structure in controlling the vertical distributions of phytoplankton, and raising questions about the relative importance of temperature, nutrient, and light limitation and the effects of density stratification.

Widespread visible blooms (red tides) of *N. scintillans* were recorded from January 22, consistent with the end stages of the bloom when senescent cells become buoyant and accumulate along surface zones of convergence (Ajani et al. 2000b).

Clearly, remote sensed ocean color together with SST supported by some *in situ* observations provide the means to forecast algal bloom risk and diagnose initiation sites, which in this case were distant from major anthropogenic nutrient discharges off Sydney. Indeed during the summer of 1998

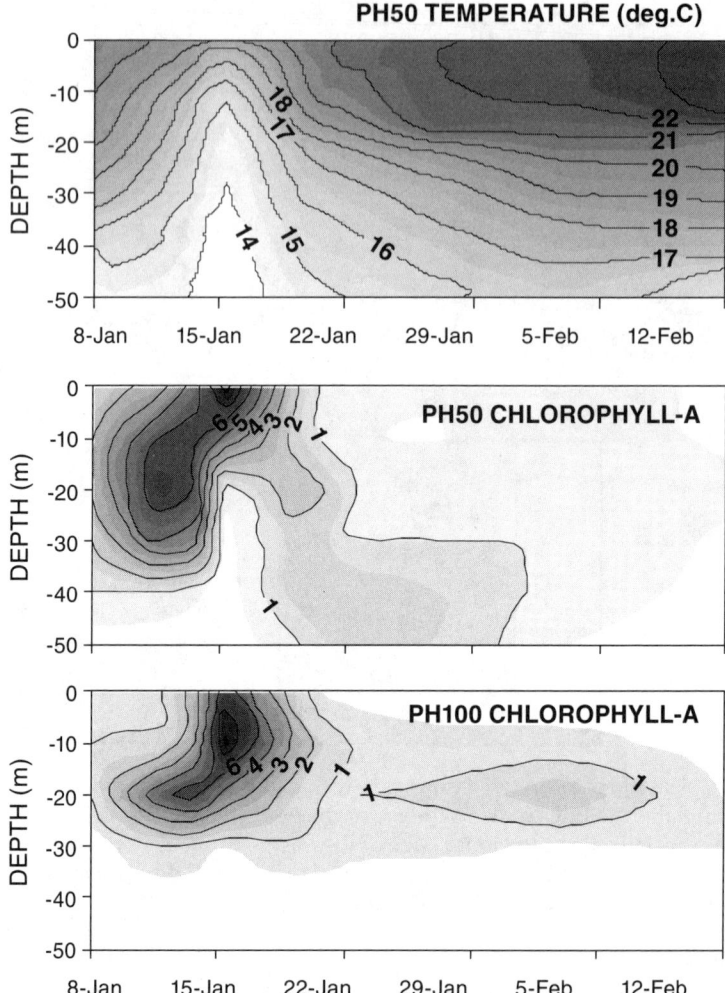

Figure 6.10 Contoured time series CTD temperature data (°C) and in situ chlorophyll-a data (mg·m⁻³) off southern Sydney at PH50 (2km offshore in 55 m of water)
and chlorophyll-a at PH100 (5km offshore in 105m of water) - based on sampling at
10 m depth intervals on 8,13,15 & 20 January and 3 & 12 February 1998.

all major visible blooms reported in the NSW marine waters were preceded
by predictions of high algal bloom risk, based mainly on remote sensed data.

Trichodesmium bloom: March and April 1998

A large *T. erythraeum* bloom developed at Batemans Bay on the south coast
of NSW in early April 1998. The cyanobacterium *T. erythraeum* is a common
red tide organism in NSW coastal waters, transported there from northern

tropical waters by the EAC. The annual distribution of this species monitored off Port Hacking shows peak concentrations in the coastal waters off
Sydney in mid-April when surface waters were more than 22°C (Ajani et
al. 2001a).

One week before the bloom was reported, AVHHR imagery for March
28, 1998, showed unusually warm water throughout the NSW south coast
area associated with a strong manifestation of the EAC (Figure 6.11).
Corresponding SeaWiFS data showed low levels of chlorophyll-a within the
EAC filament but high levels of productivity accumulated and entrained
along the inner edge of EAC water. The zone of high productivity moved
southward to Batemans Bay (on April 5), where the resulting *T. erythraeum*
bloom caused oysters from the estuary to be withdrawn from markets over
Easter. Toxicity testing using a mouse bioassay technique revealed a present,
but unknown, toxin. Previous reports (Hahn and Capra 1992; Endean et al.
1993) also suggest that *T. erythraeum* can produce compounds with mouse
intraperitoneal potency, but this requires further investigation. No human
health impacts were reported.

This case study provides a powerful example of the ability of remote
sensed synoptic data to diagnose the origins and suggest the likely prevalence of algal blooms.

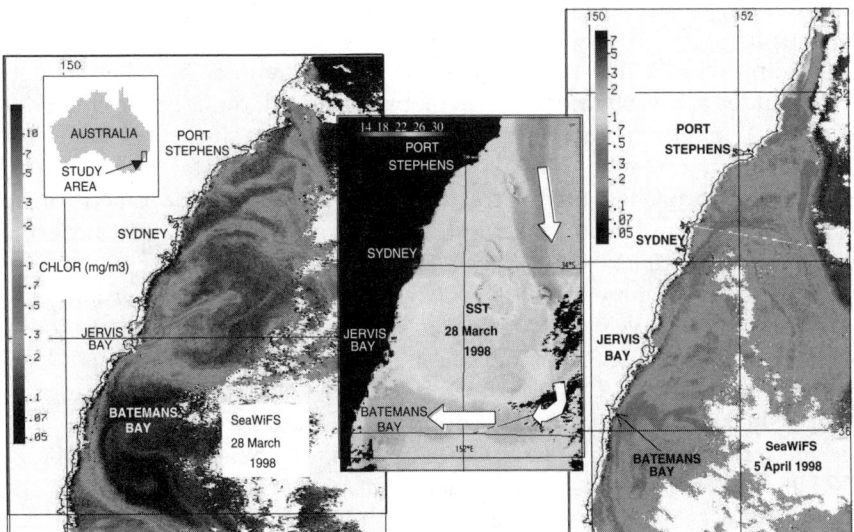

Figure 6.11 East Australian Current waters depicted by warm sea surface temperature (SST in °C) carried Trichodesmium erythraeum with high chlorophyll waters
on the EAC front to Batemans Bay (depicted by SeaWiFS chlorophyll-a in mg·m^{-3})
where oyster fisheries were disrupted during Easter 1998. Images courtesy of
CSIRO Marine Laboratory.

Conclusions

The purpose of this chapter was to demonstrate the utility of remote sensed ocean color data in order to expose opportunities for future marine ecosystem assessments.

Remotely sensed data have been critical in developing mechanistic connections among meteorological and climate change, biological productivity, carbon sequestration, and oceanic ecosystem health. Satellite-mounted ocean color sensors deliver a range of products, including chlorophyll estimates, that provide a synoptic (and global) view of phytoplankton distributions in near real time. A myriad of applications to coastal ecosystems have been spawned by the current generation of ocean color sensors. Together these studies show that a great deal of mesoscale variability can only be observed using satellite remote sensing.

The main limitations in the use of ocean color are cloud cover, confounding optical effects, and limited penetration in cases where maximum phytoplankton biomass occurs at depth. Algorithms for open ocean (case 1) waters are reasonably robust, while algorithms for coastal (case 2) waters are less reliable. Precise multispectral radiances, with contemporary optical and concentration measurements of the water constituents, are required to further develop and validate these algorithms.

There is a concerted effort to correlate the data collected by different scanners to realize the combined coverage offered by various ocean color sensors currently in orbit. Furthermore, new algorithms have been developed to provide greater consistency between new and archived ocean color data in order to investigate trends in global ocean chlorophyll since the 1980s.

Most current research using ocean color data includes synergistic analysis of a range of remote sensed and *in situ* data, often through modeling approaches. Ocean color data are increasingly applied for initialization, assimilation, calibration, and verification of physical/bio-geochemical models.

Further developments are expected for monitoring marine primary production (and its role in sequestering atmospheric carbon), algal blooms, impacts of human activities on coastal waters, and to support wild and aquaculture fisheries. Opportunities exist and will continue to emerge for synergistic analysis of multiple synoptic data sensed from space.

Free and open access of ocean color data such as that from NASA's MODIS sensors and access to merged data products promises to launch a new era of accelerated ocean color research with broad applications in ecosystem assessments.

Acknowledgements

Jocelyn Dela Cruz assisted with an extensive literature search and suggested approaches to support the literature review. The NSW Department of Environment and Conservation (formerly Environment Protection Authority) funded much of the research that led to the case studies and Commonwealth Scientific and Industrial Research Organisation Marine Research provided many of the remote sensed images.

References

Acker, J.C., Shen, S., Leptoukh, G., Serafino, G. Feldman, G, and McClain, C., 2002. SeaWiFS ocean color data archive and distribution system: assessment of system performance. *IEEE Trans. Geoscience Remote Sensing* 40 (1), 90-103.

Agostini, V.N. and Bakun, A., 2002. "Ocean triads" in the Mediterranean Sea: physical mechanisms potentially structuring reproductive habitat suitability (with example application to European anchovy, *Engraulis encrasicolus*). *Fish. Oceanogr.* 11:3, 129-142.

Aiken, J., Moore, G.F., and Holligan, P.M., 1992. Remote sensing of oceanic biology in relation to global climate change. *J. Phycology* 28, 579-590.

Aiken, J., Rees, N., Hooker, S., Holligan, P., Bale, A., Robins, D., Moore, G., Harris, R., and Pilgrim, D., 2000. The Atlantic Meridinal Transect: overview and synthesis of data. *Prog. Oceanogr.* 45, 257–312.

Ajani, P.A., Lee, R.S., Pritchard, T.R., and Krogh, M., 2001a. Phytoplankton patterns at CSIRO's long-term coastal station off Sydney, *J. Coastal Res.* 34, 60–73.

Ajani, P.A., Hallegreaff, G.M., Pritchard, T.R., 2001b. Historic overview of algal blooms in marine and estuarine waters of New South Wales, Australia, *Proc. Linnean Soc. NSW* 123, 1–22.

Anderson, D. M., 1995. Toxic red tides and harmful algal blooms: a practical challenge in coastal oceanography. *Review of Geophysics,* Supplement to the U.S. National Report to the International Union of Geodesy and Geophysics, 1991–1994, 1189–1200.

Andrefouet, S., Mumby, P.J., McField, M., Hu, C., and Muller, F.E., 2002. Revisiting coral reef connectivity. *Coral Reefs* 21, 43–48.

Antoine, D., Morel, A., Gentili, B., Gordon, H.R., Branzon, V.F., Evans, R.H., Brown, J.W., Walsh, S., Baringer, W., and Li, A., 2003. In search of long-term trends in ocean color. *EOS* 84 (32), 301–309.

Antonenko, I., McClain, C., and Feldman, G., 2001. Coastal applications of SeaWiFS data. *Proceedings of the 2nd biennial coastal geotools conference*, Charlestown, SC, January 8–11, 2001.

Armstrong, R.A., 1994. Oceanic fronts in the northeastern Caribbean revealed by satellite ocean color imaging. *Int. J. Remote Sensing* 15 (6), 1169–1994.

Barale, V., 1999. Ocean color comes of age. *Int. J. Remote Sensing* 20 (7), 1197–1199.

Bardey, P., Garnesson, P., Moussu, G., and Wald, L., 1999. Joint analysis of temperature and ocean color satellite images for mesoscale activities in the Gulf of Biscay. *Int. J. Remote Sensing* 20 (7), 1329–1341.

Barlow, R.G., Aiken, J., Sessions, H.E., Lavender, S. and Mantel, J., 2001. Phytoplankton pigment, absorption and ocean color characterization in the southern Benguela ecosystem. *South African J. Sci.* 97, 230–238.

Blain, S., Treguer, P., Belviso, S., Bucciarelli, E., Denis, M., Desabre, S., Fiala, M., Jezequel, V.M., Le Fevre, J., Mayzaud, P., Marty, J.C., and Razoul, S., 2001. A biogeochemical study of the island mass effect in the context of the iron hypothesis: Kerguelen Islands, Southern Ocean. *Deep-Sea Res. I* 48, 163–187.

Boyd, P.W. and Law, C.S., 2001. The Southern Ocean Iron Release Experiment (SOI-REE) — introduction and summary. *Deep-Sea Res. II* 48, 2425–2438.

Budd, J.W., Drummer, T.D., Nalepa, T.F., and Fahnenstiel, G.L., 2001. Remote sensing of biotic effects: zebra mussels (*Dreissena polymorpha*) influence on water clarity in Saginaw Bay, Lake Huron. *Limnol. Oceanogr.* 46 (2), 213–223.

Bukata, R.P., Jerome, J.H., Kondratyev, K.Y., and Pozdnyakov, D.V., 1995. *Optical properties and remote sensing of inland and coastal waters.* CRC Press, Boca Raton, FL.

Caldeira, R.M.A., Groom, S., Miller, P., Pilgrim, D., and Nezlin, N.P., 2002. Sea-surface signatures of the mass effect phenomena around Madeira Island, Northeast Atlantic. *Remote Sensing Environ.* 80, 336–360.

Coale, K.H., Johnson, K.S., Fitzwater, S.E., Gordon, R.M., Tanner, S., Chavez, F.P., Ferioli, L., Sakamoto, C., Rogers, P., Millero, F., Steinberg, P., Nightingale, P., Cooper, D., Cochlan, W.P., and Kudela, R., 1996. A massive phytoplankton bloom induced by an ecosystem scale iron fertilisation experiment in the equatorial Pacific Ocean. *Nature* 383, 495–501.

Cresswell, G., 1994. Nutrient enrichment of the Sydney continental shelf. *Aust. J. Mar. Freshwater Res.* 45, 677–691.

Dakin, W. J. and Colifax, A., 1933. The marine plankton of the coastal waters of New South Wales. 1. The chief planktonic forms and their seasonal distribution. *Proc. Linnean Soc. NSW* 58, 186–222.

Davis, C.S., Gallager, S.M., and Solow, A.R., 1992. Microaggregations of oceanic plankton observed by towed video microscopy. *Science* 257, 230–232.

Del Castillo, C.E., 2001. Multispectral *in situ* measurements of organic matter and chlorophyll fluorescence in seawater: documenting the intrusion of the Mississippi River plume in the West Florida Shelf. *Limnol. Oceanogr.* 46 (7), 1836–1843.

Doerffer, R., Sorensen, K. and Aiken, J., 1999. MERIS potential for coastal zone applications. *Int. J. Remote Sensing* 20(9): 1809–1818.

Endean, R., Monks, S.A., Griffith, J.K., and Llewellyn, L.E., 1993. Apparent relationships between toxins elaborated by the cyanobacterium *Trichodesmium erythraeum* and those present in the flesh of the narrow-barred Spanish mackerel *Scomberomorus commersoni. Toxicol.* 31(9):1155–1165.

Engelsen, O., Hegset, E.N., Hop, H., Hansen, E., and Falk-Petersen, S., 2002. Spatial variability of chlorophyll-a in the Marginal Ice Zone of the Barents Sea, with relations to sea ice and oceanographic conditions. *J. Mar. Syst.* 35, 79–97.

Follows, M. and Dutkiewicz, S., 2002. Meteorological modulation of the North Atlantic spring bloom. *Deep-Sea Res. II* 49, 321–344.

Froidefond, J.M., Lavender, S., Laborde, P., Herbland, P., and Lafon, V. 2002. SeaWiFS data interpretation in a coastal area in the Bay of Biscay. *Int. J. Remote Sensing* 23 (5), 881–904.

Gibbs, M.T., Marchesiello, P., and Middleton, J.H., 1997. Nutrient enrichment of Jervis Bay during the massive 1992 coccolithophorid bloom. *Mar. Freshwater Res.* 48, 473–478.

Gomes, H.R., Goes, J.I., and Saino, T., 2000. Influence of physical processes and freshwater discharge on the seasonality of phytoplankton regime in the Bay of Bengal. *Continental Shelf Res.* 20, 313–330.

Green, E.P., Mumby, P.J., Edwards, A.J., Clarke, C.D., 2000. *Remote sensing handbook for tropical coastal management.* Coastal Management Sourcebooks 3, UNESCO, Paris.

Gregg, W.W., 2002. Tracking the SeaWiFS record with a coupled physical/bio-geochemical/radiative model of the global oceans. *Deep-Sea Res. II* 49, 81–105.

Gregg, W.W., Conkright, M.E., O'Reilly, J.E, Patt, F.S., Wang, M.H.H, Yoder, J.A., and Casey, N.W. 2002. NOAA-NASA coastal zone color scanner reanalysis effort. *Appl. Opt.* 41 (9), 1615–1628.

Hahn, S.T. and Capra, M.F., 1992. The cyanobacterium *Oscillatoria erythraea* — a potential source of toxin in the ciguatera food-chain. *Food Additives Contaminants* 9 351–355.

Hastings, D.A. and Emery, W.J., 1992. The advanced very high resolution radiometer (AVHRR): a brief reference guide. *Photogrammetric Eng. Remote Sensing* 58 (8), 1183–1188.

Hallegraeff, G.M., 1995. Algal blooms in Australian waters. *Water* July/August, 19–23.

Hallegraeff, G.M. and Reid, D.D., 1986. Phytoplankton species successions and their hydrological environment at a coastal station off Sydney. *Aust. J. Mar. Freshwater. Res.* 37, 361–377.

Harris, G.P., Feldman, G.C., and Griffiths, F.B., in Barale, V. and Schlittenhardt, P.M., (Eds), 1993, *Ocean color: theory and applications in a decade of CZCS experience.* Kluwer Academic Publishers, 237–270.

Hood, R.H., Subramaniam, A., May, L.R., Carpenter, E.J., and Capone, D.G., 2002. Remote estimation of nitrogen fixation by *Trichodesmium*. *Deep-Sea Res. II* 49, 123–147.

Hooker, S.B. and McClain, C.R., 2000. The calibration and validation of SeaWiFS data. *Prog. Oceanogr.* 45, 427–465.

IOCCG, 2000. Remote sensing ocean color in coastal, and other optically-complex, waters, in Sathyendranath, S., (Ed.), *Reports of the International Ocean-Color Coordinating Group, No.3.* IOCCG, Dartmouth, Nova Scotia, Canada.

Jaquet, N., Whitehead, H., and Lewis, M., 1996. Coherence between 19th century sperm whale distributions and satellite-derived pigments in the tropical Pacific. *Mar. Ecol. Prog. Ser.* 145, 1–10.

Joint, I. and Groom, S.B., 2000. Estimation of phytoplankton production from space: current status and future potential of satellite remote sensing. *J. Exp. Mar. Biol. Ecol.* 250, 233–255.

Karabashev, G., Evdoshenko, M., and Sheberstove, S., 2002. Penetration of coastal waters into the Eastern Mediterranean using SeaWiFS data. *Oceanologica Acta* 25 (1), 31–38.

Kempe, S., 1995. Coastal seas: a net source or sink of atmospheric carbon dioxide? *Land-ocean interactions in the coastal zone (LOICZ) reports and studies,* Number 1, 27.

Kirk, J.T.O., 1994. *Light and photosynthesis in aquatic ecosystems,* 2nd ed. Cambridge University Press, England.

Kondratyev, K.Y., Pozdnyakov, D.V., and Pettersson, L.H., 1998. Water quality remote sensing in the visible spectrum. *Int. J. Remote Sensing* 129:5, 957–979.

Lee, R.S. and Pritchard, T.R., 1999. Extreme discharges into the coastal ocean: a case study of August 1998, flooding on the Hawkesbury and Hunter rivers. *Pacific Coasts and Ports '99 Proceedings,* Institute of Engineers, Australia 341–346.

Lee, R.S. Ajani, P., Wallace, S., Pritchard, T. and Black, K. 2001. Anomalous upwelling along Australia's east coast. *J. Coastal Res.* 34, 96–109.

Lin, I.-I., Khoo, V., Holmes, M., Teo, S., Koh, S.T., and Gin, K., 1999. Tropical algal bloom monitoring by sea truth, spectral and simulated satellite data. *IEEE,* 931–933.

Lockhart, R., 1994. *Remote sensing of ocean bio-mass productivity.* Curtain University of Technology, Department of Applied Physics, Rep No. UG249/1994.

McClain, C.R., Christian, J.R., Signorini, S.R., Lewis, M.R., Asanuma, I., Turk, D., and Dupouy-Douchement, C.D., 2002. Satellite ocean color observations of the tropical Pacific Ocean. *Deep-Sea Res. II* 49, 2533–2560.

McCreary, J.P., Jr., Lee, H.S., and Enfield, D.B., 1989. The response of the coastal ocean to strong offshore winds; with application to circulations in the Gulfs of Tehuantepec and Papagayo. *J.Mar. Res.* 47. 81–109.

Mertes, L.A.K. and Warrick, J.A., 2001. Measuring flood output from 110 coastal watersheds in California with field measurements and SeaWiFS. *Geology* 29:7, 659–662.

Mobley, C.D., 1994. *Light and water, radiative transfer in natural waters.* Academic Press.

Monger, B., McClain, C., and Murtugudde, R., 1997. Seasonal phytoplankton dynamics in the eastern tropical Atlantic. *J. Geophys. Res.* 102, 12389–12411.

Murray, S., and Suthers, I.M., 1999. Population ecology of *Noctiluca scintillans* Macartney, a red-tide–forming dinoflagellate. *Mar. Freshwater Res.* 50, 243–252.

Nihoul, J.C.L., (Ed.), 1984. Remote sensing of shelf sea hydrodynamics. *Elsevier Oceanogr. Ser.* 38.

Oke, P.R. and Middleton, J.H., 1999. Nutrient enrichment off Port Stephens: the role of the East Australian Current. *Continental Shelf Res.* 21, 587–606.

Oke, P.R. and Middleton, J.H., 2000. Topographically induced upwelling off Eastern Australia. *J. Phys. Oceanogr.* 30, 3, 512–531.

Paerl, H., 1997. Coastal eutrophication and harmful algal blooms: importance of atmospheric deposition and groundwater as "new" nitrogen and other nutrient sources. *Limnology and Oceanography.* 42, 5-2, 1154–1165.

Parslow, J.S., Hoepffner, N., Doerffer, R., Campbell, J.W., Schlittenhardt, P., and Sathyendranath, S., 2000. Case 2 ocean color applications, in Sathyendranath, S. (Ed.), *Remote sensing ocean color in coastal, and other optically-complex, waters.* Reports of the International Ocean-Color Coordinating Group, No.3, IOCCG, Dartmouth, Nova Scotia, Canada.

Pelley, J., 1998. Is coastal eutrophication out of control? *Environ. Sci.Technol.* October 1998, 462–466.

Platt, T., Fuentes-Yako, C., and Frank, K.T., 2003. Spring algal bloom and larval fish survival. *Nature* 423, 398–399.

Platt, T. and Sathyendranath, S., 1988. Oceanic primary production: estimation by remote sensing at local and regional scales. *Science* 241, No. 4873, 1613–1620.

Platt, T., Sathyendranath, S., Caverhill, C.M., and Lewis, M.R., 1988. Ocean primary production and available light: further algorithms for remote sensing. *Deep-Sea Res.* 135, 855–897.

Polovina, J.J., Kobayashi, D.R., Parker, D.M., Seki, M.P., and Balazs, G.H., 2000. Turtles on the edge: movement of loggerhead turtles (*Caretta caretta*) along oceanic fronts, spanning longline fishing grounds in the central North Pacific, 1997–1998. *Fish. Oceanogr.* 9, 71–82.

Pritchard, T., Lee, R., and Ajani, P., 1999. *Anthropogenic and oceanic nutrients in NSW's dynamic coastal waters and their effect on phytoplankton populations.* 14th Australian Coastal and Ocean Engineering Conference and the 7th Australasian Port and Harbour Conference – Coasts and Ports 1999 Institute of Engineers, Australia 537–543.

Pritchard, T.R, Lee, R.S., Ajani, P.A., Rendell, P.S., Black, K., and Koop, K., 2003. Phytoptoplankton responses to nutrient sources in coastal waters off southeastern Australia. *Aquatic Ecosystem Health Manage.* 6 (2), 105–117.

Rochford, P.A., Kara, A.B., Wallcraft, A.J., and Arnone, R.A. 2002. Importance of solar subsurface heating in ocean general circulation models. *J. Geophys. Res. Oceans* 106 (C12), 30923–30938.

Rud, O. and Gade, M., 2000. Using multi-sensor data for algal bloom monitoring. *IEEE* 1714–1716.

Ruddick, K.G., Ovidio, F., and Rijkeboer, M., 2000. Atmospheric correction of SeaWiFS imagery for turbid coastal and inland waters. *Appl. Opt.* 39 (6), 897–912.

Sathyendranath, S., Platt, T., Home, E.P.W., Harrison, W.G., Ulloa, O., Outerbridge, R., and Hoepffner, N., 1991. Estimation of new production in the ocean by compound remote sensing. *Nature* 353, 129–133.

Semovski, S.V., Dowell, M.D., Hapter, R., Szczucka, J., Beszcynska-Moller, A., and Darecki, M., 1999. The integration of remotely sensed, sea truth and modeling data in the investigation of mesoscale features in the Baltic coastal phytoplankton field. *Int. J. Remote Sensing* 20 (7), 1265–1287.

Shannon, L.V., (Ed.), 1985. *South African ocean color and upwelling experiment.* Sea Fisheries Research Institute, Cape Town.

Siddorn, J.R., Bowers, D.G., and Hoguane, A.M., 2001. Detecting the Zambezi River Plume using observed optical properties. *Mar. Pollut. Bull.* 42:10, 942–950.

Smith, S.V. and Hollibaugh, J.T., 1993. Coastal metabolism and the oceanic carbon balance. *Rev. Geophys.* 31: 75–89.

Solanki, H.U., Dwivedi, R.M., and Nayak, S.R., 2001. Synergistic analysis of SeaWiFS chlorophyll concentration and NOAA-AVHRR SST features for exploring marine living resources. *Int. J. Remote Sensing* 22 (18), 377–3882.

Stumpf, R.P., 2001. Applications of satellite ocean color sensors for monitoring and predicting harmful algal blooms. *Human Health Ecol. Risk Assessment* 7 (5), 1363–U15.

Subramaniam, A., Brown, C.W., Hood, R.R., Carpenter, E.J., and Capone, D.G., 2002. Detecting trichodesmium blooms in SeaWiFS imagery. *Deep-Sea Res. II* 49, 107–121.

Tranter, D.J., Carpenter, D.J., and Leech, G.S. 1986 The coastal enrichment effect of the East Australian Current eddy field. *Deep-Sea Res.* 33, 1705–1728.

Turley, C.M., Bianchi, M., Christaki, U., Conan, P., Harris, J.R.W., Psarra, S., Ruddy, G., Stutt, E.D., Tselepides, A., and Van Wambeke, F., 2000. Relationship between primary producers and bacteria in an oligotrophic sea — the Mediterranean and biogeochemical implications. *Mar. Ecol. Prog. Ser.* 193, 11–18.

Ufermann, S., Robinson, I.S., and Da Silva, J.C.B., 2001. Synergy between synthetic aperture radar and other sensors for the sensing of the ocean. *Ann. Telecommunications* 56 (11–12), 672–681.

Vance, T.C., Schumacher, J.D., and Stabeno, P.J., 1998. Aquamarine waters recorded for first time in eastern Bering Sea. *EOS Trans. Am. Geophys. Union* 79 (10), 121.

Van Der Piepen, H., Amman, V., and Barrot, K.W., 1999. Distinction of different water masses by means of remote sensing data collection during the Alboran Sea Experiment. *Int. J. Remote Sens.* 20 (7), 1319–1327.

Vasilkov, A., Krotov, N., Herman, J., McClain, C., Arrigo, K., and Robinson, W.T. 2001. Global mapping of underwater UV irradiances and DNA-weighted exposure using total ozone mapping spectrophotometer and sea-viewing wide field-of-view sensor data products. *J. Geophys. Res.* 106 (C11), 27205–27219.

Woodruff, D.L., Stumpf, R.P., Scope, J.A., and Pearl, H.W., 1999. Remote estimation of water clarity in optically complex estuarine waters. *Remote Sensing Environ.* 68, 41–52.

Yakov, D.A., Nikolay, P.N., and Kostianoy, A.G., 2001. Patterns of seasonal dynamics of remotely sensed chlorophyll and physical environment in the Newfoundland region. *Remote Sensing Environ.* 76, 268–282.

Zann, L., 1995. *Our sea, our future.* Major findings of the State of The Marine Environment Report for Australia, Great Barrier Reef Marine Park Authority (Australia) for Ocean Rescue, 2000.

chapter seven

Risk perception and public communication of aquatic ecosystem assessment information

M.R. Reiss and L. Pelstring

Contents

Introduction 229
Risk perception 230
Risk perception and aquatic ecosystem assessment 232
Aquatic ecosystem assessment communication 234
    Audience analysis 235
    Interacting with the public 239
    Communicating results of aquatic ecosystem assessments 240
    Pretesting message effectiveness 240
        Emphasizing the relevance of results 241
        Data framing 241
        Graphic and visual representations of data 242
        Uncertainty discussion 243
Conclusions 244
Summary 244
References 245

Introduction

Aquatic ecosystem assessments provide technical information about ecosystem health and integrity and inform recommendations to preserve, enhance, or restore ecosystem functions. Nontechnical experts (such as elected officials) in consultation with the public often make decisions regarding the commitment of political or resource expenditures. These decision-makers are often unfamiliar with data and techniques used to assess aquatic ecosystems. As such, it is important that assessment results be effectively communicated in comprehensible terms and language to ensure that decision-makers and the public are adequately informed.

The preceding chapters described advancements in aquatic bioassessment tools and techniques. Experts use the data obtained from studies employing these techniques in mathematical models, such as ecological risk assessments, to evaluate ecosystem health and integrity. While aquatic scientists may find the results of these models persuasive or indeed conclusive, policy-makers and the general public often remain unconvinced.

The seeming inability or unwillingness of the public to associate "appropriate" levels of risk with specific activities, technologies, and events is often frustrating to those conducting the assessments. Literature noting the disparity between risk judgments of technical and lay groups has been reported in many fields, including the environment, public health, and technology sectors (Kraus et al. 1992; Harrington 1998; Flynn et al. 1993; Wright et al. 2000). Technical experts often consider this disparity as symptomatic of a lack of education or of obstinacy on the part of the public (Slovic 1987; Kraus et al. 1992). Such a simplistic view, however, discounts the complexities of how risk attitudes are actually formed.

Clearer communication based on a better understanding of how nonexperts perceive ecological risk may close this disparity. This chapter provides the aquatic ecosystem assessor with an appreciation of the variety of factors that contribute to public perceptions of risk, an understanding of the impact of these factors on the communication of assessment results, and some specific strategies for fostering credibility and trust with public stakeholders, establishing avenues for meaningful public involvement, and communicating assessment results.

Risk perception

There are fundamental psychological, socioeconomic, and cultural dimensions to risk perception. Two dominant lines of research exploring risk attitudes are the psychometric and cultural approaches. Psychometric theory hypothesizes that risk perceptions reflect the inherent characteristics or nature of the hazard associated with a given situation (Slovic 1987). Cultural theory proposes that risk perceptions reflect an individual's life perspective or worldview (Douglas and Wildavsky 1982).

Increasingly, there is a convergence of the psychometric and cultural approaches in explaining risk attitudes. Experts acknowledge that while the specific characteristics of a situation are undoubtedly important contributors to its perceived risk, consideration of sociodemographic and cultural contexts explains much of the variability in the risk attitudes of individuals (and groups).

Surveys using the psychometric approach pose a series of questions designed to assess the perceived characteristics of potentially risky situations, and ask respondents to quantitatively rank their level of concern associated with each situation. Rank scores from each of these questions are then considered in multivariate factorial analyses and the situations are mapped in factorial space. The shared characteristics of situations occupying similar positions in factorial space can then be used to characterize the nature of the psychological factors underlying the way that risks are perceived and assigned by the respondents.

The psychometric approach described above can help identify the complex and rich assortment of underlying factors that contribute to a situation's perceived riskiness by the public. However, not everyone perceives or assigns relative risk in exactly the same manner. An individual's personal history and circumstances (such as previous accidents or illnesses, or parental status) also contribute to perception and allocation of risk (Marris et al. 1997). Differences in risk attitudes across gender, racial, and demographic lines have also been reported (Flynn et al. 1994).

Variability in risk perception is a function of the social, political, geographic (proximity to risk situation), and economic circumstances of individuals and groups. Recognition of the importance of these extra-situational factors is the impetus for cultural research. Cultural risk perception research is conducted along sociological and anthropological lines of inquiry to explain the variability in human allocation of risk.

Cultural theory suggests that an individual's worldview is supported by a set of biases that color perceptions of risk. Individuals subconsciously choose to adopt perceptions of risk that reinforce their perspective and way of life (Douglas and Wildavsky 1982). While an individual's perceptions of risk are expected to be more or less stable, a degree of evolution in an individual's outlook occurs based on life experience, social interactions, and changes in surrounding conditions (Boholm 1996). Adherents to a given worldview tend to selectively accommodate information that reinforces their worldview. Therefore it is difficult to win over skeptics solely by seeking to educate them with more or better technical information.

Adherence to a particular worldview cannot be predicted solely on the basis of social group; nevertheless, demographics and prevalence of specific worldviews are not independent (Brenot et al. 1998; Gustafson 1998; Marris et al. 1998). Cultural theorists suggest that worldviews influence attitudes toward many social issues, extending well beyond perception of risk. Differences in risk perceptions that have been reported for different social groups, such as racial and gender differences, may actually be manifestations of cultural differences in other areas, such as attitudes regarding trust, empowerment, and equity (Flynn et al. 1994).

Both psychometric and cultural risk research have shown that the degree of public trust in the institution, organization, or individual responsible for assessment and communication plays a critical role in public attitudes toward risks managed by that entity (Siegrist and Cvetkovich 2000; Bord and O'Connor 1992). Numerous surveys have been conducted to identify the important factors that the public considers when judging the trustworthiness and credibility of risk-management entities. Institutions with demonstrated records of honesty and openness, knowledge and expertise, care and concern, and commitment to public or ecological health enjoy greater levels of public trust (Frewer et al. 1996; Peters et al. 1997; Bord and O'Connor 1992). To build public trust, it is necessary to establish the credibility, integrity, and accountability of those performing key risk analysis, management, and communication functions and to demonstrate that adequate resources and technologies are available to fully address the risk situation.

Demonstrations of technical knowledge alone (presenting the public with more or better technical data) do not significantly alter trust and public risk attitudes (Slovic 1993). Therefore, investment in outreach and involvement strategies that fosters trusting relationships with public stakeholders may be a more promising direction for changing risk attitudes than simply improving technical assessment methodologies. Effective communication with public stakeholders is important in building these relationships.

Risk perception and aquatic ecosystem assessment

Attaining consensus on what should be achieved when assessing and managing aquatic ecosystems presents a challenge that is not shared by other (such as human health) risk-assessment and management scenarios — there is no single definition of ecological health or integrity that is widely accepted or that is applicable across ecosystems (McDaniels 1998). Moral, value, and ethical judgments about the system are often made in selecting a particular state for the ecosystem to be considered healthy (Fisher 1998; Kapustka and Landis 1998). When an assessment fails to address those aspects of the system that are valued by the public, there is increased potential for conflict. Making these types of judgments (establishing ecosystem goals) is an inherently societal function.

In addition, ecosystems are complex and dynamic and assessors must make numerous judgments in technical areas of the assessment (such as selection of assays, exposure assumptions, dose-response curves, and measures of fitness) to obtain maximally relevant lines of evidence (Kapustka and Landis 1998; Otway and von Winterfeldt 1992). Although the public grants assessors a degree of latitude to exercise judgment in technical areas of assessments (Fisher 1998), the public may not understand how these choices relate to its ecosystem concerns. This can present difficulties in communicating assessment results or in attaining consensus on assessment findings.

Research has shown that the public has more confidence in assessments that employ formal processes for making judgments in key areas of assessments (Otway and von Winterfeldt 1992), such as selecting goals and endpoints. Using formal processes, such as citizen advisory groups or scientific peer review panels, to inform judgments made in key areas improves public acceptance of those judgments (Otway and von Winterfeldt 1992) and can enhance perceptions of openness and trust by providing a mechanism for ongoing dialogue with the affected public (Lynn and Busenberg 1995).

Communication challenges can be minimized if public values and concerns regarding the aquatic system are evaluated as part of the study scoping process, and considered in designing and conducting assessments. Assessments should target the prevailing values and attitudes of the majority of the audience while being sensitive and responsive to minority values and views.

Relatively little research has specifically investigated risk attitudes related to perceived hazards to the environment. However, one study that has particular relevance to aquatic ecosystem assessors is that of McDaniels et al. (1997). In this study, psychometric techniques were used to evaluate how residents from three communities (suburban, rural, and mixed urban and rural) in a watershed (the Fraser River Basin in British Columbia, Canada) perceived risks to the aquatic ecosystem associated with 33 situations. Respondents were also asked questions regarding their worldviews on the environment.

The situations posed to Fraser River Basin residents by McDaniels et al. ranged broadly in nature and in potential for ecological impact to the system. Situations included activities having a direct impact on the ecosystem (such as commercial fishing, urban development, and waste disposal), human activities (such as irrigation withdrawal), indirect environmental consequences of those activities (agricultural runoff and landfill leaching), natural phenomena (drought), and recreational activities (such as canoeing and sport fishing). A survey of expert opinions (aquatic scientists and environmental managers) was also conducted to contrast and compare expert and public perceptions of risks posed by these situations. General relationships among situations revealed in this study are presented in Figure 7.1. The results of this survey are used to illustrate how to maximize effective public communication of aquatic assessment information.

Expert and lay judgments of ecological hazard were similar for most situations posed to survey respondents; there were, however, notable differences for certain situations. For example, experts associated higher levels of risk with introduced species, hydrodevelopment, and population growth, and they assigned lower risks to natural phenomena than did the lay public. These differences in risk allocations suggest that the lay public has a more limited understanding of causal relationships in ecological systems, and tends to emphasize impacts to species (including humans) in ecological-risk allocations.

The McDaniels et al. study revealed only modest differences among lay groups' perceptions of aquatic ecosystem risk. Differences in risk allocation among lay groups were correlated with differences in the level of human benefits that each group associated with the posed situation. For example, urban residents rated withdrawal of water for irrigation as riskier than did residents of rural, farming communities that would receive the benefits of

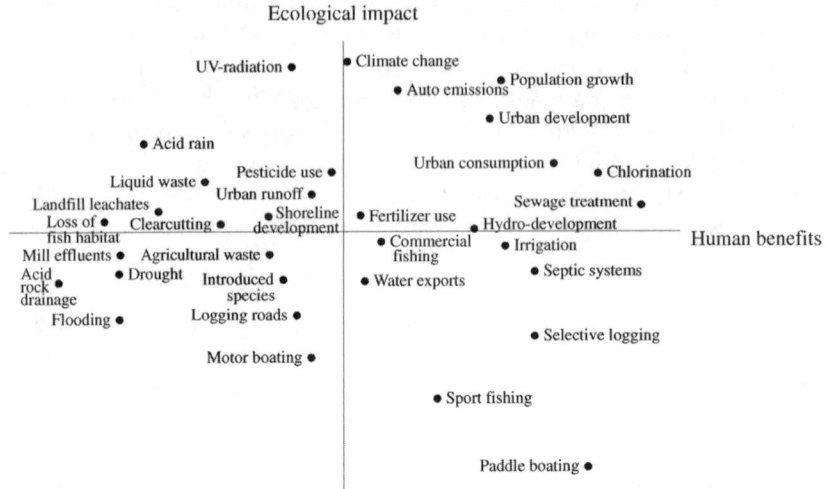

Figure 7.1 Location of 33 potential hazards to aquatic ecosystem derived from relationships among 17 risk characteristics. Situations in the upper left quadrant (high perceived impact and low perceived benefit) of the factorial space were perceived as posing the greatest risk to the river ecosystem. (Reprinted from McDaniels, T.L., Axelrod, L.J., Cavanagh, N.S., and Slovic, P., *Risk Anal.* 17, 341–352, 1997. With permission of author and Blackwell Publishers, Inc.)

such withdrawal. The inverse relationship between perceived risk and benefit has been consistently reported in studies attempting to capture this interaction (e.g., Alhakami and Slovic 1994; Gregory and Mendelsohn 1993).

Therefore, although perceived benefit is independent from technical assessments of hazard, it clearly plays an extremely important role in the psychological calculation that an individual makes in attributing "net" risk to a given situation (Gregory and Mendelsohn 1993). This suggests that effective communication might be enhanced by specifically relating assessment results to impacts on species (including humans) and derived or lost human benefits.

Aquatic ecosystem assessment communication

The intent of aquatic ecosystem assessment communication programs should *not* be to persuade or convince the public, but rather to provide them with the information necessary to understand assessment goals, methods, and findings, and the implications for those attributes of the system that they value. Aquatic ecosystem assessment communication programs require frequent interaction with the public to assist them in forming balanced opinions about the assessment and its recommendations, to identify areas of agreement and disagreement, and to solicit their input as to how any differences might be resolved.

Given the importance of public support in aquatic ecosystem manage-ment, public outreach and communication must be an integral part of overall assessment project planning. Poorly constructed public involvement and communication strategies can heighten the public's sense of mistrust and lead to conflict that can derail or delay the assessment or implementation of its recommendations (Box 7.1).

The remainder of this chapter provides an overview of risk communi-cation basics and emphasizes obtaining and using information about audi-ence concerns, values, abilities, and information preferences to improve com-munication about aquatic ecosystem assessments. Recommendations made are based on the authors' experiences in communicating aquatic ecosystem information. Some useful Web sites for obtaining additional information about public outreach and risk communication are provided in Box 7.2.

Audience analysis

Aquatic ecosystem assessors and managers typically have a very good understanding of the information they wish to convey, but have only a cursory understanding of their target audiences and the media or channels that will reach those audiences. Obtaining public attention and participation in aquatic ecosystem assessments depends on a communication plan that incorporates audience values, abilities, and preferences (Bath 1996; EPA 1995; Lundgren and McMakin 1998). Audience analysis, often referred to as human dimensions research (Decker et al. 1987), should be performed as early as possible in the assessment and communication planning process to obtain this information. Human dimensions information allows financial and staff resources to be focused on communications strategies that are tailored to effectively reach and resonate with the audience.

A communication program for a specific aquatic ecosystem assessment may target stakeholders from a relatively restricted geographic region or stakeholders that share a somewhat homogeneous demographic profile (such as farming communities within a specific watershed). In this case, it may be possible to rely on a focused range of communication formats and channels. Communicating with stakeholders from a broad geographic area or who may have diverse backgrounds and attitudes may require the use of multiple formats and channels to reach these different audiences. Appropri-ate methods for obtaining representative human dimensions information for these situations also differ.

In cases where the target population is relatively small and homoge-neous or the communicator is in the initial stages of developing an outreach strategy, small group meetings, focus groups, or interpersonal, face-to-face communication (such as interviews) are useful methods for obtaining human dimensions information to support public outreach and communication planning. For example, the U.S. Environmental Protection Agency (EPA) recommends 15 to 25 community interviews when developing a community

One of New York's largest watersheds, the Hudson River, spans 500 kilometers from its lake source in the Adirondacks to New York Harbor. It is also one of the largest hazardous waste sites in the U.S. Until 1977, the General Electric Company (GE) legally discharged approximately 600,000 kilograms of polychlorinated biphenyls (PCBs) from two capacitor-manufacturing facilities into the upper Hudson River. In the mid-1980s, the U.S. Environmental Protection Agency (EPA) designated 325 kilometers of the river as a Superfund site and embarked on a lengthy remedial investigation. GE strongly opposed dredging as a cleanup option and maintained that it was unnecessary because the river was ridding itself of PCBs through natural attenuation.

In addition to the technical and scientific complexities of the remedial investigation, the Hudson River Superfund site generated unprecedented political and public controversy. Interest groups and GE waged media battles in attempts to sway public opinion. Fearful of what dredging would do to local economies and the impact it would have on the river, many upriver residents sided with GE. Citizens living further downriver, however, tended to support dredging of the river as the preferred remedy. Gaining the public's trust was a critical issue for both GE and EPA.

To build support for its theory of natural attenuation as a cleanup strategy and discourage EPA from selecting dredging as a remedy, GE waged a massive public relations campaign. The company placed multiple television, radio, and full-page newspaper advertisements, issued colorful, glossy newsletters, established a Web site, and conducted surveys of citizens. EPA received thousands of postcards from residents opposing dredging; the postcards were provided by GE. The company spent an estimated $15 to $30 million on its public relations campaign. As the media began reporting about the millions that GE was devoting to public relations, many citizens questioned whether this money would be better spent cleaning the river.

EPA's early outreach plan included mailings about the investigation and cleanup process to roughly 1500 interested citizens, meetings with local elected officials, and establishing four stakeholder groups comprising a range of interests, including scientists, academics, interested and affected citizens, representatives from interest groups, GE employees, and state agency officials. EPA also held multiple public meetings at cities along the river, with as many as 400 citizens attending. One EPA staff member was devoted full-time to orchestrating these activities, with support from contractors.

The agency's early outreach program was subject to significant public criticism. Some complained about the balance of representation in the stakeholder advisory groups. Advocates from both the pro- and anti-dredging camps complained that EPA was not providing adequate or timely information and that its decision-making process was not transparent. These critics declared that EPA had already decided on its cleanup plan and was merely going through the motions of public involvement.

Box 7.1 Evolution of an EPA community involvement program

EPA issued a cleanup decision in early 2002 requiring the dredging of two million cubic meters of sediment to remove an estimated 68,000 kilograms of PCBs from a 250-kilometer stretch of the upper Hudson. Since its announcement to dredge, EPA has taken actions to address shortcomings in its public outreach program and to rebuild public trust. The agency discontinued the earlier stakeholder groups and replaced most formal public meetings with public availability sessions. Public availability sessions use a meeting format that allows agency officials to interact with attendees on an informal, one-on-one basis. EPA now provides timely information about the cleanup and other activities on its Web site, mails periodic newsletters that contain text and graphic presentation of information and data, and started an e-mail listserve.

A critical move demonstrating that the agency was serious about improving public involvement was the establishment of an on-site field office in Hudson Falls, NY, where criticism of the agency's efforts was often the loudest. The upriver office enables residents to obtain information quickly and agency officials to be more in tune with local concerns. EPA has also devoted significantly more financial and personnel resources to outreach — approximately three full-time staff with internal administrative support.

Finally, in early 2002, EPA contracted an independent consulting company with expertise in facilitation. The contractor helped EPA develop a community involvement program to ensure active public participation during the design and implementation of the dredging project. EPA interviewed hundreds of individuals and held local workshops to develop the community involvement plan. The final plan identifies tools and activities the agency will use to address community concerns, providing the public with multiple opportunities for involvement. The plan also contains a glossary, references, and a series of appendices designed to serve as resources for both EPA and the community. Specific sections include contact information for the EPA and the project team, local government, and media, and information on how to obtain additional information.

EPA's outreach plan for the Hudson has evolved from a largely one–staff-person effort to a comprehensive program. Increasing the agency budget and the number of personnel for outreach, using contractors with experience in facilitation, expanding the avenues by which citizens may obtain information, and developing a public involvement plan shaped by community input demonstrates an agency commitment to ensuring that citizens are able to provide informed input throughout a complex aquatic remediation project.

Box 7.1 (continued) Evolution of an EPA community involvement program

relations plan (EPA 1992). Care must be taken, however, to ensure that the public is fairly represented in these forums (Lynn and Busenberg 1995).

For those issues involving a large geographic area and potentially a more diverse range of citizen opinions, there may be a need for larger-scale investigations (such as administering questionnaires by mailings, telephone, or polling stations) to determine audiences' sociocultural attitudes and trusted information sources. The planning, execution, and results analysis for these surveys may entail resource investments that are disproportionate to the scale or controversy of many aquatic ecosystem assessment programs. Conducting such surveys, however, is not the only means to obtain information

Below is a brief list of Web sites that provide additional information about risk communication in a variety of contexts. Additional documents and sites may be obtained by typing "risk communication" in search engines at the home Web sites for the organizations. This list is not intended to provide a complete overview of Web resources, but rather to direct the reader to several particularly useful sites. Many of these sites have additional links that the reader may also find helpful.

Government

www.atsdr.cdc.gov/HEC/primer.html
This site provides a primer on health risk communication by the Agency for Toxic Substances and Disease Registry.

www.inspection.gc.ca/english/corpaffr/publications/riscomm/riscomme.shtml
This Canadian site provides an excellent overview of risk communication, including a review of recent risk communication theories.

www.epa.gov/oerrpage/superfund/tools/index.htm
This site provides guidance for promoting successful community participation, specifically for hazardous waste cleanup programs. The site contains 46 tools, each of which describes activities that the EPA has used successfully, or provides information on available resources.

www.epa.gov/waterscience/fish/guidance.html
This site provides guidance for assessing and managing health risks associated with the consumption of chemically contaminated fish. The EPA developed the guidance documents to help state, local, regional, and tribal environmental health officials who are responsible for establishing fish consumption advisories. The fourth volume of the guidance is specifically devoted to risk communication.

www.who.int/whr/2002/en/
Chapter 3 of the World Health Organization's 2002 *World Health Report* provides information about risk perception, presenting data, the importance of risk communication, the role of the media in risk perceptions, and the social and cultural interpretations of risk.

www.who.int/water_sanitation_health/Documents/IWA/iwachap14.pdf
This World Health Organization site provides a chapter on risk communication in the context of threats to water supplies. It specifically provides information on developing a risk communication program and managing the overall communication effort, including audience identification, message development, and crisis management.

Organizations

www.sra.org/
The Society for Risk Analysis (SRA) provides an open forum for all those who are interested in risk analysis. Risk analysis is broadly defined to include risk assessment, risk characterization, risk communication, risk management, and policy relating to risk.

www.riskworld.com/
RiskWorld is a comprehensive collection of links to risk-related news, events, and societies.

Box 7.2 Risk communication Internet sites

necessary to characterize the communication requirements for large or diverse audiences.

Government census databases may also provide useful demographic information (such as education level, ethnicity, income level, familial status, and age) for communication planning. Public opinion polls, such as electoral efforts, may also be available. In many countries, useful data on regional demographics and social attitudes (environmental and political values) can also be identified through commercially available databases or through consumer market data (e.g., subscription lists).

Communication strategies should never be based exclusively on information from census and market databases. Generally, these data are not sufficiently site- and issue-specific to develop an effective public outreach and communication strategy. Prevailing public attitudes revealed by these data, however, may be useful for identifying candidate strategies for conducting outreach.

Attitudes, knowledge levels, concerns, and information preferences can and often do change during the course of a communication campaign (Fischhoff 1995; Peters et al. 1997). Attitudes can change based on evolution of trust dynamics during the campaign (Peters et al. 1997) or can shift abruptly as a result of events that are outside the communicator's control, such as media reports of environmental calamities (Lindell and Perry 1990; Liu et al. 1998). Therefore, audience analysis and communication planning should be viewed as continuous programs, rather than as one-time efforts, to allow the individual and collective effectiveness of communication elements and the overall strategy to be gauged and adapted as necessary (Bradbury 1994).

Interacting with the public

Many communication efforts fail because they use ineffective methods to reach audiences. For example, government agencies typically rely on public meetings (Fiorino 1990) to convey environmental information to the public. While public meetings are often required by law, they may not always effectively inform and involve citizens (EPA 1995). Public meetings transmit information to the public but often do not enable information exchange or dialogue between speaker and audience.

Communication strategies such as interviews, small group meetings, and focus groups that allow for two-way information exchange should be emphasized in aquatic ecosystem assessment programs. These forums allow the communicator to convey a message and learn more about audience concerns. Furthermore, members of the public may share insights, experience, or expertise that can be invaluable to aquatic ecosystem assessors (for example, farmers may be uniquely qualified to comment on assumptions regarding rates of irrigation, or fertilization and pesticide application). Therefore, forums that allow for information exchange may help establish trust between the institution and the recipient, as well as enable the communicator to identify areas where improved information or alternate approaches may improve the assessment or enhance public understanding.

In addition to direct interaction with the public, there are many other options for conveying information to the public. The public is not indiscriminate in its preferences for receiving information (Jungermann et al. 1996). Ideally, a combination of media, forums, or intermediaries (such as academics or community leaders) that the audience relies on and trusts should be used to disseminate messages.

Communicating results of aquatic ecosystem assessments

Aquatic ecosystem assessments often generate a great deal of technical information. In the interest of demonstrating openness or full disclosure, or to demonstrate the thoroughness of an assessment, aquatic scientists often feel compelled to present the public with large quantities of technical data. However, presenting too much data can overwhelm and distract the public from the most essential components of the assessment (Fischhoff 1995). Only those data that are essential to convey key aspects of an assessment should be included in communication materials. The audience should also, however, be provided with information on how to obtain additional data, should they desire it.

Differences between expert and public risk perceptions relating to aquatic ecosystems are not overcome simply by crafting an effective message; well-crafted messages, however, can serve to fill knowledge gaps, reinforce public beliefs, or correct misconceptions (Fischhoff 1995). Messages should be objective and balanced — presenting factual material about all sides of the issue (Lundgren and McMakin 1998). Message content must be economically, socially, and culturally responsive to the needs, interests, and values of the audience (EPA 1995). Crafting such a message is difficult, if not impossible, without early and adequate audience analysis.

Pretesting message effectiveness

A wide variety of message formats and contents are possible for presenting aquatic ecosystem information to the public. Because not all formats are equally effective in communicating to different audiences, it is important to pretest the effectiveness and clarity of multiple information formats and contents. The effectiveness of candidate messages can be pretested by presenting them to representative samples of the audience in small, interactive forums, such as stakeholder focus groups. Pretesting in this manner ensures that concepts, language, and graphics used in messages are clear, comprehensive, and unbiased (Lundgren and McMakin 1998).

Communicating aquatic ecosystem assessment information may be more effective if attention is devoted to three areas in crafting messages: emphasizing the relevance of results, using graphics and framing to convey data, and addressing uncertainty. A brief discussion of these points is provided below.

Emphasizing the relevance of results

As discussed above, technical assessments of environmental quality of an aquatic system are necessarily focused on measurable endpoints. The rationale for these endpoints may be self-evident to the aquatic assessor. However, the public may have an entirely different set of endpoints or concerns and may not understand how assessment findings relate to their concerns. Similarly, while comparing data to established numerical standards can help convey the relevance and importance of assessment findings to the public (Lundgren and McMakin 1998), such comparisons are of limited assistance if the public does not understand how those standards relate to their particular endpoints or concerns.

Because the public employs ecological risk constructs that emphasize impacts to species (including humans) and perceived human benefits (McDaniels et al. 1997), implications of assessment results for the public's ecosystem goals can be emphasized in communicating with the public by specifically relating assessment results to impacts to species (including humans) and derived or lost human benefits.

For example, an assessor may learn that an audience has identified recreational angling as an important value for the ecosystem. The assessor may then extend a water-quality model to explicitly consider the effect of changes in water quality on game fish populations. Assessment results can then be communicated in terms that more fully resonate with public values and concerns (the importance of angling and impacts to species) while accommodating potential limits in the public's knowledge of ecological relationships and linkages.

Data framing

An important consideration in presenting assessment results is the effects of framing of data (the specific manner in which data is presented) on public reactions. For example, research has shown that the public may respond differently to the same proportion or probability result, depending on whether it is expressed as probability (p) of an adverse effect occurring or as the inverse probability ($1 - p$) of no adverse effect occurring (Slovic 2001). For example, the public may judge a bioassay very differently based solely on whether the results are presented as a percentage of survival or a percentage of mortality.

There are appropriate and useful applications for data framing. Use of comparisons to familiar concepts can help communicate numerical data that the public may not otherwise understand (Fisher 1998). Quantities (such as rates, volumes, areas) involved in aspects of aquatic ecosystem assessments (such as groundwater discharge to water bodies, areal extent of impact) may be difficult for the public to grasp. These challenges may be minimized by using comparisons to situations that are familiar to the public to provide senses of magnitude or proportion. For example, the public may better understand the rate of groundwater discharge to a water body if it is compared with a more familiar concept, such as a typical year's rainfall, the volume of water used in a shower, or the flow of a local stream. Alternatively, adding a cup of water to a bathtub could be used to illustrate the proportion of daily groundwater discharge to the total volume of the water body. Geographic areas might be effectively illustrated by comparisons to city blocks or local parks.

While comparing data to situations of a similar or neutral nature can be useful for public communication, using comparisons of concepts that differ significantly in nature should be avoided (Freudenberg and Rursch 1994). Psychometric research has shown that differences in the characteristics and nature of individual situations can invoke very different public risk perceptions (Slovic 1987). For example, using the amount of oil released in a recent spill to describe the volume of oil introduced to a water body from urban runoff would not be useful. The acute and visual impacts of spills evoke feelings of outrage in the public. Using such a comparison could result in the public misjudging the nature of the urban runoff problem.

Graphic and visual representations of data

Displaying data in a graphic format organizes and reduces voluminous information and may effectively increase public comprehension. Graphic and visual representations of data may also enhance the media's judgment of aquatic ecosystem information as newsworthy and increase their willingness to report it (Greenberg et al. 1989). Visual and graphic representation of data must effectively support key aspects of the message and be as clear and objective as possible. Communicators of aquatic assessment information are encouraged to be creative in developing visual data representations to support public communication. However, because of the powerful impact of graphic and visual data representations, it is very important that audience reactions to the representations be carefully evaluated before they are incorporated into public communication materials.

Direct mail pieces and other printed communications generally allow the audience greater time to digest, and therefore allow for inclusion of more complex representations (such as tabulated data). Spoken presentations or communications intended to be shown on television allow less time for public consumption; accompanying graphics must be correspondingly less complex (Lundgren and McMakin 1998).

Certain visual representations are better suited for communicating specific kinds of information. Where providing full or accurate descriptions is difficult using text alone, photographs or illustrations (or video and animation, where practical) may effectively support messages and provide visual relief from text (Lundgren and McMakin 1998). Communication situations that might benefit from picture-aided text include descriptions of abnormal environmental conditions (such as red tides), pathologies (such as lesions or tumors in biota), or changes in conditions (before-and-after comparisons). Pictures, illustrations, and animations may also be useful aids for effectively communicating methods and tools that are foreign to the public (EPA 1995). Graph and chart displays are particularly effective in facilitating comparison of data among sites, timeframes, or management options (Lundgren and McMakin 1998). Appropriately scaled maps and charts aid the communication of geographic boundaries or relationships. Care must be taken when using maps, however, as the public may make inappropriate inferences regarding the relationships of geographically proximate map features (Moen and Ale 1998). For example, responsibility for ecosystem impairments could mistakenly be attributed to adjacent industries based solely on geographic proximity.

Scientists are very familiar and comfortable with graphic representations of data. It is important to remember, however, that graphics that are familiar and comprehensible to scientists may be misleading or bewildering to the public. As a result, the public often misinterprets certain graphics (such as cumulative probability plots) that are routinely used by scientists (Ibrekk and Morgan 1987). In addition, positioning, scaling, and coloring of graphics have been shown to affect the public's interpretation of the meaning of graphically displayed data (Sandman et al. 1994; Moen and Ale 1998).

Uncertainty discussion

Formal recognition and analysis of areas of uncertainty has become an integral part of the ecosystem assessment process. Assessment uncertainty is one of the more important pieces of information that managers consider when evaluating management recommendations that might arise from an assessment. As the public becomes more involved in aquatic ecosystem issues, they increasingly demand to be informed about uncertainties associated with assessments.

Failing to discuss the limitations of assessments may erode the public's perception of the honesty and integrity of the assessment process and of those conducting the assessment. However, disclosure of uncertainty does not necessarily result in improved public perceptions. Acknowledging assessment uncertainty can either enhance the public's perception of those responsible for conducting the assessment as honest and forthcoming, or be interpreted as evidence that the assessors are incapable or unqualified (Johnson and Slovic 1995). Ultimately, the degree of trust that is present in the assessor's relationship with the public determines how uncertainty disclosures are perceived.

Discussing quantitative uncertainty associated with assessment data to the public can be difficult. The public is generally unfamiliar with representations of uncertainty (Johnson and Slovic 1995) and tends to view guidelines and decision points as dichotomous thresholds (Lundgren and McMakin 1998). Probabilistic techniques used to derive goals are difficult to explain and percentiles used to set goals (such as a 95% probability) may be misconstrued by the public as resulting in the loss of some level of ecological function (Roberts 1999). When discussing quantitative uncertainty with the public it is important to discuss the nature of the uncertainty, why it exists, and steps (if any are possible) that will be taken to reduce uncertainty (Lundgren and McMakin 1998).

Conclusions

Moral, value, and ethical judgments are often made by the public in selecting preferred ecological states for aquatic ecosystems. While primarily technical in nature, the ecosystem assessment should seek to be responsive to societal values and concerns. Therefore, it is important that public preferences and values for the system be evaluated and considered in the design and conduct of aquatic ecosystem assessments.

Managers are often reluctant to allocate the required resources and commit to meaningful public involvement in ecosystem assessment and management, fearing delays or a loss of institutional control over the process (Lundgren and McMakin 1998). However, active and early participation by the public in resource-management issues is more likely to result in decisions and actions that incorporate a broader range of public values (Fiorino 1990) and thereby enjoy greater public acceptance (Landre and Knuth 1993). Therefore, budgeting the time and funding necessary to promote meaningful public participation in the assessment process should be incorporated into overall project planning.

Substantial research has shown that public perceptions of the credibility, concern, and commitment of those conducting technical assessments typically have more impact on public attitudes toward the findings than do the technical aspects of the assessment itself (Siegrist and Cvetkovich 2000; Bord and O'Connor 1992). An important benefit of a well-designed and executed public outreach and involvement plan is that it can significantly enhance public perception of the institution's credibility and its concern for societal values and goals for aquatic ecosystems. It is likely that the trust gained by adopting more democratic approaches to public involvement will translate to better public acceptance of aquatic ecosystem assessments as the basis for formulating management recommendations.

Summary

Attaining consensus on what should be achieved when assessing and managing aquatic ecosystems presents certain challenges that are not shared by other (such as human health) risk assessment and management scenarios (McDaniels 1998). Ecological goals reflect underlying value and ethical judgments regarding the system. Communication challenges can be minimized if public values and concerns regarding the aquatic system are identified and considered when establishing goals and designing assessments for aquatic ecosystems. Public perceptions of assessor credibility, concern, and commitment have more impact than do technical aspects on public attitudes toward assessment findings. Therefore, public outreach and involvement strategies that enable dialogue and build trusting relationships with stakeholders should be emphasized in communicating with the public about aquatic ecosystems.

This paper has not been subjected to Agency review. Therefore, it does not necessarily reflect the views of the National Oceanic and Atmospheric Administration or the U.S. Environmental Protection Agency.

References

Alhakami, A.S. amd Slovic, P., 1994. A psychological study of the inverse relationship between perceived risk and perceived benefit. *Risk Anal.* 14(6), 1085–1096.

Bath, A.J., 1996. Increasing the applicability of human dimensions research to large predators. *J. Wildlife Res.* 1(2), 215–219.

Boholm, A., 1996. Risk perception and social anthropology: critique of cultural theory. *Ethnos* 61, 64–84.

Bord, R.J. and O'Connor, R.E., 1992. Determinants of risk perceptions of a hazardous waste site. *Risk Anal.* 12(3), 411–416.

Brenot, J., Bonnefous, S., and Marris, C., 1998. Testing the cultural theory of risk in France. *Risk Anal.* 18(6), 729–739.

Bradbury, J.A., 1994. Risk communication in environmental restoration programs. *Risk Anal.* 14(3), 357–363.

Decker, D.J., Brown, T.L., Driver, B.L., and Brown, P.J., 1987. Theoretical developments in assessing social values of wildlife: toward a comprehensive understanding of wildlife recreation involvement, in D.J. Decker and G.R. Goff, (Eds.), *Valuing wildlife — economic and social perspectives*, 76–95. Westview Press, Boulder, CO.

Douglas, M. and Wildavsky, A., 1982. *Risk and culture*. University of California Press, Inc., Berkeley, CA.

EPA (U.S. Environmental Protection Agency). 1992. *Community relations in Superfund: a handbook*. Office of Emergency and Remedial Response, U.S. EPA, Washington, D.C.

EPA (U.S. Environmental Protection Agency). 1995. *Guidance for assessing chemical contaminant data for use in fish advisories: volume 4: risk communication*. Office of Science and Technology, U.S. EPA, Washington, D.C.

Fiorino, D.J., 1990. Citizen participation and environmental risk: a survey of institutional mechanisms. *Sci. Technol. Hum. Val.* 15(2), 226–243.

Fischhoff, B., 1995. Risk perception and communication unplugged: twenty years of process. *Risk Anal.* 15(2), 137–145.

Fisher, A., 1998. The challenges of communicating health and ecological risks. *Hum. Ecol. Risk Assess.* 4(3), 623–626.

Flynn, J., Slovic, P., and Mertz, C.K., 1993. Decidedly different: expert and public views of risks from a radioactive waste repository. *Risk Anal.* 13(6), 643–648.

Flynn, J., Slovic, P., and Mertz, C.K., 1994. Gender, race, and perception of environmental health risks. *Risk Anal.* 14(6), 1101–1108.

Freudenberg, W.R. and Rursch, J.A., 1994. The risks of "putting the numbers in context": a cautionary tale. *Risk Anal.* 14(6), 949–958.

Frewer, L.J., Howard, C., Hedderley, D., and Shepherd, R., 1996. What determines trust in information about food-related risks? Underlying psychological constructs. *Risk Anal.* 16(4), 473–486.

Greenberg, M.R., Sachsman, D.B., Sandman, P.M., and Salomone, C.L., 1989. Network evening news coverage of environmental risk. *Risk Anal* 9(1), 119–126.

Gregory, R. and Mendelsohn, R., 1993. Perceived risk, dread, and benefits. *Risk Anal.* 13(3), 259–264.

Gustafson, P.E., 1998. Gender differences in risk perception: theoretical and methodological perspectives. *Risk Anal.* 18(6), 805–811.

Harrington, J.M., 1998. Facts, fallacies, and fears: the public and the health professionals at odds. *Ann. Occup. Hyg.* 42(4), 227–232.

Ibrekk, H. and Morgan, M.G., 1987. Graphical communication of uncertain quantities to nontechnical people. *Risk Anal.* 7(4), 515–529.

Johnson, B.B. and Slovic, P., 1995. Presenting uncertainty in health risk assessment: initial studies of its effects on risk perception and trust. *Risk Anal.* 15(4), 485–494.

Jungermann, H., Pfister, H.-R., and Fischer, K., 1996. Credibility, information preferences, and information interests. *Risk Anal.* 16(2), 251–261.

Kapustka, L.A. and Landis, W.G., 1998. Ecology: the science versus the myth. *Hum. Ecol. Risk Assess.* 4(4), 829–838.

Kraus, N., Malmfors, T., and Slovic, P., 1992. Intuitive toxicology: expert and lay judgments of chemical risks. *Risk Anal.* 12(2), 215–232.

Landre, B.K. and Knuth, B.A., 1993. The role of agency goals and local context in Great Lakes water resources public involvement programs. *Environ. Manage.* 17(2), 153–165.

Lindell, M.K. and Perry, R.W., 1990. Effects of the Chernobyl accident on public perceptions of nuclear plant accident risks. *Risk Anal.* 10(3), 393–399.

Liu, S., Huang, J-C., and Brown, G.L., 1998. Information and risk perception: a dynamic adjustment process. *Risk Anal.* 18(6), 689–699.

Lundgren, R. and McMakin, A., 1998. *Risk communication: a handbook for communicating environmental, safety, and health risks.* 2nd ed. Battelle Press, Inc. Columbus, OH.

Lynn, F.M. and Busenberg, G.J., 1995. Citizen advisory committees and environmental policy: what we know, what's left to discover. *Risk Anal.* 15(2), 147–162.

Marris, C., Langford, I., Saunderson, T., and O'Riordan, T., 1997. Exploring the "psychometric paradigm": comparisons between aggregate and individual analyses. *Risk Anal.* 17(3), 303–312.

Marris, C., Langford, I.H. and O'Riordan, T., 1998. A quantitative test of the cultural theory of risk perceptions: comparison with the psychometric paradigm. *Risk Anal.* 18(5), 635–647.

McDaniels, T.L., 1998. Systemic blind spots: implications for communicating ecological risk. *Hum. Ecol. Risk Assess.* 4(3), 633–638.

McDaniels, T.L., Axelrod, L.J., Cavanagh, N.S. and Slovic, P., 1997. Perception of ecological risk to water environments. *Risk Anal.* 17(3):341–352.

Moen, J.E.T. and Ale, B.J.M., 1998. Risk maps and communication. *J. Hazardous Mater.* 61, 271–278.

Otway, H. and von Winterfeldt, D., 1992. Expert judgment in risk analysis and management: process, context, and pitfalls. *Risk Anal.* 12(1), 83–93.

Peters, R.G., Covello, V.T., and McCallum, D.B., 1997. The determinants of trust and credibility in environmental risk communication: an empirical study. *Risk Anal.* 17(1), 43–54.

Roberts, S.M., 1999. Practical issues in the use of probabilistic risk assessment. *Hum. Ecol. Risk Assess.* 5(4), 729–736.

Sandman, P.M., Weinstein, N.D., and Miller, P., 1994. High risk or low: how location on a "risk ladder" affects perceived risk. *Risk Anal.* 14(1), 35–45.

Siegrist, M. and Cvetkovich, G., 2000. Perception of hazards: the role of social trust and knowledge. *Risk Anal.* 20(5), 713–719.

Slovic, P., 1987. Perception of risk. *Science.* 236, 280–285.

Slovic, P., 1993. Perceived risk, trust, and democracy. *Risk Anal.* 13(6), 675–682.

Slovic, P., 2001. The risk game. *J. Hazardous Mater.* 86, 17–24.

Wright, G., Pearman, A., and Yardley, K., 2000. Risk perception in the U.K. oil and gas production industry: are expert loss-prevention managers' perceptions different from those of members of the public? *Risk Anal.* 20(5), 681–690.

chapter eight

Ecotoxicological testing of marine and freshwater ecosystems: synthesis and recommendations

P.J. den Besten and M. Munawar

Contents

Application of toxicity tests 250
Application of biomarkers 251
    Biomarkers in combination with bioassays 251
    Biomarkers in tiered approaches 252
    Biomarkers linked with chemical analysis 253
    Biomarkers as diagnostic tools 254
New technologies 254
Remote sensing 256
Risk perception 256
Conclusions and emerging research needs 256
Final remarks 258
Acknowledgments 258
References 259

Over the past 25 years major developments have been made in the field of ecotoxicology. Traditional testing methods have improved in robustness, representativeness, and in their integration in decision support systems such as whole effluent assessment. Furthermore, a number of new techniques have been presented in the literature for which important applications are foreseen in the quality assessment of surface water, drinking water, wastewater, sediment (*in situ*), and dredged material. This chapter provides a synthesis of these developments and discusses further research requirements.

Application of toxicity tests

Chapters 1 and 2 provide details of standardized toxicity tests (or bioassays) that have been developed for specific purposes, such as screening, high-tiered risk assessment, or toxicity identification evaluation procedures. In addition to these standardized tests, new ones are being developed using species of ecological relevance. Standardized tests are a logical choice when they are used in early-warning assessments or in a screening battery of tests. For site-specific risk assessment, however, there is a clear need for tests with species that are present in the environment being investigated. In many projects, decisions can more easily be made when they are based on data with high relevance to the field situation. In some countries there is a growing trend to develop targets for water-quality and sediment-quality improvement based on location or region-specific scales. This will also stimulate the use of tests with ecologically relevant species.

Multispecies strategies are also being developed. Interactions between species are important factors that influence the degree of impact on individual species. Risk-assessment work should also account for possible indirect effects, such as the results of changes in food availability or in the pressure of predators on the population size. Multispecies tests can be effective in identifying such effects. These tests can also allow the focus of toxicity studies to be changed from endpoints in single species to parameters that relate better to the functioning of ecosystems, such as biomass production.

A large gap exists between results of laboratory tests and the effects occurring in the field. The extrapolation of results from biotesting in the laboratory to estimates of the actual risks caused by contaminants under field conditions is hampered by many factors that cannot easily be quantified, such as:

- Route of exposure
- Exposure to complex mixtures of chemicals
- (Bio)transformation of the chemical, resulting in enhanced or decreased toxicity
- Change in concentration at which organisms are exposed to the compound, due to the chemical binding to the solid phase in sediment
- Failure to use ecologically relevant species in laboratory experiments
- Nutritional and physiological status of the test organism
- Multistress situations
- Variation in the exposure intensity over time
- Relation between indirect effects and the endpoints measured in laboratory toxicity tests
- Physiological or genetic adaptation
- Relation between changes in ecosystem structure and function

Field bioassays or *in situ* exposure tests may help to address some of the issues listed above. Considerable progress has been made in the application of *in situ* exposure bioassays (Chappie and Burton 2000; Burton et al. 2003; Den Besten et al. 2003). Field bioassays can be valuable in situations where it is difficult or undesirable to collect animals directly from the field. For those situations, *in situ* bioassays can be used for surrogate ecological measurements.

Application of biomarkers

An important, ongoing advancement in biotesting techniques is the shift from broad-spectrum tests to receptor-based assays with high specificity. This will result in the development of diagnostic approaches where toxicity is only one of the stressors present in the field. Biomarkers are useful tools in this respect. There are different concepts for the use of biomarkers (Depledge and Fossi 1994; Den Besten 1998):

- Biomarkers in combination with bioassays as parameters in water- or sediment-quality monitoring (trend analysis)
- Biomarkers that lead the investigations from screening to detailed assessment (tiered approaches or weight-of-evidence approaches)
- Biomarkers linked with chemical analysis (hyphenated approaches or toxicity identification evaluation [TIE])
- Biomarkers as diagnostic tools

Biomarkers in combination with bioassays

For many environmental quality assessments, bioassays and biomarkers can be used together. Having a battery of bioassays and biomarkers enables coverage of a broad spectrum of chemicals and provides better representation of the species present in the field. On the other hand, concepts can be chosen for which biomarkers clearly give additional information. For example, bioassays are selected for their ability to detect adverse toxic effects on ecosystem components, whereas biomarkers are included as measures of health and fitness of selected species (from bioassays or from the field). Biomarkers often provide an avenue for studying combination effects and enable in-depth analysis of toxic mechanisms on molecular and cellular levels, thus allowing insight into causal and adaptive responses. In some cases, biomarkers are integrated in bioassays, as is the case for the fluorescent bacterium *Vibrio fischeri* (fluorescence production is the biomarker for energy metabolism). Standard bioassays are widely used because they are designed to fulfill regulatory purposes in a reliable way. Practical demand comes to the fore compared to scientific demand. However, the European Water Framework Directive (Anonymous 2000) requires "good ecological quality" far beyond established trigger values that call for increased scientific demand. Therefore, more sensitive and more specific approaches have to be used.

Biomarker responses integrate toxicokinetics and toxic interactions if exposed to mixtures. The rapid responses provided by biomarkers allow an early-warning system of longer-term effects. Biomarker approaches also overcome the problem of extrapolation of *in vitro* measurements to *in vivo* responses by their potential application in laboratory tests as well as in field monitoring. *In vitro* tests provide insights in toxicological mechanisms, a thorough balance of protection and susceptibility factors, comparisons of organ and species sensitivity, and links to chemical analysis and causative agents. On the other hand, biomarker measurements in the field integrate exposure of different routes over time and ideally over a range of species.

For trend monitoring (both in time and space), it is important to translate quality objectives for the environment (often chemically oriented) to criteria for biomarker responses (for example, defining a range for a biomarker value that is characteristic for an unpolluted environment). Since it is always problematic to define an unpolluted and clean or completely natural state of an ecosystem, it may be more advantageous to track gradually changing biomarker responses in relation to increasing or decreasing pollution over time or space. The *in situ* bioassays (field exposure of caged organisms) mentioned earlier could provide material for biomarker measurements. In the case of animals collected from the field, sessile organisms such as clams and mussels could be used to identify "hot spots" and locally specialized organisms can provide a geographical resolution of pollution and risk (Shugart et al. 1992).

Biomarkers in tiered approaches

Tiered approaches provide a step-by-step application of different bioassays and biomarkers that can be very effective for estimating water quality and environmental health in field areas suitable for regulatory and standard monitoring. In the case of the first screening step, the bioassay or biomarker may be used as a first and cost-effective measurement in a stepwise approach intended to signal the presence of or the effects caused by pollutants (early-warning system; Den Besten 1998). Biomarkers used for screening may be markers of exposure (with specificity for certain contaminants) or markers of toxic effect. Their function is to trigger further research, based on an indication that the organism is exposed to pollutants at levels exceeding the capacity of normal detoxification or repair systems (Shugart et al. 1992). Following the use of biomarkers (or bioassays) to indicate toxicity in the initial assessment, the second step is to refine those responses by using more specific biomarkers so that more comprehensive results can be obtained. For this purpose several methods are available (Hoppe, 1991; Münster 1991; Obst et al. 1995). Eukaryotic organisms such as invertebrates may be used as a link between biochemical and subcellular responses and effects on populations and communities. Lysosomal responses may act as general biomarkers for stress, whereas more specific responses such as cholinesterase, phase I biotransformation, and metallothioneins give insight to toxic mechanisms and perhaps to causative agents (see Chapter 3 on biomarkers).

Tiered risk assessments often are synonymous with weight-of-evidence (WOE) approaches. Biomarkers may also be used in higher tiers. In this case, biomarkers can be important supplementary tools. WOE approaches combine information from different sources and disciplines in order to build lines of evidence (Burton et al. 2002). For instance, if in the field negative effects are observed in fish populations, and bioassays with fish larvae also indicate effects of water-borne contaminants, biomarker measurements in fish collected from the field would complement the field and laboratory observations, and enhance the consistency of the risk assessment. When within a line of evidence there is consistency in results, and when different lines of evidence build up a consistent assessment of environmental risks, the risk manager can be advised to take certain actions.

Chapter 3 also discusses differences in the response of a specific type of biomarker in different species. Differences in the sensitivity of biomarkers among species can be used to estimate ecosystem damage as shown in Figure 8.1 (see also Den Besten 1998). A biomarker response in a species known for its sensitivity would, according to the concept in Figure 8.1, give the risk assessor an indication of limited risk. Conversely, responses of biomarkers in keystone species or known insensitive species is a signal of high risk. Such a concept could be refined by making a distinction between markers of exposure and markers of effect. More research is needed to clarify the interaction between effects caused by contaminants and other environmental threats. An example is the virus-associated mass mortality among harbor seals due to immunotoxic effects of contaminants such as PCBs, PCCDs, and others accumulated by the food chain (Van Loveren et al. 2000). Bioaccumulative properties, however, are not necessarily related to an enhanced toxicity under prolonged exposure (Segner and Braunbeck 1998). The application of higher-level biomarkers such as histological, immunological, or bioenergetic parameters to indicate cumulative stress may be a contribution to the solution to these questions (Shugart et al. 1992).

Biomarkers linked with chemical analysis

Since at least some biomarkers give greater insight into the effect mechanism, they represent a linkage between cause and effect more strongly than do bioassays. This creates the possibility of integrating biomarkers with chemical analysis and using this as a first screening step (Den Besten 1998). The *in vitro* (bioassay) techniques are an especially growing field (see Chapter 5 on bioassays and biosensors). The combination of biological responses detected by biomarkers with chemical fractionation and analysis is one of the approaches that can help identify causative agents, and provides the basis for closing sources of pollution as well as for remediation procedures (Segner and Braunbeck 1998). This approach is realized in toxicity identification evaluation and in the bioassay-directed determination of toxic agents

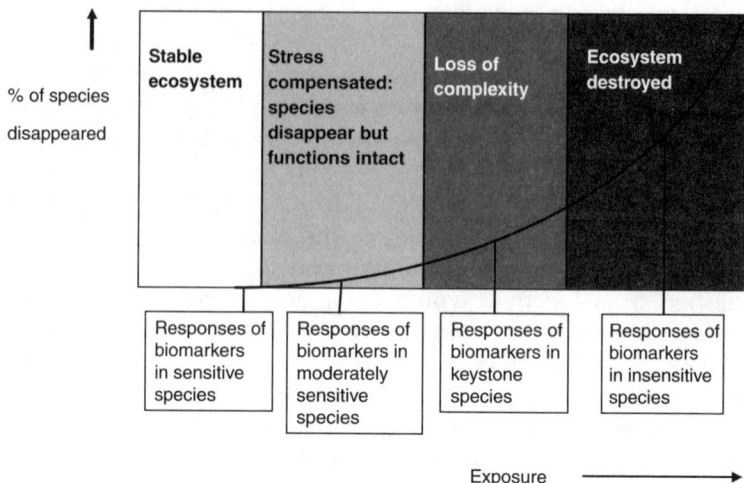

Figure 8.1 Interpretation of species sensitivity differences for the use of biomarkers for ecosystem health assessment.

in environmental samples (Schuetzle and Lewtas 1986; Ankley et al. 1992; Burgess et al. 1995). A general scheme of this hyphenated approach is given in Figure 8.2.

Biomarkers as diagnostic tools

While biomarkers of exposure can be linked (hyphenated) with chemical analysis, biomarkers of effect can be used as diagnostic tools. The term diagnosis refers to the application of a suite of biomarkers that can signal specific effects in wildlife (comparable to the application of biomarkers in human medicine, where biomarkers are used to determine whether or not an individual is physiologically "normal"). Biomarkers on different levels of biological organization can reflect progressive toxic interactions (Walker 1998). To apply biomarkers in this context, it is necessary to know at what point a departure from the normal and healthy state (homeostasis) is likely to affect the performance of an organism (survival, growth, or reproduction). Biomarkers related to the performance or fitness of an organism can be used to detect deviations from homeostasis and may serve as early-warning signals for effects on the population level that are not yet imminent (Walker 1998). The ideal application for these diagnostic biomarkers is *in vivo* measurements, such as in animals collected from the field. With respect to this, noninvasive biomarker techniques (Fossi et al. 1993; Fossi and Marsili 1997) are of great importance.

New technologies

Environmental toxicology is now expanding to new molecular biological methods such as genomics, transcriptomics, and proteomics. Genomics encompasses many different technologies that are related to the content and

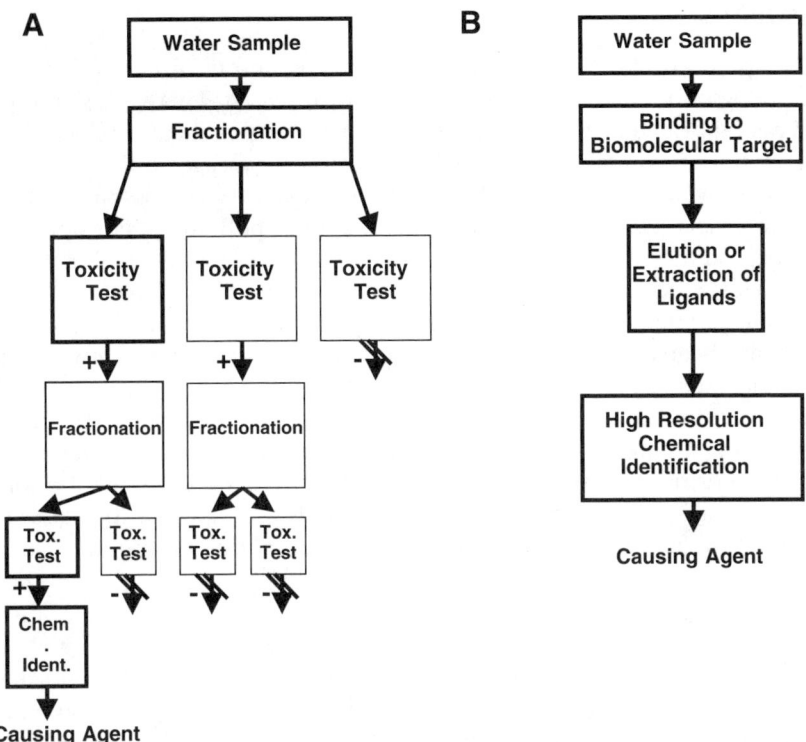

Figure 8.2 Hyphenated approaches. A: bioassay-directed chemical analysis or toxicity identification evaluation; B: bioresponse-linked instrumental analysis.

function of DNA and RNA in a cell or organism (Eisenbrand et al. 2002). For toxicological purposes, two main approaches can be used: (1) the generation of mRNA expression maps (transcriptomics), and (2) the analysis of the expression profile of proteins (proteomics) (Eisenbrand et al. 2002).

Recent developments in the use of polymerase chain reaction techniques for the analysis of mRNA expression patterns after reverse transcription were described in Chapter 4. These techniques will allow researchers to unravel early cellular or individual responses to chemical stress on the genetic level. The analysis of genetic expression at the protein level (proteomics) may be used in toxicology for predictive toxicology and rapid screening, especially in lower doses, by establishing relationships between toxic effects and protein patterns or protein markers (Kennedy 2002). Moreover, identification of new biomarkers may be done by comparing the protein expression of control and exposed cells or organisms. Likewise, new target molecules for the biological selection step in bioresponse-linked instrumental analysis may be found. There are many preclinical and clinical applications of pharmacaproteomics (Moyses 1999) that could also be modified for use in ecotoxicology. These techniques would be a breakthrough in diagnostic studies in situations with multiple stressors.

Remote sensing

In Chapter 6 it was shown that remote-sensing and information-processing technologies are also fast evolving areas of research. There are major environmental problems that become apparent at the global scale. Global warming, flooding events in river catchments (in many cases due to decreased upstream water retention capacity) and in coastal zones, discharge of effluents in coastal zones, atmospheric deposition of pollutants, eutrophication, overexploitation of ecosystems (such as fish stocks), loss of habitat, and spread of introduced species are issues that require risk-assessment and risk-management tools on different geographical scales. Remotely sensed data have been critical in developing mechanistic connections between meteorological/climate change, biological productivity, and carbon sequestration thus providing a better insight in oceanic ecosystem health. An accurate monitoring of mesoscale variations can only be achieved using satellite remote sensing, as was shown for studies of phytoplankton distributions in coastal areas and oceans. Further developments are expected for monitoring marine primary production, algal blooms, and marine pollution.

Risk perception

Chapter 7 on risk perception and communication showed that no matter what the choice of techniques used in monitoring or risk assessment, the value of the data from those techniques depends to a large extent on how the results are communicated to the public and stakeholders. Molecular techniques may have the advantage of providing rapid signals that indicate early effects, but their acceptance for decision-making frameworks might be problematic when investigators fail to show linkage with effects on species or on the functioning of the ecosystem.

Especially with large-scale efforts such as cleanup projects, communication with the public is often carried out on a somewhat *ad-hoc* basis, and systematic analysis of stakeholders is not done. Problems arise in such projects due to the failure to communicate, or due to badly timed or poorly organized attempts to do so. Another frequent mistake is failing to react adequately to signals from interested local groups (Terlien and Bentum 2002). For these reasons it is necessary to make a systematic analysis of local interests at the earliest possible stage, and to develop a communication plan that brings a clear message about the objectives of the work and shows stakeholders how they can influence the process.

Conclusions and emerging research needs

From the discussion above, a number of focal points in ecotoxicology become clear. In comparison to a few decades ago, there are now more effect-based approaches for the assessment of water and sediment quality that can be used in addition to classical chemical analyses. When the quality assessment of surface water, drinking water, wastewater, sediment (*in situ*), and dredged material is also based (in higher tiers) on ecotoxicological data, the resulting decisions will better relate to the actual problem. Seen from this viewpoint, it can be expected that the ecological relevance of ecotoxicological techniques (validation) will become a crucial factor in frameworks when the assessment of damage to the local ecoystem is the main focus. The use of keystone species in bioassays therefore will become even more important in the future. Field exposures (*in situ* bioassays) can also help to demonstrate the ecological relevance of the techniques. Also very important will be the causal relationships between effect and presence of contaminants. More data on the sensitivity of bioassays for specific chemicals are needed to build databases that can be used for finding those relationships (Den Besten et al. 1995). Furthermore, TIE (Ankley and Schubauer-Berigan 1995; Norberg-King et al. 1992) procedures need to be integrated in multitiered risk assessments.

For linking effects with causing agents, information about the bioavailability of contaminants is essential (Peeters et al. 2001). Chemical measurements have also developed over the past decade. At present, very sophisticated methods are available that can characterize the bioavailability of contaminants (Vink 2000; Cornelissen et al. 2001; Burgess et al. 2003). Metal levels in the pore water from the aerobic sediment top layer have shown a better relation with bioaccumulation than total levels in sediment (Vink 2000). Likewise, for organics, mild extractions with Tenax or acetyl acetate have proved to give better results than total extraction (Burgess et al. 2003; Ten Hulscher et al. 2003). Therefore, analysis of the bioavailable fraction of contaminants seems important for finding cause-effect relationships and building lines of evidence in WOE approaches.

For screening bioassays or biomarkers and for biosensors, ecological relevance is usually less important than the sensitivity range and complementarity of the techniques. For these applications it seems much more important to gain knowledge of the specificity and sensitivity range of tests for a broad array of chemicals. Here the challenge is to develop a battery of tests that covers all relevant modes of action. Not only acute toxicity should be included, but also sublethal modes of toxicity (effects on fecundity, growth, immuno-competence, and so on) need to be included in tests used for getting early-warning signals.

In vitro toxicity on the cellular and molecular levels, genomics, and proteomics are promising developments, but many questions are left open. The development of these techniques should be accompanied by thorough investigations of toxicity profiles, including toxicokinetics/biotransformation and barrier and transporter functions, and of differences among species, within one species, and among tissues. New endpoints of toxicity are urgently needed to provide more detailed insight into the fate of hazardous chemicals and into the responses of aquatic populations.

Much more attention should be focused on quality assurance of ecotoxicological techniques. Effect-based quality-assessment approaches provide more information about the actual risks for ecosystems than do the classical chemical approaches. Even if bioavailable fractions are measured, chances are that (many) toxic compounds are overlooked and combination effects are difficult to predict. This step forward also creates concern about the quality assurance of the techniques. Chapters 1 and 2 described in detail what has been achieved with the standardization of techniques and the validity criteria for the acceptance of test results for the decision-makers. The selection of a reference that is meaningful for the site under consideration is important when using ecotoxicological tests in decision support systems. In most countries, the development of different water-quality and sediment-quality assessment approaches includes different choices of references as well. Water-quality and sediment-quality management in coastal zones or in river catchments can be difficult as a result of differences in the choice of reference and use of statistics. Therefore, more harmonization, especially with respect to this part of assessment approaches, is clearly necessary.

The final challenge in ecotoxicology is to combine all existing and new techniques into a number of transparent risk-assessment strategies. Ecosystem health management requires predictive (for early warning), diagnostic (for risk characterisation), or monitoring frameworks with clear steps that lead the responsible managers to the right decisions. The integration of ecotoxicological techniques in such frameworks will continue to be a challenge in the coming years.

Final remarks

In environmental management, aquatic ecosystem health is a key issue, but not the only one. Furthermore, it should be realized that water pollution, which has been the primary focus of this book, may not be the main water-quality driver in many parts of the world. Where human populations are dense, bacteriological status may be the most urgent problem. Many countries also have to deal with water-quantity issues, such as limited drinking water reserves, flooding events, or themes related to other environmental compartments such as soil and air pollution. Because of the great diversity in environmental matters, there will be a continuing need for simple techniques that help prioritize the issues. This book may help inform those responsible for managing risk and for designing water and sediment monitoring programs.

Acknowledgments

The authors are indebted to Dr. Ursula Obst, who contributed to the discussion on the application of biomarkers and bioassay-directed chemical analysis. We would also like to thank Dr. Sharon Lawrence for her constructive editing of the manuscript.

References

Ankley, G.T and Schubauer-Berigan, M.K., 1995. Background and overview of current sediment toxicity identification evaluation procedures. *J. Aquatic Ecosystem Health* 4, 133–149.

Ankley, G.T., Schubauer-Berigan, M.K., and Hoke, R.A., 1992. Use of toxicity identification evaluation techniques to identify dredged material disposal options: a proposed approach. *Environ. Manage.* 16, 1–6.

Anonymous, 2000. *The EU water framework directive.* Off. J. Europ. Commun., December 22th, 2000.

Burgess, R.M., Ho, K.T., Tagliabue, M.D., Kuhn, A., Comeleo, R., Comeleo, P., Modica, G., and Morrison, G.E., 1995. Toxicity characterisation of an industrial and a municipal effluent discharging to the marine environment. *Mar. Pollut. Bull.* 30, 524–535.

Burgess, R.M., Ahrens, M.J., Hickey, C.W., Den Besten, P.J., Ten Hulscher, T.E.M., Van Hattum, B., Meador, J.P., and Douben, P.E.T., 2003. An overview of the partitioning and bioavailability of PAHs in sediments and soils, in P.E.T. Douben, (Ed.), *Polyaromatic hydrocarbons: an ecological perspective*, 99–126. Wiley, West Sussex, England.

Burton, G.A., Jr., Batley, G.E., Chapman, P.M., Forbes, V.E., Smith, E.P., Reynoldson, T., Schlekat, C.E., Den Besten, P.J., Bailer, J., Green, A.S., and Dwyer, R.L., 2002. A weight-of-evidence framework for assessing ecosystem impairment: improving certainty in the decision-making process. *Human Ecol. Risk Assess.* 8, 1675–1696.

Burton, G.A., Jr., Rowland, C.D., Greenberg, M.S., Lavoie, D.R., Nordstrom, J.F., and Eggert, L.M., 2003. A tiered, weight-of-evidence approach for evaluating aquatic ecosystems, in M. Munawar, (Ed.), *Sediment quality assessment and management: insights and progress*, 3–22. Ecovision World Monograph Series, Goodword Books Pvt. Ltd., New Delhi, India.

Chappie, D.J. and Burton, G.A., Jr., 2000. Applications of aquatic and sediment toxicity testing *in situ. J. Soil Sediment Contamination* 9, 219–246.

Cornelissen, G., Rigterink, H., Ten Hulscher, T.E.M., Vrind, B.A., and Van Noort, P.C.M., 2001. A simple Tenax extraction method to determine the availability of sediment-sorbed organic compounds. *Environ. Toxicol. Chem.* 20, 706–711.

Den Besten, P.J., 1998. Concepts for the implementation of biomarkers in environmental monitoring. *Mar. Environ. Res.* 46, 253–256.

Den Besten, P.J., Naber, A., Grootelaar, E.M.M., and Van de Guchte, C., 2003. *In situ* bioassays with *Chironomus riparius*: laboratory-field comparisons of sediment toxicity and effects during wintering. *Aquatic Ecosystem Health Manage.* 6, 217–228.

Den Besten, P.J., Schmidt, C.A., Ohm, M., Ruys, M.M., Van Berghem, J.W., and Van de Guchte, C., 1995. Sediment quality assessment in the delta of the rivers Rhine and Meuse based on field observations, bioassays and food chain implications. *J. Aquatic Ecosystem Health* 4, 257–270.

Depledge, M.H. and Fossi, M.C., 1994. The role of biomarkers in environmental assessment (2). *Ecotoxicology* 3, 161–172.

Eisenbrand, G., Pool-Zobel, B., Baker, V., Balls, M., Blaauboer, B.J., Boobis, A., Carere, A., Kevekordes, S., Lhuguenot, J.-C., Pieters, R., and Kleiner, J., 2002. Methods of *in vitro* toxicology. *Food Chem. Toxicol.* 40, 193–236.

Fossi, M.C. and Marsili, L., 1997. The use of non-destructive biomarkers in the study of marine mammals. *Biomarkers* 2, 205–216.

Fossi, M.C., Leonzio, C., and Peakall, D., 1993. The use of non-destructive biomarkers in the hazard assessment of vertebrate populations, in M.C. Fossi and C. Leonzio, (Eds.), *Nondestructive biomarkers in vertebrates*, 3–36. CRC/Lewis Publishers, Boca Raton FL.

Hoppe, H.-G., 1991. Microbial extracellular enzyme activity: a new key parameter in aquatic ecology, in R.J. Chróst, (Ed.), *Microbial enzymes in aquatic environments*, 60–80. Springer Verlag, Heidelberg.

Kennedy, S., 2002. The role of proteomics in toxicology: identification of biomarkers of toxicity by protein expression analysis. *Biomarkers* 7, 269–290.

Moyses, C., 1999. Pharmacogenetics, genomics, proteomics: the new frontiers in drug development. Int. J. Pharm. Med. 13, 197–202. Referred to in Kennedy, S., 2002. The role of proteomics in toxicology: identification of biomarkers of toxicity by protein expression analysis. *Biomarkers* 7, 269–290.

Münster, U., 1991. Extracellular enzyme activity in eutrophic and polyhumic lakes, in R.J. Chróst, (Ed.), *Microbial enzymes in aquatic environments*, 96–122. Springer Verlag, Heidelberg.

Norberg-King, T.J., Mount, D.I., Amato J.R., Jensen, D.A., and Thompson, J.A., 1992. *Methods for aquatic toxicity identification evaluations: phase 1 toxicity characterisation procedures (second edition).* EPA/600/6-91/003, U.S. Environmental Protection Agency, Washington, D.C.

Obst, U., Holzapfel-Pschorn, A., Weßler, A., and Wiegand-Rosinus, M., 1995. Enzymatische Tests für die Wasseranalytik. 2. Auflage. R. Oldenbourg Verlag, München.

Peeters, E.T.H.M., Dewitte, A., Koelmans, A.A., Van der Velden, J.A., and Den Besten, P.J., 2001. Evaluation of bioassays versus contaminant concentrations in explaining the macroinvertebrate community structure in the Rhine-Meuse Delta, the Netherlands. *Environ. Toxicol. Chem.* 20, 2883–2891.

Schuetzle, D. and Lewtas, J., 1986. Bioassay-directed chemical analysis in environmental research. *Anal. Chem.* 58, 1060A–1075A.

Segner, H. and Braunbeck, T., 1998. Cellular response profile to chemical stress, in G. Schüürmann and B. Markert, (Eds.), *Ecotoxicology*, 521–569. Wiley und Spektrum Akad, Verlag, Heidelberg.

Shugart, L.R., McCarthy, J.F., and Halbrook, R.S., 1992. Biological markers of environmental and ecological contamination: an overview. *Risk Anal.* 12, 353–360.

Ten Hulscher, T.E.M., Postma, J., Den Besten, P.J., Stroomberg, G.J., Belfroid, A., Wegener, J.W., Faber, J.H., Van der Pol, J.J.C., Hendriks, A.J., and Van Noort, P.C.M., 2003. Tenax extraction mimics benthic and terrestrial bioavailability of organic compounds. *Env. Tox. Chem.* 22, 2258–2265.

Terlien, D. and Bentum, S., 2002. *Taking local interests seriously.* Report of the Dutch Aquatic Sediment Expert Centre (AKWA), Rp 02.001.

Van Loveren, H., Ross, P.S., Osterhaus, A.D.M.E., and Vos, J.G., 2000. Contaminant-induced immunosuppression and mass mortalities among harbour seals. *Toxicol. Lett.* 112–113, 319–324.

Vink, J.P.M., 2000. SOFIE®: An integrated test-system for the determination of chemical-toxicological transfer-functions for heavy metals in sediments. *Environ. Pollut. Res.* 1, 49.

Walker, C.H., 1998. Biomarker strategies to evaluate the environmental effects of chemicals. *Environ. Health Perspect.* 106, Suppl. 2, 613–620.

Index

A

Acartia tonsa, 61, 76

Acceptance and sediment quality assessments, test, 16, 38, 40

Acetylcholinesterase (ACHE) as biomarker
amphibians, 120
fish, 109
mammals/birds, 117
methods used to determine biomarker assays, 140
overview, 99–100
standard operating procedures, 148–151

Acetylthiocholin (ACTC), 148

ACHE, see Acetylcholinesterase

Activated sludge and monitoring effluents, 71, 73–74, 82

Acute fish toxicity tests, 60–61, 69–70, 76

Acute invertebrate toxicity tests, 70–72, 76–77

Adsorbable organic halides (AOX), 46

Adsorption kinetics and isotherms, 82

Advanced very-high-resolution radiometers (AVHRRs), 197, 212–213, 221; see also Satellite remote sensing

Aeration and pretreatment of effluents, 54

Alanine aminotransferase (ALT), 139

Aldehydes, 107

Algae and monitoring effluents, 61, 74, 78

Algal blooms and satellite remote sensing, 207–208, 213; see also under Satellite remote sensing

Alkoxyresorufin-O-dealkylase (AROD) biomarkers
aminolevulinic acid dehydratase compared to, 112
amphibians, 120
mammals/birds, 115
phase I biotransformation enzymes, 96

American Society for Testing and Materials (ASTM)
effluents, monitoring
bivalve embryo-larval development toxicity test, 77–78
Ceriodaphnia dubia, 76
protocols, test, 49
rotifers, toxicity tests with, 77
sediment quality assessments
Chironomus riparius, 38
design, sample, 9
dredged materials, 8
storing samples, 11

Ames assay, 80

Aminolevulinic acid dehydratase (ALAD) as a biomarker
alkoxyresorufin-O-dealkylase compared to, 112
amphibians, 112, 120
mammals/birds, 116
methods used to determine biomarker assays, 139
overview, 98

Ampelisca abdita, 39–40

Amphibians and biomarkers, 110–114, 120

Amphipods, see Sediment quality assessments, toxicity tests for

Anaerobic bacteria inhibition test, 73

Antibodies and biosensors, 185

Apoptosis and real-time reverse-transcription PCR, 165

Army Corps of Engineers, U.S. (USACE), 6, 23

Aroclor 1254, 165, 167

Aromatase, 116

Aryl hydrocarbon receptor (AhR), 96–97, 159

Aspartate aminotransferase (AST), 139

Association Francaise de Normalisation (AFNOR), 74

Audience analysis and communicating
    assessment information, 235, 237,
    239
Australia, 7–8, 211–212; *see also* algal blooms
    under Satellite remote sensing
Australian and New Zealand Guidelines for
    Fresh and Marine Water Quality
    (ANZECC/ARMCANZ), 8
Automatization and bioassays, 189

B

Bacteria, acute/chronic tests for, 73, 77
    b5 and biomarkers, 144–145
Bay of Biscay, 211
BEEP, European, 122
Benefits and risk perception, 233
BEQUALM (biological-effects quality
    assurance in monitoring
    programs), 122
Best available technology (BAT)
    Germany, 58, 59
    Netherlands, 56
    tiered approaches, 55
    United Kingdom, 62
Bile levels used as biomarkers, 107
Bill deformations in birds, 117–118
Bioaccumulation, 51
Bioassays and biosensors; *see also* Effluents/
    surface water quality
    bioassays, 180–184
    biosensors, 184–186
    complementary/integrative technologies,
        187–188
    defining terms, 179–180
    endpoints, biological, 187
    future perspectives, 188–190
    history of, 178–179
    in vitro bioassays, 180–182
    in vivo bioassays, 180
    keystone species used, 257
    recognition elements, biological, 184–186
    summary/conclusions, 190
    synthesis/recommendations, 251–252
    transgenic animals, 182–183
    validation and application, 188
Biogeochemical cycles and satellite remote
    sensing, 208
Biological oxygen demand (BOD), 54, 81
Biomarkers in environmental assessment; *see
    also* Molecular methods for gene
    expression analysis
    amphibians, 110–114
    chemical monitoring, advantages over,
        90–91

criteria for candidate biomarkers, 93
    defining terms, 89–91
early-warning biomarker signals, 92
fish, 105–110
improper application/interpretation of
    responses, 92
integrated monitoring programs, 93–94
invertebrates, 101–105
mammals and birds, 114–118
methods used to determine, 138–140
microarray analysis, 166
perspectives and recommendations,
    121–122
sequential order of responses to pollutant
    stress, 91–92
standard operating procedures
    ACHE activity in tissue homogenates,
        148–151
    EROD activity in liver microsomes,
        142–144
    GST activity of liver cytosol, 146–147
    lysosomal membrane stability in cells,
        151–152
    microsomes/cytosol isolated from
        liver tissues, 141–142
    P450 and b5 contents of liver
        microsomes, 144–145
    SOD activity in liver cytosol, 147–148
summary/conclusions
    amphibians, 120
    fish, 119–120
    invertebrates, 119
    mammals/birds, 120–121
    overview, 118–119
synthesis/recommendations, 251–254
types currently being used
    biotransformation products, 97–98
    genotoxic parameters, 100
    hematological parameters, 98–99
    immunological parameters, 99
    neuromuscular parameters, 99–100
    overview, 94
    oxidative stress parameters, 97
    phase I biotransformation enzymes,
        95–96
    phase II biotransformation enzymes,
        96–97
    physiological/morphological
        parameters, 100–101
    proteomics and genomics, 101
    reproductive/endocrine parameters,
        99
    stress proteins/metallothioneins/
        MXR, 98

Biosensors, 179–180, 184–186; *see also* Bioassays and biosensors
Biotransformation index (BTI), 106
Biotransformation products and biomarkers, 97–98, 107
Bird eggs and contaminated sediments, 3
Birds/mammals and biomarkers, 114–118, 120–121
Bivalve embryo-larval development toxicity test, 77–78
Bottom-up description of stress response of an organism, 154
Brachionus calyciflorus, 70, 73
Breast cancer, 181

C

Cadmium, 114
Caenorhabditis elegans, 184
CALUX bioassay systems, 181–183, 185
Canada, 7, 11
Carbamates (CAs), 117
Carbon and satellite remote sensing, 203
Carcinogenicity/mutagenicity/reproductive (CMR) toxicity, 187
Case 2 waters and satellite remote sensing, 201–202
Catalase (CAT) and biomarkers
    fish, 107
    methods used to determine biomarker assays, 139
    overview, 97
    oxidative stress parameters, 102
Ceriodaphnia dubia, 70, 72–73, 76
Chemical analysis, biomarkers linked with, 253–254
Chemical-oriented approach, limitations of, 46, 55, 90–91, 177–178
Chemical oxygen demand (COD), 46, 54, 81, 83
Chinese moderate resolution imaging spectroradiometer (CMODIS), 199
Chironomus riparius, 38–39
Chlorophyll and satellite remote sensing, 200–201, 203, 206, 209–210, 218
Chromosomal aberration in eukaryotic cells, 81; *see also* DNA *listings*
Chronic tests
    Environmental Protection Agency, 47
    long-term freshwater, 75–76
    long-term marine, 79
    short-term freshwater, 72–75
    short-term marine, 77–79
    United Kingdom, 61

CITY FISH, 122
Cluster analysis, hierarchical, 167
Coastal Zone Color Scanner (CZCS), 197, 210
Coastal zones and satellite remote sensing, 211–213
Color, ocean, *see under* Satellite remote sensing
Comet assay, 100, 140
Communication of assessment information
    audience analysis, 235, 237, 239
    community involvement program, 236–237
    data framing, 241–242
    graphic/visual representations, 242–243
    interacting with the public, 239–240
    overview, 234
    pretesting message effectiveness, 240
    relevance of results, emphasizing the, 241
    results, 240, 241
    risk perception, 232–234, 238
    summary/conclusions, 244
    uncertainty discussion, 243
Community involvement and communicating assessment information, 236–237
Composting and sediment quality assessments, 10
Condition factor (CF), 100–101, 110, 140
Control genes and molecular methods for gene expression analysis, 171
Convention for the Protection of the Marine Environment of the North-East Atlantic (OSPAR) in 1997, 50, 84
Copepod toxicity test, marine, 76
Core sediment sampling device, 11
Cultural theory and risk perception, 231
CYP1A (P450-1A)
    biomarkers
        amphibians, 111
        dioxins and planar PCBs, 95
        fish, 105–106, 119
        ligand binding of contaminants to receptors, 96
        methods used to determine biomarker assays, 138
        molecular methods for gene expression analysis, 164, 167
    Cyprinus carpio, 61

D

Daphnia, 60, 61, 70, 75
Data framing and communicating assessment information, 241–242
DDT, 107, 113, 114

Denmark, 73
Detroit River, 3
Differential display technique, 155
Dinophysis acuminata, 215–216
Dioxins and planar PCBs, 95
Direct toxicity assessment (DTA), 55, 59–62, 121
Dissolved organic carbon (DOC), 81, 83
Dithiobisnitrobenzoate (DTNB), 148
DNA adducts/transcription level and biomarkers; *see also* Genotoxicity
 biotransformation products, 98
 fish, 109–110, 119
 genomics, 101
 invertebrates, 119, 120
 methods used to determine biomarker assays, 140
 oxidative stress parameters, 103, 107
 phase I biotransformation enzymes, 96, 97
DNA arrays, 154–155; *see also* Molecular methods for gene expression analysis
Dredged materials and contaminated sediments, 3, 5–8
Dredge samplers, 11

E

Early life stage (ELS) toxicity test, 52, 75, 78
East Australian Current (EAC), 217, 220–221
Ecological Risk Assessments (ERAs), 8–9
Economic impacts of contaminated sediments, 3–4
Ecotoxicology: synthesis/recommendations; *see also individual subject headings*
 biomarkers, application of, 251–254
 new technologies, 254–255
 remote sensing, 256
 risk perception, 256
 summary/conclusions and emerging research needs, 256–258
 toxicity tests, application of, 250–251
Effect-based quality-assessment approaches, 257–258
"Effluent Ecotoxicology: A European Perspective," 50
Effluents/surface water quality, bioassays/ tiered approaches for monitoring
 acute tests using fish, freshwater, 69–70
 acute tests using fish, marine, 76
 acute tests using invertebrates, freshwater, 70–72
 acute tests using invertebrates, marine, 76–77

bioassay types for effluent monitoring/ assessment
 bioaccumulation, 51
 complement to chemical/ecological measures, 47
 criteria for test selection, 48–49
 defining terms, 46
 genotoxicity/mutagenicity, 51
 holistic nature of, 47
 overview, 49–50
 pretreatment of effluents, 54–55
 toxicity, 51–54
biodegradation and sorption tests
 biological oxygen demand, 81
 elimination of biological effects, 84
 evaporation, removal by, 82–83
 sludge, sorption to activated, 82
 solids and sediments, sorption to, 82
 treatment plant simulation model, 83
 Zahn-Wellens test, 83
chemical-oriented approach, limitations of, 46
 genotoxicity tests
  Ames assay, 80
  chromosomal aberration, 81
  overview, 79–80
  UmuC assay, 80–81
introduction, 45
long-term chronic tests, freshwater, 75–76
long-term chronic tests, marine, 79
regulatory test batteries, 69
sampling, effluent, 55
short-term chronic tests, freshwater, 72–75
short-term chronic tests, marine, 77–79
summary, 45
summary/conclusions, 64–66
surface water quality, assessing, 48–49
tiered approaches
 Germany, 58–59
 Netherlands, 56–58
 overview, 55–56
 United Kingdom, 59–62
 United States, 62–64
Eggs of birds/turtles and contaminated sediments, 3
El Niño and satellite remote sensing, 206–208
Elutriates and sediment quality assessments, 15, 16
Embryo-larval survival and teratogenicity test, 78–79
Emission-based approach for assessing effluents, 48
Endocrine parameters and biomarkers, 116–117; *see also* Reproductive/

endocrine parameters and
biomarkers
Endpoints, bioassays/biosensors and
biological, 187
Environment Agency for England and Wales
(UKEA), 48
effluents, monitoring
algal growth inhibition test, 74
direct toxicity assessment, 61
pretreatment of effluents, 54
prioritizing effluents based on hazard,
62
short-term chronic tests, freshwater, 73
Tisbe battagliai, 79
Environmental Monitoring and Assessment
(EMAP), 22–23
Environmental Protection Agency, U.S.
(USEPA)
communicating assessment information,
235
effluents, monitoring
acute toxicity tests with marine fish, 76
bivalve embryo-larval development
toxicity test, 77
Ceriodaphnia dubia, 70, 72, 76
gammarid toxicity test, 70
integrated strategy for water-quality
standards, 63
larval survival and teratogenicity test,
78
minnow, fathead, 72
mysid shrimp toxicity test, 77
oyster toxicity test, 77
pretreatment of effluents, 54
treatment plant simulation model, 83
Hudson River, 236–237
responsibilities of, 6
sediment quality assessments
Chironomus riparius, 39
design, sample, 9
ecological risk assessments, 8
future research recommendations, 23
laboratory and field results, 22–23
short/long-term tests, 38
storing samples, 11
short-term toxicity tests, 47
Environment Canada, 11
Enzyme-linked immunosorbent assay
(ELISA), 138, 164
Enzymes and biomarkers, *see* Biomarkers in
environmental assessment;
CYP1A; P450
Enzymes and biosensors, 185
Eohaustorius estuarius, 22
Epoxide hydrolase (EH), 115

EROD, *see* Ethoxyresorufin-O-dealkylase
E-SCREEN, 181
Estradiol, 108, 109
Estrogenic effects, 108–109, 181, 183; *see also*
Reproductive/endocrine
parameters and biomarkers
Ethinyloestradiol, 108
Ethoxyresorufin-O-dealkylase (EROD) as
biomarker
amphibians, 111
fish, 106, 119
invertebrates, 102
mammals/birds, 117
methods used to determine biomarker
assays, 138
overview, 95
standard operating procedures, 142–144
European medium resolution imaging
spectrometer (MERIS), 199
European Union (EU), 55, 122, 187
*Evaluation of Dredged Material Proposed for
Discharge in Waters of the
U.S.-Testing Manual*, 6
*Evaluation of Dredged Material Proposed for
Ocean Disposal-Testing Manual*, 6
Evaporation, wastewater samples removed
by, 82–83
Expressed sequence tags (ESTs), 165

F

Federal Environment Agency in Germany
(UBA), 49
Fenitrothion, 113
Field exposures *vs.* laboratory results, 17,
22–23
Fish
acute tests, freshwater, 60–61, 69–70
acute tests, marine, 76
bioassays, 183
biomarkers, 105–110, 119–120
early life stage and toxicity test, 78
long-term chronic tests, freshwater, 75–76
satellite remote sensing and fisheries,
213–215
sediments, contaminated, 2, 3
short-term chronic tests, freshwater, 72
Flame retardants, brominated, 114
Flow-through tests, 16
Fluorescence resonance energy transfer
(FRET), 161
Fluorescent aromatic compound levels
(FAC), 107
Fold inductions and molecular methods for
gene expression analysis, 168

Food chain and contaminated sediments, 3
Fraser River, 233

G

Gammarid toxicity test, 70
Gammarus pulex, 61
Gas chromatography/mass spectrometry
    (GC-MS), 51, 181–182
General Electric, 236
Genomics, 101, 254–255; *see also* DNA *listings*
Genotoxicity; *see also under* Effluents/surface
    water quality
    biomarkers
        comet assay, 100
        fish, 109–110
        invertebrates, 104
        P-postlabeling technique, 100
    protocols, test, 51
Germany, 56, 58–59, 69, 83, 84
Glutamate-cysteine ligase (GLCL), 165
Glutathione (GSH), 139
Glutathione-dependent peroxidase (GPOX)
    as biomarker, 97, 102, 107
Glutathione:oxidized glutathione ratio
    (GSH:GSSG) as biomarker, 107,
    112, 139
Glutathione reductase (GRED) and
    biomarkers, 97, 107
Glutathion-S-transferases (GSTs)
    biomarkers
        amphibians, 111
        fish, 106
        mammals/birds, 115
        methods used to determine biomarker
          assays, 139
        phase II biotransformation enzymes,
          96
        standard operating procedures,
          146–147
    molecular methods for gene expression
        analysis, 165
Glyceraldehyde-3-phosphate dehydrogenase
    (GAPDH), 171
Grab sediment sampling device, 11
Graphic/visual representations and
    communicating assessment
    information, 242–243
GSSG, *see* Glutathione:oxidized glutathione
    ratio
GST, *see* Glutathion-S-transferases

H

Heat-shock proteins (HSPs), *see* Stress
    proteins/metallothioneins/MXR
    as biomarkers
Heavy metals, 114–115
Hematological parameters and biomarkers,
    98–99, 112, 116
Hierarchical cluster analysis, 167
Hierarchical division of (sub)organismal
    responses, 154
High-performance liquid chromatography
    (HPLC), 51
Hormones, *see* Reproductive/endocrine
    parameters and biomarkers
Hudson River, 236–237
Hyalella azteca, 10, 22, 36–38

I

Immunological parameters and biomarkers,
    99
Imposex, 104
Indiana Harbor Ship Canal, 3
Inland Testing Manual (ITM), 6–7, 12
Inland waters and satellite remote sensing,
    211–213
Insecticides, 113
In situ toxicity test methods, 23, 49; *see also*
    Bioassays and Biosensors;
    Biomarkers in environmental
    assessments
Institute for Inland Water Management and
    Waste Water Treatment (RIZA),
    57
Integrated monitoring programs and
    biomarker responses, 93–94
Integration and bioassays, 189–190
International Ocean Color Ocean
    Coordination Group (IOCCG),
    197, 201–202
International Organization for
    Standardization (ISO), 49
    effluents, monitoring
        activated sludge respiration inhibition
          test, 71
        acute toxicity tests with marine fish, 76
        copepod toxicity test, marine, 76
        Daphnia, 75
        nitrification inhibition test, 71
        Pseudomonas putida, 73
        sampling, effluent, 55
        sludge microorganisms, growth
          inhibition of activated, 73–74

standards (test) based on guidelines
of, 52
Internet and risk communication, 238
Interstitial water (ITW)
sediment quality assessments
food chain, 3
laboratory/field results, differences
between, 17
oxidation state, changes in, 15
sieving, 12–14
storage, 11
whole-sediment matrix *vs.* ITW test,
15–16
Invertebrates
acute tests, freshwater, 70–72
acute tests, marine, 76–77
biomarkers, 101–105, 119
IPPC (Integrated Pollution Prevention and
Control) Directive, 55
Iraq, 213
Iron hydroxide and interstitial water, 15
ITW, *see* Interstitial water

K

Keystone species and bioassays/biosensors,
257
Knock-out technologies, 182–183
Kuwait, 213

L

Laboratory *vs.* field exposures, 17, 22–23
La Niña and satellite remote sensing, 206–208
Larval development and biomarkers, 114
Larval survival and teratogenicity test, 78
Legislation
Australia
Environment Protection and
Biodiversity Conservation Act of
1999, 7–8
Environment Protection (Sea
Dumping) Act of 1981, 7
Canada
Canadian Environmental Protection
Act (CEPA) of 1999, 7
Germany
German Federal Water Act (WHG), 58
United States
Clean Water Act (CWA), 6, 63
Comprehensive Environmental
Response, Compensation and
Liability Act (CERCLA) of 1980, 8
Marine Protection, Research and
Sanctuaries Act (MPRSA), 6

Superfund Amendments and
Reauthorization Act (SARA) of
1986, 8, 10, 236
Lemna minor, 61, 74–75
Leptocheirus plumulosus, 23
Lipid peroxidation (LPO), 107
Liver somatic index (LSI), 100–101, 110; *see
also* standard operating
procedures *under* Biomarkers in
environmental assessment
Local species and sediment quality
assessments, 15
London Convention (1996), 7
Long-term chronic tests, freshwater, 75–76
Long-term chronic tests, marine, 79
Long-term sediment test, 38
Lysosomal changes/stability and
biomarkers, 105, 140, 151–152

M

Maagd de, P. G.-J., 51
Madeira Island, 210
Mammals/birds and biomarkers, 114–118,
120–121
Masculinization phenomena and
biomarkers, 104
Melanogrammus aeglefinus, 214
Menidia, 76
Mesoscale processes and satellite remote
sensing, 208–211
Metabolite levels and biomarkers, 97–98, 107
Metallothioneins (MTs), 98, 139; *see also* Stress
proteins/metallothioneins/MXR
as biomarkers
Metals, heavy, 114–115
Methoxyresurufin-Odealkylase (MROD), 95
Methylcholanthrene, 167
Methylmercury, 165
Microarray analysis, 156, 165–169
Microsomal monooxygenase (MO) enzymes,
95
Microtox test, 10, 41
Miniaturization and bioassays, 189
Minnows, fathead, 72
Mississippi River, 211
Mixed-function oxidase (MFO) system, 95
Moderate resolution imaging
spectroradiometer (MODIS), 197,
199, 222
Molecular methods for gene expression
analysis
control genes, 171
data, lack of genomic data, 169–170

future developments: pitfalls/
    recommendations, 169–172
microarray analysis, 165–169
overview, 153–156
real-time reverse-transcription PCR,
    160–165
suppression subtractive hybridization,
    156–160
Monguagon Creek, 3
Morphological/histological parameters and
    biomarkers, 117–118; *see also*
    Physiological/morphological
    parameters and biomarkers
Multixenobiotic resistance (MXR) and
    biomarkers, 98; *see also* Stress
    proteins/metallothioneins/MXR
    as biomarkers
Mutagenicity, 51
Mysid shrimp toxicity test, 77, 79

N

National Ocean Disposal Guidelines for
    Dredged Material, 7, 8
National Oceanic and Atmospheric
    Administration (NOAA), 197
National Oil and Hazardous Substances
    Pollution Contingency Plan
    (NCP), 8
National Priorities List (NPL), Superfund, 8
Natural resource damage assessments
    (NRDAs), 5
Necturus maculosus, 3
Netherlands, 56–58, 81, 84
N-ethyl perfluorooctanesulfonamido ethanol
    (N-EtFOSE), 165
Neuromuscular parameters and biomarkers
    fish, 109
    invertebrates, 104
    mammals/birds, 117
    overview, 99–100
Nicotinamide adenine dinucleotide
    phosphate oxidase (NADPH),
    138
Nitocra spinipes, 76
Nitrification inhibition test, 71
Noctiluca scintillans, 215–220

O

Ocean general-circulation models (OGCMs),
    208
Ocean Testing Manual, 6–7, 12
Oncorhynchus mykiss, 60

Organization for Economic Cooperation and
    Development (OECD)
    effluents, monitoring
        activated sludge respiration inhibition
            test, 71
        algal growth inhibition test, 74
        biodegradation, 81
        Daphnia, 75
        early life stage toxicity test, 75
        Lemna minor, 74
        protocols, test, 49
        protozoans, toxicity tests with, 71
        Pseudomonas putida, 73
        treatment plant simulation model, 83
    standards (test) based on guidelines of, 52
    toxicity tests recommended by, 53
Organochlorines, 114–116; *see also* DDT;
    Polychlorinated biphenyls
Organophosphorus compounds (OPs), 117
OSPAR WEA Demonstration Program (2003),
    84
Oxidative stress parameters and biomarkers
    amphibians, 112
    fish, 106–107
    invertebrates, 102–103
    overview, 97
Oyster toxicity test, 77
Ozone depletion, 208

P

P450 as a biomarker; *see also* CYP1A
    invertebrates, 102
    mammals/birds, 115
    methods used to determine biomarker
        assays, 138
    reproductive/endocrine parameters, 99
    standard operating procedures, 144–145
PCB pattern analysis (PPA), 115; *see also*
    Polychlorinated biphenyls
Perfluorooctanesulfonate (PFOS), 165
Peripheral blood mononuclear cells
    (PBMCs), 164
Persian Gulf, 213
Pesticides, *see* DDT; Polychlorinated
    biphenyls
P-glycoproteins (PGPs) as biomarkers, 98,
    103–104, 108
pH adjustment and pretreatment of effluents,
    54–55
Phase I biotransformation enzymes as
    biomarkers
    amphibians, 111
    fish, 105–106, 110
    invertebrates, 102

mammals/birds, 115
overview, 95–96
Phase II biotransformation enzymes as
          biomarkers
    amphibians, 111–112
    fish, 106
    mammals/birds, 115
    overview, 96–97
Phenobarbital (PB), 96, 111, 166
Photosynthetically available radiation (PAR),
          212
Physiological/morphological parameters
          and biomarkers
    amphibians, 113–114
    invertebrates, 104–105
    mammals/birds, 117–118
    overview, 100–101
Pimephales promelas, 72
Planar PCBs, dioxins and, 95
Polluter pays principle (PPP), 59
Pollution in environment, unreliability of risk
          assessment based solely on, 91
Polychlorinated biphenyls (PCBs)
    biomarkers
        biotransformation products, 107
        dioxins and planar PCBs, 95
        phase I biotransformation enzymes,
              102, 115
        reproductive/endocrine parameters,
              108
    effects of chronic exposure to, 3
Polycyclic aromatic hydrocarbons (PAHs) as
          biomarkers
    fish, 107
    methods used to determine biomarker
          assays, 140
    overview, 116–117
    oxidative metabolism, 97–98
    thyroid hormones, 116–117
Polymerase chain reaction (PCR), *see*
          Real-time reverse-transcription
          PCR
Pore water, *see* Interstitial water and
          contaminated sediments
Potter homogenizer, 138
Precautionary principle, 56, 59
Pretreatment of effluents, 54
Principal-component analysis, 166
Probability-based sampling designs, 9–10
Proteomics and biomarkers, 101
Protocol to the Convention on Prevention of
          Marine Pollution by Dumping of
          Wastes and Other Matter (1996),
          7
Protozoans, toxicity tests with, 71, 77

Pseudomonas putida, 73
Psychometric approach and risk perception,
          231

R

Reactive oxygen intermediates (ROIs), 97
Reactive oxygen species (ROS), 97
Real-time reverse-transcription PCR
          (RT-PCR), 156, 160–165
Remote sensing technologies, 196, 256; *see
          also* Satellite remote sensing
Replication and sediment quality
          assessments, 10, 16, 40
Reporter genes and bioassays, 180–181
Reproductive/endocrine parameters and
          biomarkers
    amphibians, 112–113
    fish, 108–109
    invertebrates, 104
    overview, 99
Retinoic acid, 114
Rhepoxynius abronius, 22
Ribonucleic acid (RNA), 98, 255; *see also*
          Molecular methods for gene
          expression analysis
Risk perception, 230–234, 238, 256
Rotifers, toxicity tests with, 70, 73, 77

S

Satellite remote sensing
    algal blooms off southeast Australia
        management issues, 215–216
        Noctiluca scintillans, 217–220
        predictive/diagnostic capability,
              216–217
        Trichodesmium erythraeum, 220–221
    color, history/relevance of ocean
        Case 2 waters, 201–202
        chlorophyll and primary productivity,
              200–201
        key satellite-mounted sensors,
              197–199
        limitations of sensors, 199–200
        overview, 196–197
        products, remote sensed, 199–200
    environmental issues/applications
        algal blooms, 207–208
        biogeochemical cycles, 208
        carbon, 203
        chlorophyll, 203, 206
        coastal zones, 211–213
        El Niño/La Niña, 206–208
        fisheries, 213–215

global scale phenomena, 203–208
overview, 202
regional seas: mesoscale processes,
208–211
summary/conclusions, 222
Scope for growth (SFG) as a biomarker, 101,
104–105, 140
Scophthalmus maximus, 76
Sea surface temperature (SST), *see* Satellite
remote sensing
Sea-viewing wide field-of-view sensor
(SeaWiFS), 197; *see also* Satellite
remote sensing
Sediment quality assessments, toxicity tests
for
applications of sediment toxicity tests, 5–9
chemistry, sediment, 4
disadvantages of using, 4
future research recommendations, 23
interpretation, 17, 22–23
introduction, 2
need for, 2–5
organisms, freshwater test
Chironomus riparius, 38–39
Hyalella azteca, 36–38
organisms, marine test
Ampelisca abdita, 39–40
Microtox test, 41
recommended procedures for
freshwater/marine test
organisms, 14–21
sampling
collection/processing/transport/
storage, 10–12
design, sample, 9–10
manipulation of sediments, 12–14
summary/conclusions, 24
tiered testing, 5
Sequential order of responses to pollutant
stress, 91–92
Sex characteristics and biomarkers, 104,
112–113, 116
Shipping (water commerce) industry and
contaminated sediments, 3–4
Short-term tests
chronic freshwater, 72–75
chronic marine, 77–79
Environmental Protection Agency, U.S.,
47
sediment, 37–38
Shrimp toxicity test, 77, 79
Sieving and sediment quality assessments,
12–13, 37
Single-cell gel electrophoresis (SCGE), 100

Sludge and monitoring effluents, 71, 73–74,
82
Society of Environmental Toxicology and
Chemistry (SETAC), 49, 50
SOD, *see* Superoxide dismutase
Solid-phase microextraction (SPME), 51
Sorption, *see* biodegradation and sorption
tests *under* Effluents/surface
water quality
Spermatogenesis, 116
Static toxicity tests, 16, 17
Steroids and *in vitro* bioassays, 181
Storage and sediment quality assessments,
11–12
Stress proteins/metallothioneins/MXR as
biomarkers
amphibians, 112
fish, 108
invertebrates, 103–104
mammals/birds, 116
overview, 98
Substance-oriented approach, limitations of,
46
Superfund, 8, 10, 236
Superoxide dismutase (SOD) and biomarkers
fish, 107
methods used to determine biomarker
assays, 139
overview, 97
oxidative stress parameters, 102
standard operating procedures, 147–148
Suppression subtractive hybridization
polymerase chain reaction
(SSH-PCR), 155–160
Surface water quality, assessing, 48–49; *see
also* Effluents/surface water
quality
Surrogates and sediment quality
assessments, 15
Sweden, 81
SYBR-Green-based detection, 161, 163

T

TaqMan assay, 161–162
Temperature, sea surface, *see* Satellite remote
sensing
Testis change/volume, 114
Testosterone, 116
Tetrachlorodibenzo-p-dioxin (TCDD), 96,
116, 168
Thyroid hormones and biomarkers, 112–114,
116

Tiered testing approaches, 5, 7–8, 252–253; *see also under* Effluents/surface water quality
Tisbe battagliai, 60, 61, 76, 79
Total maximum daily life loads (TMDLs), 5
Total organic carbon (TOC), 46
Total oxidant scavenging capacity (TOSC), 103, 107
Toxicity identification evaluation (TIE), 23, 61, 63, 84, 257
Toxicity-reduction evaluations (TREs), 63
Toxicity tests; *see also* Bioassays and biosensors; Biomarkers in environmental assessment; Chronic tests; Effluents/surface water quality; Long-term tests; Short-term tests
  direct toxicity assessment, 55, 59–62, 121
  nonstandardized, 52–53
  standardized, 51–52
  synthesis/recommendations, 250–251
  validity criteria, 53–54
Transducers used in biosensors, 186
Transgenic animals and bioassays, 182–183
Transport and sediment quality assessments, 10–11
Treatment plant simulation model, 83, 84
Triad approach, 93–94
Tributyltin (TBT), 104
Trichodesmium erythraeum, 213, 217, 220–221
Triclopyr, 113
Triphenyltin (TPT), 104
Turbidity and pretreatment of effluents, 54
Turtle eggs and contaminated sediments, 3

U

Ultracentrifuge, 138
Ultra-Turrax homogenizer, 138
UmuC assay, 80–81
Uncertainty discussion and communicating assessment information, 243
United Kingdom, 56, 59–62, 69
United Nations Economic Commission for Europe (UN/ECE), 48
United States, 56, 62–64, 69; *see also individual subject headings*
Uridine diphosphate glucuronyl-transferases (UDPGTs) as biomarkers
  fish, 106
  mammals/birds, 115

methods used to determine biomarker assays, 138–139
phase II biotransformation enzymes, 96

V

Validity criteria and testing standards, 52, 53–54
Vibrio fischeri, 10, 61, 71–73, 77, 84
Virilization of females and biomarkers, 104
Vitellogenin (VTG) and biomarkers
  amphibians, 112, 113, 120
  fish, 108
  methods used to determine biomarker assays, 140
  reproductive/endocrine parameters, 99
Volume per sample and sediment quality assessments, 10

W

Wastewater discharges, monitoring, 50; *see also* Effluents/surface water quality
Water Framework Directive, 55
Water-quality-based approach for assessing effluents, 48
WEA, *see* Whole-effluent assessment
Weight-of-evidence (WOE) approaches, 253, 257
Whole-effluent assessment (WEA)
  aim of, 55–56
  degradability of biological effects, 84
  Netherlands, 57–58
  United States, 62
Whole effluent toxicity (WET) testing, 62–64, 70, 71
Whole-sediment matrix, interstitial water tests compared to, 15–16
Whole-sediment toxicity tests, 37, 40
World Health Organization (WHO), 91

Z

Zahn-Wellens test, 82–84
Zambezi River, 211
Zebrafish and bioassays, 183
Zona radiate protein (ZRP) and biomarkers
  fish, 108, 109
  methods used to determine biomarker assays, 140
  reproductive/endocrine parameters, 99